Boundary Value Problems

Second Edition

Boundary
Value
Problems

Second
Edition

David L. Powers
Clarkson College

ACADEMIC PRESS
New York
San Francisco
London
A Subsidiary of Harcourt Brace Jovanovich, Publishers

Academic Press, Inc.
111 Fifth Avenue,
New York, New York 10003

United Kingdom Edition published by
Academic Press, Inc. (London) Ltd.
24/28 Oval Road,
London NW1

ISBN: 0-12-563760-8
Library of Congress Catalog Card Number: 78-57142

Printed in the United States of America

Contents

Preface

This text is designed for a one-semester or two-quarter course in partial differential equations given to third- and fourth-year students of engineering and science. It may also be used as the basis for an introductory course for graduate students. Mathematical prerequisites have been kept to a minimum —calculus and differential equations. No vector calculus or linear algebra, other than 2×2 determinants, is necessary. The reader is assumed to have enough background in physics to follow the derivations of the heat and wave equations.

The principal objective of the book is solving boundary value problems. Separation of variables receives the greatest attention because of its frequent appearance elsewhere and because it provides a uniform method for solving important cases of the heat, wave, and potential equations. One technique is not enough, of course. D'Alembert's solution of the wave equation is developed in parallel with the series solution, and the distributed-source solution is constructed for the heat equation. In addition, there are chapters on Laplace transform techniques and on numerical methods.

The secondary objective is to tie together the mathematics developed and the student's physical intuition. This is accomplished by deriving the mathematical model in a number of cases, by using physical reasoning in the mathematical development from time to time, by interpreting mathematical results in physical terms, and by studying the heat, wave, and potential equations separately.

In the service of both objectives, there are many fully worked examples and over 550 exercises, including miscellaneous exercises at the end of each chapter. The level of difficulty ranges from drill and verification of details to development of new material. Answers are supplied for all odd-numbered exercises.

There are many ways of choosing and arranging topics from the book so as to provide an interesting and meaningful course. The following sections form the core, requiring at least 14 hours of lecture: Chapter 1, Sections 1–3; Chapter 2, Sections 1–5; Chapter 3, Sections 1–3; Chapter 4, Sections 1, 2, and 4. These cover the basics of Fourier series and the solutions of heat, wave, and potential equations on a finite interval. My choice for the next most important block of material is the Fourier integral and the solution of problems on unbounded regions: Chapter 1, Section 8; Chapter 2, Sections 10 and 11; Chapter 3, Section 6; Chapter 4, Section 3. These require at least six more lectures.

The taste of the instructor and the needs of his audience will govern the choice of further material. A rather theoretical flavor results from including: Chapter 1, Sections 4–6 on Fourier series; Chapter 2, Sections 7–9 on Sturm–Liouville problems, and the sequel, Chapter 3, Section 4; and the more difficult parts of Chapter 5, Sections 5–9 on Bessel functions and Legendre polynomials. On the other hand, inclusion of Chapter 7, Numerical Methods, gives a very applied flavor, especially if students write programs and run them on a computer.

Most ordinary differential equations courses make little or no mention of boundary value problems. Moreover, many students have had ample time to forget many of the techniques learned there. For these reasons, Chapter 0 on ordinary differential equations and boundary value problems has been added to this edition. Equilibrium forms of the heat and wave equations are derived here; the derivations in Chapters 2 and 3 still stand on their own if Chapter 0 is not presented but may go more quickly if it is. Section 2, Variation of Parameters, is useful if the Laplace transform (Chapter 6) is to be studied. Other new material in this edition is Chapter 1, Section 5, Operations on Fourier Series; an additional example on the heat equation in Chapter 2, Section 5; a solution of the heat equation in a circular cylinder in Chapter 5, Section 6; and numerical methods for the wave equation in Chapter 7, Section 6. Further changes in the second edition are rearrangement of some material, addition of over 200 exercises, and a few deletions and corrections.

Now it is my pleasure to thank publicly some friends and colleagues. Victor Lovass-Nagy has suggested numerous improvements and supplied some exercises; Charles Cullen has consistently given good advice; and many colleagues, especially Abdul J. Jerri, Gustave Rabson, Hayley Shen, Anthony Zeppetella, Heino Ainso, Charles W. Haines, Harvey Segur, Charles Marshall, Robert Feinberg, and Paul Bugl, have given advice, suggestions, encouragement, and help in various forms. Finally, I wish to acknowledge the assistance rendered by the reviewers, George E. Abdo of the Florida Institute of Technology, Chris Rorres of Drexel University, and Monty Strauss of Texas Tech University, for their constructive criticisms and suggestions.

Boundary Value Problems

Second Edition

chapter 0
Ordinary Differential Equations

1   Ordinary differential equations

The subject of most of this book is partial differential equations: their physical meaning, problems in which they appear, and their solutions. Our principal solution technique will involve separating a partial differential equation into ordinary differential equations. Therefore we begin by reviewing some facts about ordinary differential equations and their solutions.

We are interested mainly in linear differential equations of first and second orders, as shown in Eqs. (1) and (2):

$$\frac{du}{dt} = k(t)u + f(t), \tag{1}$$

$$\frac{d^2u}{dt^2} + b(t)\frac{du}{dt} + p(t)u = f(t). \tag{2}$$

In either equation, if $f(t)$ is 0, the equation is *homogeneous*. (Another test: if the constant function $u(t) \equiv 0$ is a solution, the equation is homogeneous.)

1

The most important feature of a linear, homogeneous equation is that any sum of constant multiples of solutions is also a solution. Thus, if $u_1(t)$ and $u_2(t)$ are solutions of a linear, homogeneous differential equation, so is $c_1 u_1(t) + c_2 u_2(t)$, for any choice of the constants c_1 and c_2. (A sum of constant multiples of two or more functions is called a *linear combination* of them.)

The *general* solution of an ordinary differential equation contains as many arbitrary parameters as the order of the equation. The general solution of a nonhomogeneous linear differential equation is formed as the sum of a particular solution (that is, any solution) of the nonhomogeneous equation and the general solution of the corresponding homogeneous equation (obtained by replacing $f(t)$ with 0).

In each of the numbered items that follow, a differential equation and its solution are given. These will be especially important in the rest of the book. Throughout this section we use u and t as the dependent and independent variables.

1. $\dfrac{du}{dt} = f(t); \quad u(t) = c + \displaystyle\int f(t)\, dt$

This may be thought of as the simplest of all first-order differential equations. In fact, the definition of the indefinite integral in the solution is any function whose derivative is $f(t)$. For example, the differential equation

$$\frac{du}{dt} = \sin t$$

has as its solution $u(t) = c - \cos t$, since the derivative of $-\cos t$ is $\sin t$. Another legitimate expression for the solution is $u(t) = c' + (1 - \cos t)$, since the derivative of $1 - \cos t$ is also $\sin t$. Of course, the arbitrary constants c and c' would have to differ by 1 in order for values of the two functions to match.

2. $\dfrac{du}{dt} = ku; \quad u(t) = ce^{kt}$

It is easy to verify that the function shown does indeed satisfy this differential equation. The multiplier c may be any number. The solution $u(t)$ approaches 0 as t increases, provided that k is a negative constant. In the opposite case, when k is positive, the solution increases in magnitude as t increases. (This exponential growth often signals disaster in physical situations, since it cannot be sustained indefinitely.)

3. $\dfrac{d^2u}{dt^2} = g(t); \quad u(t) = c_2 + c_1 t + \displaystyle\iint g(t)\, dt\, dt$

In this simplest of second-order differential equations, notice that one integration of both sides of the equation leads to

$$\frac{du}{dt} = c_1 + G(t)$$

where $G(t)$ is an indefinite integral of $g(t)$. This equation fits the pattern of the equation in item 1 with $f(t) = c_1 + G(t)$. Since the solution of the differential equation involves a total of two integrations, two arbitrary constants appear.

mt

Try $u(t) = e$

4. $\dfrac{d^2u}{dt^2} - \lambda^2 u = 0; \quad u(t) = c_1 e^{\lambda t} + c_2 e^{-\lambda t}$

A useful technique for solving homogeneous linear equations with constant coefficients is to assume a solution of the form $u(t) = e^{mt}$. Substituting into the differential equation and canceling the common factor e^{mt} from all terms leads to a polynomial equation called the *indicial equation* for m. In the case at hand, assume $u(t) = e^{mt}$; then we have

$$\frac{du}{dt} = m e^{mt}, \qquad \frac{d^2u}{dt^2} = m^2 e^{mt}.$$

Substituting in the differential equation gives

$$m^2 e^{mt} - \lambda^2 e^{mt} = 0 \quad \text{or} \quad m^2 - \lambda^2 = 0$$

after cancellation of e^{mt}. Acceptable values of m are λ and $-\lambda$. Thus $e^{\lambda t}$ and $e^{-\lambda t}$ are both solutions of the differential equation. Since it is linear and homogeneous, any linear combination (sum of constant multiples) of these solutions is again a solution. By this route we arrive at the solution noted at the beginning.

It is sometimes helpful to write the solution of this differential equation in another form. The hyperbolic sine and cosine are defined by

$$\sinh A = \tfrac{1}{2}(e^A - e^{-A}), \qquad \cosh A = \tfrac{1}{2}(e^A + e^{-A}).$$

Therefore, both $\sinh \lambda t$ and $\cosh \lambda t$, being linear combinations of $e^{\lambda t}$ and $e^{-\lambda t}$, are solutions of this differential equation. Hence we can express the general solution as

$$u(t) = b_1 \cosh \lambda t + b_2 \sinh \lambda t.$$

Try

$u(t) = e^{mt}$

(m complex)

5. $\dfrac{d^2u}{dt^2} + \lambda^2 u = 0; \quad u(t) = c_1 \cos \lambda t + c_2 \sin \lambda t$

This differential equation is linear and homogeneous, with constant coefficients. If we attempt to find solutions by the substitution $u(t) = e^{mt}$, as suggested in item 4, we find that $m^2 + \lambda^2 = 0$ is the indicial equation. Since the roots are $m = \pm i\lambda$ (where $i^2 = -1$), we are led to consider the complex exponentials $e^{i\lambda t}$ and $e^{-i\lambda t}$ as solutions. In order to find a convenient real form of the solution, we use Euler's formula,

$$e^{iA} = \cos A + i \sin A,$$

which relates the imaginary exponential to the real sine and cosine functions. It may be verified then that the equations

$$\tfrac{1}{2}(e^{i\lambda t} + e^{-i\lambda t}) = \cos \lambda t$$

$$(1/2i)(e^{i\lambda t} - e^{-i\lambda t}) = \sin \lambda t$$

are valid. The two trigonometric functions, being linear combinations of solutions, are themselves solutions of the differential equation. (Of course, this may be verified independently.)

6. $t^2 \dfrac{d^2u}{dt^2} + t \dfrac{du}{dt} - \lambda^2 u = 0; \quad u = c_1 t^\lambda + c_2 t^{-\lambda}$

This equation is special but important. It is a particular case of the second-order Cauchy–Euler equation,

Try

$$t^2 \dfrac{d^2u}{dt^2} + bt \dfrac{du}{dt} + pu = 0, \qquad u(t) = t^m$$

where b and p are constants. An nth-order Cauchy–Euler equation can be identified by the facts that it is linear and that the coefficient of the kth derivative is a constant multiple of t^k. A technique for solving the Cauchy–Euler equation is to assume that $u(t) = t^m$ and substitute in the differential equation to find an acceptable constant m.

Following this suggestion for the equation

$$t^2 \dfrac{d^2u}{dt^2} + t \dfrac{du}{dt} - \lambda^2 u = 0$$

leads to the equation

$$t^2 m(m-1) t^{m-2} + tm t^{m-1} - \lambda^2 t^m = 0,$$

which becomes, after cancellation of the common factor t^m,

$$m(m-1) + m - \lambda^2 = 0.$$

Evidently $m = \pm\lambda$ are the only values of m for which t^m is a solution of the differential equation. The solution shown in the heading is a linear combination of t^λ and $t^{-\lambda}$. In those cases where m is complex, it is convenient to use the identity

$$t^m = \exp(m \ln t)$$

together with Euler's formula from item 5 to obtain real solutions.

Exercises

1. Solve each of the following. Be careful to identify the dependent and independent variables.

   a. $\dfrac{d}{dx}\left((b + px)\dfrac{dv}{dx}\right) = 0$

   b. $\dfrac{d}{dx}\left(e^x \dfrac{d\phi}{dx}\right) + \lambda^2 e^x \phi = 0$

   c. $\dfrac{d}{d\rho}\left(\rho^2 \dfrac{dR}{d\rho}\right) + \lambda^2 \rho^2 R = 0$ (Hint: let $R = u/\rho$.)

   d. $\dfrac{1}{r}\dfrac{d}{dr}\left(r\dfrac{dw}{dr}\right) - \dfrac{\lambda^2}{r^2}w = 0$

   e. $\rho^2 \dfrac{d^2 R}{d\rho^2} + 2\rho \dfrac{dR}{d\rho} - n(n+1)R = 0$

2. Solve each of the following:

   a. $\dfrac{d^2 u}{dt^2} = 0$

   b. $\dfrac{d^2 \phi}{dx^2} + \lambda^2 \phi = 0$

   c. $\dfrac{dT}{dt} + \lambda^2 kT = 0$

   d. $\dfrac{d^2 \phi}{dx^2} - \mu^2 \phi = 0.$

3. Solve each of the following:

   a. $\dfrac{d^2u}{dx^2} = b$

   b. $\dfrac{d^2u}{dx^2} - k^2u = 1.$

4. Solve the differential equation

   $$\dfrac{d^2u}{dx^2} = pu$$

   and compare the behavior of the solution in the three cases

   a. $p = \mu^2$

   b. $p = 0$

   c. $p = -\lambda^2.$

5. Solve the fourth-order differential equations

   a. $\dfrac{d^4u}{dx^4} - \lambda^4 u = 0$

   b. $\dfrac{d^4u}{dx^4} + \lambda^4 u = 0.$

6. Find the general solution of the linear first-order equation below. Divide both sides by u, change the dependent variable, and recognize the resulting differential equation.

   $$\dfrac{du}{dt} = k(t)u$$

7. Solve the differential equations

   a. $r^2 \dfrac{d^2u}{dr^2} + r \dfrac{du}{dr} + \lambda^2 u = 0$

   b. $\dfrac{1}{r}\dfrac{d}{dr}\left(r\dfrac{du}{dr}\right) = 0.$

8. Graph the solution of the initial value problem below in the three different cases

   a. $k > 0$

   b. $k = 0$

c. $k < 0$

$$\frac{du}{dt} = ku, \qquad u(0) = 1.$$

2   Variation of parameters

Variation of parameters is a device for finding additional solutions of a linear differential equation, once one solution is known. It may also be used to determine the solution of a nonhomogeneous equation if the general solution of the corresponding homogeneous equation is known.

1.   Solution of $\dfrac{du}{dt} = ku + f(t)$

Since the homogeneous equation

$$\frac{du}{dt} = ku$$

has the solution e^{kt}, we assume that the solution of the nonhomogeneous equation has the form $u(t) = v(t)e^{kt}$. Substitution into the differential equation yields

$$\frac{dv}{dt}e^{kt} + vke^{kt} = kve^{kt} + f(t)$$

or, after some simple manipulation,

$$\frac{dv}{dt} = e^{-kt}f(t).$$

This equation can be solved by one integration to yield

$$v(t) = c + \int e^{-kt}f(t)\, dt,$$

and consequently the general solution of the original problem is

$$u(t) = \left(c + \int e^{-kt}f(t)\, dt\right)e^{kt}.$$

2.   Second solution of $\dfrac{d^2u}{dt^2} + b(t)\dfrac{du}{dt} + p(t)u = 0$

We assume that one solution of this differential equation, say $u_1(t)$, is known. We then seek a second solution in the form

$$u(t) = v(t)u_1(t).$$

When u in this form is substituted into the differential equation, the result is

$$\frac{d^2v}{dt^2}u_1 + 2\frac{dv}{dt}\frac{du_1}{dt} + v\frac{d^2u_1}{dt^2} + b\frac{dv}{dt}u_1 + bv\frac{du_1}{dt} + pvu_1 = 0.$$

The three terms in which v is a multiplier add up to 0 because they are the three terms of the homogeneous equation of which u_1 is a solution. The remaining terms must also add up to 0:

$$\frac{d^2v}{dt^2}u_1 + 2\frac{dv}{dt}\frac{du_1}{dt} + b\frac{dv}{dt}u_1 = 0$$

or

$$\frac{d^2v}{dt^2} + \frac{dv}{dt}\left(b + 2u_1^{-1}\frac{du_1}{dt}\right) = 0.$$

This last equation is of first order in dv/dt. Standard solution procedures lead to

$$\frac{dv}{dt} = \exp\left[-\int\left(b + 2u_1^{-1}\frac{du_1}{dt}\right)dt\right]$$

$$= u_1^{-2}\exp\left(-\int b(t)\,dt\right).$$

With one more integration, one arrives at the function $v(t)$ and thus at a second independent solution $u(t) = v(t)u_1(t)$.

An example will help clarify the procedure. Suppose we are attempting to solve the equation

$$t^2\frac{d^2u}{dt^2} - t\frac{du}{dt} + u = 0,$$

of which $u_1(t) = t$ is a solution. Although the equation is of the Cauchy–Euler type, the technique suggested in Section 1, item 6 produces only this one solution. If we seek a second solution in the form $u(t) = v(t) \cdot t$, we find that $v(t)$ satisfies

$$\frac{d^2v}{dt^2} + \left(\frac{2}{t} - \frac{1}{t}\right)\frac{dv}{dt} = 0.$$

$$\frac{d}{dt}\left[t^4 \frac{dv}{dt}\right] + \left[2t^2 + t^4\right] = 0$$

$$u = t^r y$$

Then it is easy to find that

$$\frac{dv}{dt} = \frac{c_1}{t}$$

and $v(t) = c_1 \ln t + c_2$. A second solution of the equation is therefore

$$u(t) = (c_1 \ln t + c_2)t.$$

In particular, one might take $u(t) = t \ln t$ as the second solution.

3.   Solution of $\dfrac{d^2 u}{dt^2} + b(t)\dfrac{du}{dt} + p(t)u = f(t)$

When two independent solutions, say u_1 and u_2, of the homogeneous equation

$$\frac{d^2 u}{dt^2} + b(t)\frac{du}{dt} + p(t)u = 0$$

are known, we can find a solution of the given nonhomogeneous equation by assuming that u has the form

$$u(t) = v_1(t)u_1(t) + v_2(t)u_2(t).$$

Here, $v_1(t)$ and $v_2(t)$ are to be found. Insertion of $u(t)$ in this form into the original differential equation leads to one complicated equation in the two unknown functions v_1 and v_2. If we require, in addition, that v_1 and v_2 satisfy

$$\frac{dv_1}{dt}u_1 + \frac{dv_2}{dt}u_2 = 0, \tag{i}$$

then the one complicated equation representing the original differential equation simplifies to

$$\frac{dv_1}{dt}\frac{du_1}{dt} + \frac{dv_2}{dt}\frac{du_2}{dt} = f(t). \tag{ii}$$

There are good reasons for imposing condition (i). For one we now have a system of two linear algebraic equations, in both of which dv_1/dt and dv_2/dt are the unknowns. Furthermore, the determinant of the coefficients of this system (called the Wronskian of u_1 and u_2) is nonzero, provided that u_1 and u_2 are independent. Thus the two simultaneous equations (i) and (ii) can be solved algebraically for dv_1/dt and dv_2/dt in terms of known functions of t. Then v_1 and v_2 are found by integration.

As an example, suppose we attempt to solve the nonhomogeneous equation

$$t^2 \frac{d^2 u}{dt^2} - t\frac{du}{dt} + u = t.$$

In item 2 we found that two independent solutions of the homogeneous equation

$$t^2 \frac{d^2u}{dt^2} - t\frac{du}{dt} + u = 0$$

are $u_1(t) = t$ and $u_2(t) = t \ln t$. Thus we look for a solution of the nonhomogeneous equation in the form

$$u(t) = v_1(t)t + v_2(t)t \ln t.$$

Equations (i) and (ii) become, in this specific case,

$$\frac{dv_1}{dt} t + \frac{dv_2}{dt} t \ln t = 0$$

$$\frac{dv_1}{dt} + \frac{dv_2}{dt}(1 + \ln t) = \frac{1}{t}.$$

(Note that $f(t) = 1/t$, not t, when the differential equation is arranged so that the highest-order term has a coefficient of unity.) When these equations are solved simultaneously for dv_1/dt and dv_2/dt, we find that

$$\frac{dv_1}{dt} = -\frac{(\ln t)}{t}, \frac{dv_2}{dt} = \frac{1}{t}.$$

Hence $v_1(t) = -\frac{1}{2}(\ln t)^2 + c_1$, $v_2(t) = \ln t + c_2$, and the (general) solution of the nonhomogeneous equation is

$$u(t) = (c_1 - \tfrac{1}{2}(\ln t)^2)t + (\ln t + c_2)t \ln t$$

$$= \tfrac{1}{2}t(\ln t)^2 + c_1 t + c_2 t \ln t.$$

Exercises

1. For each equation, find one solution by the methods of Section 1 and a second by variation of parameters.

a. $\dfrac{d^2u}{dt^2} + 2\dfrac{du}{dt} + u = 0$

b. $\dfrac{d^2u}{dt^2} - 2a\dfrac{du}{dt} + a^2u = 0$

c. $t^2\dfrac{d^2u}{dt^2} + bt\dfrac{du}{dt} + \dfrac{(b-1)^2}{4}u = 0$

2. Obtain the formula given below for a particular solution of the differential equation

$$\frac{d^2u}{dt^2} + \gamma^2 u = f(t),$$

$$u(t) = \frac{1}{\gamma} \int_0^t \sin \gamma(t - t')f(t')\, dt'.$$

(Hint: use $u_1 = \sin \gamma t$, $u_2 = \cos \gamma t$.)

3. Use variation of parameters to solve each of the following:

a. $\dfrac{du}{dt} + a(u - T) = 0$ (T a constant)

b. $\dfrac{du}{dt} + au = e^{at}$

c. $\dfrac{du}{dt} + au = e^{-at}$.

4. Find a second solution of the differential equation

$$\frac{d}{dx}\left(x\frac{d\phi}{dx}\right) + \left(x - \frac{1}{4x}\right)\phi = 0,$$

$$\phi_1(x) = \frac{\cos x}{\sqrt{x}}.$$

5. Find a second solution of the differential equation

$$\frac{d}{dx}\left(\sin x\frac{du}{dx}\right) + 2 \sin x\, u = 0,$$

$$u_1(x) = \cos x.$$

6. Find the general solution of

$$\frac{d^2u}{dt^2} + \omega^2 u = \sin \omega t.$$

3   Boundary value problems

A *boundary value problem* in one dimension is an ordinary differential equation together with conditions involving values of the solution and/or its derivatives at two or more points. The number of conditions imposed is equal to the order of the differential equation. Usually, boundary value problems of any physical relevance have these characteristics: (1) the conditions are imposed at two different points; (2) the solution is of interest only between those two points; (3) the independent variable is a space variable, which we shall represent as x. In addition, we are primarily concerned with cases where the differential equation is linear and of second order. However, problems in elasticity often involve fourth-order equations.

In contrast to initial value problems, even the most innocent-looking boundary value problem may have exactly one solution, no solution, or an infinite number of solutions. Exercise 1 illustrates these cases.

When the differential equation in a boundary value problem has a known general solution, we use the two boundary conditions to supply two equations that are to be satisfied by the two constants in the general solution. If the differential equation is linear, these are two linear equations and can be easily solved, if there is a solution.

In the rest of this section we examine some physical examples that are naturally associated with boundary value problems.

The hanging cable. A cable hung between two pegs carries a distributed load; we wish to find its shape, described by its height $u(x)$ above the horizontal, as shown in Fig. 0.1. The governing differential equation can be found by

Figure 0.1 The hanging cable.

applying Newton's second law to a segment of the cable between x and $x + \Delta x$ (see Fig. 0.2). We assume that the cable is perfectly flexible—that it offers no resistance to bending. A consequence of this assumption is that the tensions exerted on the segment in Fig. 0.2 are tangent to the cable. The sum of forces in the horizontal and vertical directions give these two equations:

$$-T(x) \cos \alpha + T(x + \Delta x) \cos \beta = 0 \tag{1}$$

$$-T(x) \sin \alpha + T(x + \Delta x) \sin \beta - f(x) \Delta x = 0. \tag{2}$$

Both right-hand sides are 0 because we assume no movement—and thus no acceleration—in any direction. In the second equation, $f(x)$ is the loading (force per unit length) on the segment of cable.

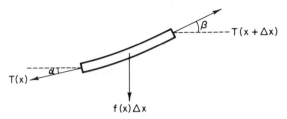

Figure 0.2   Section of cable showing forces acting on it.

We rearrange Eq. (1) to read

$$T(x) \cos \alpha = T(x + \Delta x) \cos \beta$$

and assign T to the mutual value of these two expressions. Then we can find the tensions

$$T(x) = \frac{T}{\cos \alpha}, \qquad T(x + \Delta x) = \frac{T}{\cos \beta}.$$

Substitution of these expressions into Eq. (2) gives us

$$-T \tan \alpha + T \tan \beta = f(x) \, \Delta x.$$

Now, $\tan \alpha$ is just the slope of the cable at x; that is, $\tan \alpha = u'(x)$, and $\tan \beta = u'(x + \Delta x)$. We can rewrite the last equation above as

$$T(u'(x + \Delta x) - u'(x)) = f(x) \, \Delta x.$$

Dividing through by Δx yields

$$T \frac{u'(x + \Delta x) - u'(x)}{\Delta x} = f(x).$$

In the limit, as Δx approaches 0, the difference quotient in the left member becomes the second derivative of u, and the result is the equation

$$T \frac{d^2 u}{dx^2} = f(x), \tag{3}$$

which is valid for x in the range $0 < x < a$, where the cable is. In addition, $u(x)$ must satisfy the boundary conditions

$$u(0) = h_0, \qquad u(a) = h_1. \tag{4}$$

About the load $f(x)$, we might first assume that the cable is hanging under its own weight of w units of weight per unit length of cable. Then in Eq. (2) we should put

$$f(x) \, \Delta x = w \frac{\Delta s}{\Delta x} \, \Delta x$$

where s represents arclength along the cable. In the limit, as Δx approaches 0, $\Delta s/\Delta x$ has the limit

$$\lim_{\Delta x \to 0} \frac{\Delta s}{\Delta x} = \sqrt{1 + \left(\frac{du}{dx}\right)^2}.$$

Therefore, with this assumption the boundary value problem that determines the shape of the cable is

$$\frac{d^2u}{dx^2} = \frac{w}{T}\sqrt{1 + \left(\frac{du}{dx}\right)^2}, \qquad 0 < x < a \tag{5}$$

$$u(0) = h_0, \qquad u(a) = h_1. \tag{6}$$

Notice that the differential equation is nonlinear.

Another case arises when the cable supports a load uniformly distributed in the horizontal direction, as given by

$$f(x)\,\Delta x = w\,\Delta x.$$

This is approximately true for a suspension bridge. The boundary value problem to be solved is then

$$\frac{d^2u}{dx^2} = \frac{w}{T}, \qquad 0 < x < a \tag{7}$$

$$u(0) = h_0, \qquad u(a) = h_1.$$

It is a simple matter to solve this problem. The solution is

$$u(x) = \frac{w}{2T}(x^2 - ax) + \frac{h_1 - h_0}{a}x + h_0. \tag{8}$$

Buckling of a column. A long, slender column whose bottom end is hinged carries an axial load as as shown in Fig. 0.3. The upper end of the column can move up or down but not sideways. The displacement of the column from the vertical is given by $u(x)$. If the column were cut at any point x, an upward force P and a clockwise moment $Pu(x)$ would have to be applied to the upper part to keep it in equilibrium (see Fig. 0.4). This force and moment must be supplied by the lower part of the column.

It is known that the internal bending moment (positive when counterclockwise) in a column is given by the product

$$EI\,\frac{d^2u}{dx^2}$$

where E is Young's modulus and I is the moment of inertia of the cross-

Figure 0.3   Column carrying load *P*.

sectional area. ($I = b^4/12$ for a column whose cross section is a square of side
b.) Thus equating the external moment to the internal moment gives the
differential equation

$$EI \frac{d^2u}{dx^2} = -Pu, \qquad 0 < x < a \tag{9}$$

which, together with the boundary conditions

$$u(0) = 0, \qquad u(a) = 0, \tag{10}$$

determines the function $u(x)$.

Figure 0.4   Section of column showing forces and moments.

In order to study this problem more conveniently, we set

$$P/EI = \lambda^2,$$

so that the differential equation becomes

$$\frac{d^2u}{dx^2} + \lambda^2 u = 0, \qquad 0 < x < a. \tag{11}$$

Now, the general solution of this differential equation is

$$u(x) = c_1 \cos \lambda x + c_2 \sin \lambda x.$$

Since $u(0) = 0$, we must choose $c_1 = 0$, leaving $u(x) = c_2 \sin \lambda x$. The second boundary condition requires that

$$u(a) = c_2 \sin \lambda a = 0.$$

If $\sin \lambda a$ is not 0, the only possibility is that $c_2 = 0$. In this case we find that the solution is

$$u(x) \equiv 0, \qquad 0 < x < a.$$

Physically, this means that the column stands straight and transmits the load to its support, as it was probably intended to do.

Something quite different happens if $\sin \lambda a = 0$, for then any choice of c_2 gives a solution. The physical manifestation of this case is that the column assumes a sinusoidal shape and may then collapse, or buckle, under the axial load. Mathematically, the condition $\sin \lambda a = 0$ means that λa is an integer multiple of π, since $\sin \pi = 0$, $\sin 2\pi = 0$, etc., and integer multiples of π are the only arguments for which the sine function is 0. The equation $\lambda a = \pi$, in terms of the original parameters, is

$$\sqrt{\frac{P}{EI}}\, a = \pi.$$

It is reasonable to think of E, I, and a as given quantities; thus it is the force

$$P = EI(\pi/a)^2,$$

called the critical or Euler load, which causes the buckling. The higher critical loads, corresponding to $\lambda a = 2\pi$, $\lambda a = 3\pi$, etc., are so unstable as to be of no physical interest in this problem.

Conduction of heat. A long rod of uniform material and cross section conducts heat along its axial direction (see Fig. 0.5). We assume that the temperature in the rod, $u(x)$, does not change in time. A heat balance ("what goes in must come out") applied to a slice of the rod between x and $x + \Delta x$

Figure 0.5   Cylinder of heat-conducting material.

(see Fig. 0.6) shows that the heat flow rate q, measured in units of heat per unit time per unit area, obeys the equation

$$q(x) \cdot A + r(x)A \, \Delta x = q(x + \Delta x) \cdot A \tag{12}$$

in which A is the cross-sectional area and r is the rate at which heat enters the slice by means other than conduction through the two faces. For instance, if heat is generated in the slice by an electric current I, we might have

$$r(x)A \, \Delta x = \theta I^2 R \, \Delta x \tag{13}$$

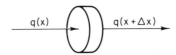

Figure 0.6   Section cut from heat-conducting cylinder showing heat flow.

where R is the resistance of the rod per unit length and θ is the factor of conversion from units of electrical energy to units of heat: for example, $\theta = 860$ cal/watt-hr. If heat is lost through the cylindrical surface of the rod by convection to a surrounded medium at temperature T, then $r(x)$ would be given by "Newton's law of cooling,"

$$r(x)A \, \Delta x = -h(u(x) - T)C \, \Delta x, \tag{14}$$

where C is the circumference of the road and h is the heat transfer coefficient. (This minus sign appears because, if $u(x) > T$, heat actually leaves the rod.)

Equation (12) may be altered algebraically to read

$$\frac{q(x + \Delta x) - q(x)}{\Delta x} = r(x).$$

and limiting process leaves

$$\frac{dq}{dx} = r(x). \tag{15}$$

The unknown function $u(x)$ does not appear in Eq. (15). However, a well-known experimental law (Fourier's law) says that the heat-flow rate through a unit area of material is directly proportional to the temperature difference and inversely proportional to thickness. In the limit this law takes the form

$$q = -\kappa \frac{du}{dx}. \tag{16}$$

The minus sign expresses the fact that heat moves from hotter toward cooler regions.

Combining Eqs. (15) and (16) gives the differential equation

$$-\kappa \frac{d^2u}{dx^2} = r(x), \qquad 0 < x < a, \tag{17}$$

where a is the length of the rod and the conductivity κ is assumed to be constant.

If the two ends of the rod are held at constant temperature, the boundary conditions on u would be

$$u(0) = T_0, \qquad u(a) = T_1. \tag{18}$$

On the other hand, if heat were supplied at $x = 0$ (by a heating coil, for instance), the boundary condition there would be

$$-\kappa A \frac{du}{dx}(0) = H, \tag{19}$$

where H is measured in units of heat per unit time.

As an example, we solve the problem

$$-\kappa \frac{d^2u}{dx^2} = -hu(x)C/A, \qquad 0 < x < a \tag{20}$$

$$u(0) = T_0, \qquad u(a) = T_0. \tag{21}$$

(Physically, the rod is losing heat to a surrounding medium at temperature 0, while both ends are held at the same temperature T_0.) If we designate $\mu^2 = hC/\kappa A$, the differential equation becomes

$$\frac{d^2u}{dx^2} - \mu^2 u = 0, \qquad 0 < x < a,$$

with solution

$$u(x) = c_1 \cosh \mu x + c_2 \sinh \mu a.$$

Application of the boundary condition at $x = 0$ gives $c_1 = T_0$; the second boundary condition requires that

$$u(a) = T_0 = T_0 \cosh \mu a + c_2 \sinh \mu a.$$

Thus $c_2 = T_0(1 - \cosh \mu a)/\sinh \mu a$ and

$$u(x) = T_0 \left(\cosh \mu x + \frac{1 - \cosh \mu a}{\sinh \mu a} \sinh \mu x \right).$$

Exercises

1. Of these three boundary value problems, one has no solution, one has exactly one solution, and one has an infinite number of solutions. Which is which?

    a. $\dfrac{d^2 u}{dx^2} + u = 0,\ u(0) = 0,\ u(\pi) = 0$ $v = A \sin x \quad (\infty)$

    b. $\dfrac{d^2 u}{dx^2} + u = 1,\ u(0) = 0,\ u(1) = 0$ $1\ \text{sol}^{\underline{n}}$

    c. $\dfrac{d^2 u}{dx^2} + u = 0,\ u(0) = 0,\ u(\pi) = 1$ $no\ \text{sol}^{\underline{n}}$

2. Find the Euler buckling load of a steel column with a 2 in. × 3 in. rectangular cross section. The parameters are $E = 30 \times 10^6$ lb/in^2, $I = 2$ in^4, $a = 10$ ft.

3. Find all values of the parameter λ for which these homogeneous boundary value problems have a solution other than $u(x) \equiv 0$.

    a. $\dfrac{d^2 u}{dx^2} + \lambda^2 u = 0,\ u(0) = 0,\ \dfrac{du}{dx}(a) = 0$

    b. $\dfrac{d^2 u}{dx^2} + \lambda^2 u = 0,\ \dfrac{du}{dx}(0) = 0,\ u(a) = 0$

    c. $\dfrac{d^2 u}{dx^2} + \lambda^2 u = 0,\ \dfrac{du}{dx}(0) = 0,\ \dfrac{du}{dx}(a) = 0$

4. Verify, by differentiating and substituting, that

    $$u(x) = c' + \frac{1}{\mu} \cosh \mu(x + c)$$

    is the general solution of the differential equation (5). (Here $\mu = w/T$. The graph of $u(x)$ is called a catenary.)

5. Find the values of c and c' for which the function $u(x)$ in Exercise 4 satisfies the conditions

$$u(0) = h, \qquad u(a) = h.$$

6. The ends of a taut string are fixed to the centers of synchronized rotating shafts (the axis of rotation of both shafts coincides with the string). If x denotes distance measured from one end along the axis of rotation and $u(x)$ denotes the radial distance from the axis to the string, then $u(x)$ satisfies the boundary value problem

$$T\frac{d^2u}{dx^2} = -\rho\omega^2 u, \qquad 0 < x < a$$

$$u(0) = 0, \qquad u(a) = 0.$$

In this problem T is tension, ω is the angular velocity, and ρ is the density of the string per unit length.

a. Derive the differential equation by balancing the inward component of tension against centrifugal force.

b. Find those values of ω for which a nonzero solution of the boundary value problem exists (critical speeds).

c. If the end at $x = a$ is attached to a spring with spring constant k, the second boundary condition would be

$$T\frac{du}{dx}(a) = -ku(a).$$

Answer the question of part (b) with this boundary condition instead of $u(a) = 0$.

7. The temperature $u(x)$ in a cooling fin satisfies the differential equation

$$\frac{d^2u}{dx^2} = \frac{hC}{\kappa A}(u - T), \qquad 0 < x < a$$

and boundary conditions

$$u(0) = T_0, \qquad -\kappa\frac{du}{dx}(a) = h(u(a) - T).$$

That is, the temperature at the left end is held at $T_0 > T$, while the surface of the rod and its right end exchange heat with a surrounding medium at temperature T. Find $u(x)$, taking C/A numerically equal to 1.

8. Calculate the limit as a tends to infinity of $u(x)$, the solution of the problem in Exercise 7. Is the result physically reasonable?

9. In an electrical heating element, the temperature satisfies the boundary value problem

$$\frac{d^2u}{dx^2} = \frac{hC}{\kappa A}(u - T) - \theta \frac{I^2 R}{\kappa A}, \qquad 0 < x < a$$

$$u(0) = T, \qquad u(a) = T.$$

Find $u(x)$.

10. Verify that the solution of the problem given in Eqs. (20) and (21) can also be written

$$u(x) = T_0 \frac{\cosh \mu(x - \frac{1}{2}a)}{\cosh(\mu a/2)}.$$

4   Singular boundary value problems

It sometimes happens that the differential equation in a boundary value problem has a singular point at one boundary. Recall that a point x_0 is a (regular) singular point of the differential equation

$$u'' + b(x)u' + p(x)u = f(x)$$

if the products

$$(x - x_0)b(x), \qquad (x - x_0)^2 p(x)$$

both have Taylor series expansions centered at x_0, but either $b(x)$ or $p(x)$ or both become infinite at x_0. For example, the point $x_0 = 1$ is a regular singular point of the differential equation

$$(1 - x)u'' + u' + xu = 0,$$

since both

$$b(x) = \frac{1}{1 - x} \quad \text{and} \quad p(x) = \frac{x}{1 - x}$$

become infinite at $x = 1$, but $(x - 1)b(x)$ and $(x - 1)^2 p(x)$ both have Taylor series expansions about the center $x = 1$. Another convenient example is provided by the Cauchy–Euler equation of Section 1, which has a regular singular point at the origin.

This situation typically arises when a boundary point is a mathematical boundary without being a physical boundary. For instance, a circular disk of radius c may be described in polar (r, θ) coordinates as occupying the region $0 \leq r \leq c$. The origin, at $r = 0$, is a mathematical boundary, yet physically this point is in the interior of the disk.

At a singular point, one cannot specify a value for $u(x_0)$, the solution of the differential equation, or for its derivative. However, it is usually necessary to require that both $u(x_0)$ and $u'(x_0)$ be finite, or bounded. Tacitly, we *always* require that the solution and its derivative be finite at *every* point of the interval where we are solving a differential equation. But when a singular point is a boundary point of that interval, we enforce the condition explicitly. In the example below we shall see how these conditions act so as to make the solution of a boundary value problem unique.

Radial heat flow. Suppose a long cylindrical bar, surrounded by a medium at temperature T, carries an electrical current. If heat flows in the radial direction much faster than in the axial direction, the temperature $u(r)$ in the rod may be described by the problem

$$\frac{1}{r} \frac{d}{dr} \left(r \frac{du}{dr} \right) = -H, \qquad 0 \leq r < c, \tag{1}$$

$$u(c) = T. \tag{2}$$

Here, c is the radius of the rod, r is a polar coordinate, and H (constant) is proportional to the electrical power being converted into heat.

In this problem, only the physical boundary condition has been noted. The mathematical boundary $r = 0$ is a singular point, as is clear from the differential equation in the form

$$\frac{d^2 u}{dr^2} + \frac{1}{r} \frac{du}{dr} = -H.$$

Thus, at this point we will require that u and du/dr be finite:

$$u(0), u'(0) \text{ finite}. \tag{3}$$

Now the differential equation (1) is easy to solve. Multiply through by r and integrate once to find that

$$r \frac{du}{dr} = -H \frac{r^2}{2} + c_1.$$

Divide through this equation by r and integrate once more to determine that

$$u(r) = -H \frac{r^2}{4} + c_1 \ln r + c_2.$$

Application of the special condition, that $u(0)$ and $u'(0)$ be finite, immediately tells us that $c_1 = 0$; for both $\ln r$ and its derivative $1/r$ become infinite as r approaches 0.

The physical boundary condition, Eq. (2), says that

$$u(c) = -H\frac{c^2}{4} + c_2 = T.$$

Hence $c_2 = Hc^2/4 + T$, and the complete solution is

$$u(r) = H\frac{(c^2 - r^2)}{4} + T. \tag{4}$$

From this example it is clear that the "artificial" boundary condition, boundedness of $u(r)$ at the singular point $r = 0$, works just the way an ordinary boundary condition works at an ordinary (not singular) point. It gives one condition to be fulfilled by the unknown constants c_1 and c_2, which are then completely determined by the second boundary condition.

Another type of singular boundary value problem is one for which the interval of interest is infinite. (Of course, this is always a mathematical abstraction that cannot be realized physically.) For instance, on the interval $0 < x < \infty$, sometimes called a semi-infinite interval, since it does have one finite endpoint, a boundary condition would normally be imposed at $x = 0$. At the other "end," no boundary condition is imposed, since no boundary exists. But we normally require that both $u(x)$ and $u'(x)$ remain bounded as x increases. In precise terms, we require that there exist constants M and M' for which

$$|u(x)| \leq M \quad \text{and} \quad |u'(x)| \leq M'$$

are both satisfied for all x, no matter how large. We never identify M or M', and the entire condition is usually written

$$u(x) \text{ and } u'(x) \qquad \text{bounded as } x \to \infty.$$

Cooling fin. A long cooling fin has one end held at a constant temperature T_0 and exchanges heat with a medium at temperature T through convection. The temperature $u(x)$ in the fin, satisfies the requirements

$$\frac{d^2 u}{dx^2} = -\frac{hC}{\kappa A}(u - T), \qquad 0 < x \tag{5}$$

$$u(0) = T_0 \tag{6}$$

(see Section 3). As the problem has been posed for a semi-infinite interval (because the fin is very long and, perhaps, to mask our ignorance of what is happening at the other physical end), we must also impose the condition

$$u(x), u'(x) \qquad \text{bounded as } x \to \infty. \tag{7}$$

Now, the general solution of the differential equation (5) is

$$u(x) = T + c_1 \cosh \mu x + c_2 \sinh \mu x,$$

where $\mu = \sqrt{hC/\kappa A}$. The boundary condition at $x = 0$ requires that

$$u(0) = T + c_1 = T_0.$$

The boundedness condition, Eq. (7), requires that

$$c_2 = -c_1.$$

The reason for this is that of all the linear combinations of cosh and sinh, the only one that is bounded as $x \to \infty$ is

$$\cosh \mu x - \sinh \mu x = e^{-\mu x}.$$

The final solution is easily found to be

$$u(x) = T + (T_0 - T)(\cosh \mu x - \sinh \mu x).$$

Satisfying the boundedness condition would have been simpler, had we expressed the general solution of the differential equation (5) as

$$u(x) = T + c_1' e^{\mu x} + c_2' e^{-\mu x}.$$

We would have seen immediately that choosing $c_1' = 0$ is the only way to satisfy the boundedness condition.

Exercises

1. Put each equation in the form

$$u'' + bu' + pu = f$$

and identify the singular point(s).

a. $\dfrac{1}{r} \dfrac{d}{dr}\left(r \dfrac{du}{dr}\right) = u$

b. $\dfrac{d}{dx}\left((1 - x^2) \dfrac{du}{dx}\right) = 0$

c. $\dfrac{d}{d\phi}\left(\sin \phi \dfrac{du}{d\phi}\right) = \sin \phi \, u$

d. $\dfrac{1}{\rho^2}\dfrac{d}{d\rho}\left(\rho^2\dfrac{du}{d\rho}\right) = -\lambda^2 u$

2. The temperature u in a large object having a hole of radius c in the middle may be said to obey the equations

$$u = c_1 \ln r + c_2$$

$$\frac{1}{r}\frac{d}{dr}\left(r\frac{du}{dr}\right) = 0, \qquad r > c$$

$$u(c) = T. \qquad U = T$$

Solve the problem adding the appropriate boundedness condition.

3. Compact kryptonite produces heat at a rate of H cal/sec cm^3. If a sphere (radius c) of this material transfers heat by convection to a surrounding medium at temperature T, the temperature $u(\rho)$ in the sphere satisfies the boundary value problem

$$\frac{1}{\rho^2}\frac{d}{d\rho}\left(\rho^2\frac{du}{d\rho}\right) = -H/\kappa, \qquad 0 < \rho < c$$

$$-\kappa\frac{du}{d\rho}(c) = h(u(c) - T).$$

Supply the proper boundedness condition and solve. What is the temperature at the center of the sphere?

4. (Critical radius) The neutron flux u in a sphere of uranium obeys the differential equation

$$\frac{\lambda}{3}\frac{1}{\rho^2}\frac{d}{d\rho}\left(\rho^2\frac{du}{d\rho}\right) + (k-1)Au = 0$$

in the range $0 < \rho < a$, where λ is the effective distance traveled by a neutron between collisions, A is called the absorption cross section, and k is the number of neutrons produced by a collision during fission. In addition, the neutron flux at the boundary of the sphere is 0. Make the substitutions $u = v/\rho$ and $3(k-1)A/\lambda = \mu^2$, and determine the differential equation satisfied by $v(\rho)$.

5. Solve the equation found in Exercise 4 and then find $u(\rho)$ that satisfies the boundary value problem (with boundedness condition) stated in Exercise 4. For what radius a is the solution not identically 0?

Miscellaneous exercises

In Exercises 1–15, solve the given boundary value problem, supplying boundedness conditions where necessary.

1. $\dfrac{d^2u}{dx^2} - \gamma^2 u = 0, \quad 0 < x < a$

$u(0) = T_0, \quad u(a) = T_1$

2. $\dfrac{d^2u}{dx^2} - r = 0, \quad 0 < x < a$

$u(0) = T_0, \quad \dfrac{du}{dx}(a) = 0$

3. $\dfrac{d^2u}{dx^2} = 0, \quad 0 < x < a$

$u(0) = T_0, \quad \dfrac{du}{dx}(a) = 0$

4. $\dfrac{d^2u}{dx^2} - \gamma^2 u = 0, \quad 0 < x < a$

$\dfrac{du}{dx}(0) = 0, \quad u(a) = T_1$

5. $\dfrac{1}{r}\dfrac{d}{dr}\left(r\dfrac{du}{dr}\right) = -p, \quad 0 < r < a$

$u(a) = 0$

6. $\dfrac{1}{r}\dfrac{d}{dr}\left(r\dfrac{du}{dr}\right) = 0, \quad a < r < b$

$u(a) = T_0, \quad u(b) = T_1$

7. $\dfrac{1}{\rho^2}\dfrac{d}{d\rho}\left(\rho^2\dfrac{du}{d\rho}\right) = -H, \quad 0 < \rho < a$

$u(a) = T_0$

8. $\dfrac{1}{r}\dfrac{d}{dr}\left(r\dfrac{du}{dr}\right) = 0, \quad a < r < \infty$

$u(a) = T$

9. $\dfrac{d^2u}{dx^2} - \gamma^2(u - T) = 0, \quad 0 < x < a$

$\dfrac{du}{dx}(0) = 0, \quad u(a) = T_1$

10. $\dfrac{d^2u}{dx^2} - \gamma^2 u = 0, \quad 0 < x < \infty$

$u(0) = T$

11. $\dfrac{d^2u}{dx^2} = \gamma^2(u - T_0), \quad 0 < x < \infty$

$u(0) = T$

12. $\dfrac{d}{dx}\left(x^3 \dfrac{du}{dx}\right) = -k, \quad a < x < b$

$u(a) = 0, \quad u(b) = 0 \quad$ (Note: $0 < a$.)

13. In this problem, h is the ground water level between two trenches in which water is held at constant levels. Note that the equation is non-linear.

$$\dfrac{d}{dx}\left(h \dfrac{dh}{dx}\right) + e = 0, \qquad 0 < x < a$$

$$h(0) = h_0, \qquad h(a) = h_1$$

14. $\dfrac{d^4u}{dx^4} = w, \quad 0 < x < a \quad$ (w a constant)

$u(0) = 0, \quad u(a) = 0, \quad \dfrac{d^2u}{dx^2}(0) = 0, \quad \dfrac{d^2u}{dx^2}(a) = 0$

15. $\dfrac{d^4u}{dx^4} + \dfrac{k}{EI} u = w, \quad 0 < x < \infty \quad$ (w a constant)

$u(0) = 0, \quad \dfrac{d^2u}{dx^2}(0) = 0$

16. Show that any two of the four functions $\sinh \lambda x$, $\sinh \lambda(a - x)$, $\cosh \lambda x$, $\cosh \lambda(a - x)$ are independent solutions of the differential equation

$$\phi'' - \lambda^2 \phi = 0.$$

17. In this problem, u is the temperature in a wall composed of two substances. Find $u(x)$.

$$\frac{d^2u}{dx^2} = 0, \qquad 0 < x < \alpha a \quad \text{and} \quad \alpha a < x < a$$

$$u(0) = T_0, \qquad u(a) = T_1$$

$$\kappa_1 \frac{du}{dx}(\alpha a -) = \kappa_2 \frac{du}{dx}(\alpha a +)$$

$$u(\alpha a -) = u(\alpha a +)$$

The last two conditions say that the heat flow rate and the temperature are both continuous across the interface at $x = \alpha a$.

18. Find the general solution of the differential equation

$$\frac{1}{x^2} \frac{d}{dx} \left(x^2 \frac{du}{dx} \right) + ku = 0$$

for the cases $k = \lambda^2$ and $k = -p^2$. (Hint: let $u(x) = v(x)/x$ and find the equation that $v(x)$ satisfies.)

19. Find the solution of the boundary value problem

$$e^x \frac{d}{dx} \left(e^x \frac{du}{dx} \right) = -1, \qquad 0 < x < a$$

$$u(0) = 0, \qquad u(a) = 0.$$

20. Solve the boundary value problem

$$\frac{1}{r} \frac{d}{dr} \left(r \frac{du}{dr} \right) = -r^k, \qquad 0 < r < a$$

$$u(0) \text{ bounded} \quad \text{and} \quad u(a) = 0.$$

21. Solve the differential equation

$$\frac{d^2u}{dx^2} = p^2u, \qquad 0 < x < a$$

subject to boundary conditions

a. $u(0) = 0, u(a) = 1$

b. $u(0) = 1, u(a) = 0$

c. $u'(0) = 0, u(a) = 1$

d. $u(0) = 1, u'(a) = 0$

e. $u'(0) = 1, u'(a) = 0$

f. $u'(0) = 0, u'(a) = 1.$

22. Solve the integro-differential boundary value problem

$$\frac{d^2u}{dx^2} = \gamma^2 \left(u - \int_0^1 u(x) \, dx \right) \qquad 0 < x < 1$$

$$\frac{du}{dx}(0) = 0, \qquad u(1) = T.$$

(Hint: Look for a solution in the form $u(x) = A \cosh \gamma x + B \sinh \gamma x + C$.)

23. Use variation of parameters to find a second independent solution of each of the following differential equations. In each case, one solution is given in parentheses.

a. $\dfrac{d^2u}{dx^2} - \dfrac{2x}{1 - x^2}\dfrac{du}{dx} + \dfrac{2}{1 - x^2}u = 0 \quad (u = x)$

b. $\dfrac{d^2u}{dx^2} + \dfrac{1 - x}{x}\dfrac{du}{dx} + \dfrac{1}{x}u = 0 \quad (u = 1 - x)$

24. By applying the method of variation of parameters, derive this formula for a particular solution of the differential equation

$$\frac{d^2u}{dx^2} - \gamma^2 u = f(x)$$

$$u(x) = \int_0^x f(x') \frac{\sinh \gamma(x - x')}{\gamma} \, dx'.$$

chapter 1
Fourier
Series
and
Integrals

1   Periodic functions and Fourier series

A function f is said to be periodic with period p if (1) $f(x)$ has been defined for all x, and (2) $f(x + p) = f(x)$ for all x. The familiar functions $\sin x$ and $\cos x$ are simple examples of periodic functions with period 2π, and the functions $\sin(2\pi x/p)$ and $\cos(2\pi x/p)$ are periodic with period p.

A periodic function has many periods, for if $f(x) = f(x + p)$ then also

$$f(x) = f(x + p) = f(x + 2p) = \cdots = f(x + np),$$

where n is any integer. Thus $\sin x$ has periods 2π, 4π, 6π, $\ldots$, $n \cdot 2\pi$. The period of a periodic function is generally taken to be positive, but the periodicity condition holds for negative as well as positive changes in the argument. That is to say, $f(x - p) = f(x)$ for all x, since $f(x) = f(x - p + p) = f(x - p)$. Also

$$f(x) = f(x - p) = f(x - 2p) = \cdots = f(x - np).$$

The definition of periodic says essentially that functional values repeat themselves. This implies that the graph of a periodic function can be drawn

30

for all x by making a template of the graph on any interval of length p and then copying the graph from the template up and down the x axis (see Fig. 1.1).

Many functions which occur in engineering and physics are periodic in space or time—for example, acoustic waves—and in order to understand them better it is often desirable to represent them in terms of the very simple periodic functions 1, $\sin x$, $\cos x$, $\sin 2x$, $\cos 2x$, and so forth. *All* of these functions have the common period 2π, although each has other periods as well.

If f is periodic with period 2π, then we attempt to represent f in the form of an infinite series

$$f(x) = a_0 + \sum_{n=1}^{\infty} (a_n \cos nx + b_n \sin nx). \qquad (1)$$

Each term of the series has period 2π, so if the sum of the series exists, it will be a function of period 2π. There are two questions to be answered: (a) What values must a_0, a_n, b_n have; and (b) If the appropriate values are assigned to the coefficients, does the series actually represent the given function $f(x)$?

On the face of it, the first question is tremendously difficult, for Eq. (1) represents an equation in an infinite number of variables. But a reasonable answer can be found easily by using the following *orthogonality** relations:

$$\int_{-\pi}^{\pi} \sin nx \, dx = 0$$

$$\int_{-\pi}^{\pi} \cos nx \, dx = \begin{cases} 0, & n \neq 0 \\ 2\pi, & n = 0 \end{cases}$$

$$\int_{-\pi}^{\pi} \sin nx \cos mx \, dx = 0 \qquad (2)$$

$$\int_{-\pi}^{\pi} \sin nx \sin mx \, dx = \begin{cases} 0, & n \neq m \\ \pi, & n = m \end{cases}$$

$$\int_{-\pi}^{\pi} \cos nx \cos mx \, dx = \begin{cases} 0, & n \neq m \\ \pi, & n = m \neq 0. \end{cases}$$

The fundamental idea is that if the equality proposed in Eq. (1) is to be a real equality, then both sides must give the same result after the same operation. The orthogonality relations then suggest operations which simplify the right-hand side of Eq. (1). Namely, we multiply both sides of the proposed equation by one of the functions that appears there and integrate from $-\pi$ to π. (We must assume that the integration of the series can be carried out term-by-term. This is sometimes difficult to justify, but we do it anyway.)

---

* The word "orthogonality" should not be thought of in the geometric sense.

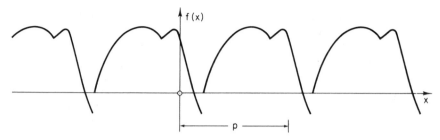

Figure 1.1 A periodic function of period p.

Multiplying both sides of Eq. (1) by the constant $1(=\cos 0x)$ and integrating from $-\pi$ to π, we find

$$\int_{-\pi}^{\pi} f(x)\, dx = a_0 \int_{-\pi}^{\pi} 1\, dx + \sum_{n=1}^{\infty} \int_{-\pi}^{\pi} (a_n \cos nx + b_n \sin nx)\, dx.$$

Each of the terms in the integrated series is zero, so the right-hand side of this equation reduces to $2\pi \cdot a_0$, giving

$$a_0 = \frac{1}{2\pi} \int_{-\pi}^{\pi} f(x)\, dx.$$

(This means that a_0 is the mean value of $f(x)$ over one period.)

Now multiplying each side of Eq. (1) by $\sin mx$, where m is a fixed integer, and integrating from $-\pi$ to π, we find

$$\int_{-\pi}^{\pi} f(x) \sin mx\, dx = \int_{-\pi}^{\pi} a_0 \sin mx\, dx + \sum_{n=1}^{\infty} \int_{-\pi}^{\pi} a_n \cos nx \sin mx\, dx$$

$$+ \sum_{n=1}^{\infty} \int_{-\pi}^{\pi} b_n \sin nx \sin mx\, dx.$$

All terms containing a_0 or a_n disappear, according to Eq. (2). Furthermore, of all those containing a b_n, the only one which is not zero is that one in which $n = m$. (Notice that n is a summation index and runs through all the integers $1, 2, \ldots$. We chose m to be a fixed integer, so $n = m$ once.) We now have the formula

$$b_m = \frac{1}{\pi} \int_{-\pi}^{\pi} f(x) \sin mx\, dx.$$

By multiplying both sides of Eq. (1) by $\cos mx$ (m is a fixed integer) and integrating, we also find

$$a_m = \frac{1}{\pi} \int_{-\pi}^{\pi} f(x) \cos mx\, dx.$$

We can now summarize our results. In order for the proposed equality

$$f(x) = a_0 + \sum_{n=1}^{\infty} (a_n \cos nx + b_n \sin nx) \qquad (1')$$

to hold, the as and bs must be chosen according to the formulas

$$a_0 = \frac{1}{2\pi} \int_{-\pi}^{\pi} f(x)\, dx \qquad (3)$$

$$a_n = \frac{1}{\pi} \int_{-\pi}^{\pi} f(x) \cos nx\, dx \qquad (4)$$

$$b_n = \frac{1}{\pi} \int_{-\pi}^{\pi} f(x) \sin nx\, dx. \qquad (5)$$

When the coefficients are chosen this way, the right-hand side of Eq. (1) is called the *Fourier series* of f. The as and bs are called *Fourier coefficients*. We have not yet answered question (b) about equality, so we write

$$f(x) \sim a_0 + \sum (a_n \cos nx + b_n \sin nx)$$

to indicate that the Fourier series corresponds to $f(x)$.

Example   Suppose that $f(x)$ is periodic with period 2π and is given by the formula $f(x) = x$ in the interval $-\pi < x < \pi$ (see Fig. 1.2). According to our formulas,

$$a_0 = \frac{1}{2\pi} \int_{-\pi}^{\pi} f(x)\, dx = \frac{1}{2\pi} \int_{-\pi}^{\pi} x\, dx = 0$$

$$a_n = \frac{1}{\pi} \int_{-\pi}^{\pi} f(x) \cos nx\, dx = \frac{1}{\pi} \int_{-\pi}^{\pi} x \cos nx\, dx$$

$$= \frac{1}{\pi} \left[\frac{\cos nx}{n^2} + \frac{x \sin nx}{n} \right]_{-\pi}^{\pi} = 0$$

$$b_n = \frac{1}{\pi} \int_{-\pi}^{\pi} f(x) \sin nx\, dx = \frac{1}{\pi} \int_{-\pi}^{\pi} x \sin nx\, dx$$

$$= \frac{1}{\pi} \left[\frac{\sin nx}{n^2} - \frac{x \cos nx}{n} \right]_{-\pi}^{\pi}$$

$$= \frac{1}{\pi} \frac{(-2\pi)\cos n\pi}{n} = \frac{2}{n}(-1)^{n+1}.$$

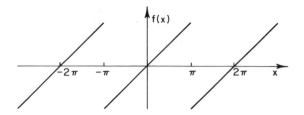

Figure 1.2 $f(x) = x$, $-\pi < x < \pi$, f periodic with period 2π.

Thus, for this function, we have

$$f(x) \sim \sum_{n=1}^{\infty} \frac{2(-1)^{n+1}}{n} \sin nx$$

$$\sim 2(\sin x - \tfrac{1}{2} \sin 2x + \tfrac{1}{3} \sin 3x - + \cdots).$$

Exercises

1. Find the Fourier coefficients of the functions given below. All are supposed to be periodic with period 2π. Sketch the graph of the function.

   a. $f(x) = x$, $\quad -\pi < x < \pi$ $\qquad$ b. $f(x) = |x|$, $\quad -\pi < x < \pi$

   c. $f(x) = \begin{cases} 0, & -\pi < x < 0 \\ 1, & 0 < x < \pi \end{cases}$ $\qquad$ d. $f(x) = |\sin x|$

2. Sketch for at least two periods the graphs of the functions defined by:

   a. $f(x) = x$, $\qquad -1 < x \le 1$, $\quad f(x+2) = f(x)$

   b. $f(x) = \begin{cases} 0, & -1 < x \le 0, \\ x, & 0 < x < 1, \end{cases}$ $\quad f(x+2) = f(x)$

   c. $f(x) = \begin{cases} 0, & -\pi < x \le 0, \\ 1, & 0 < x \le 2\pi, \end{cases}$ $\quad f(x+3\pi) = f(x)$

   d. $f(x) = \begin{cases} 0, & -\pi < x \le 0, \\ \sin x, & 0 < x \le \pi, \end{cases}$ $\quad f(x+2\pi) = f(x).$

3. Show that the constant function $f(x) = 1$ is periodic with every possible
   period $p > 0$.

4. Carry out the details of deriving the equation for a_m.

5. Suppose $f(x)$ has period p. Show that for any c,

$$\int_{c}^{c+p} f(x)\, dx = \int_{0}^{p} f(x)\, dx.$$

6. Suppose $f(x)$, $g(x)$ are periodic with a common period p. Show that $af(x) + bg(x)$ and $f(x) \cdot g(x)$ also are periodic with period p (a, b are constants).

7. Find the Fourier series of each of the following periodic functions. Integration is not necessary. $a_0 = \frac{1}{2}$

   a. $f(x) = \cos^2 x$
   b. $f(x) = \sin(x - \pi/6)$
   c. $f(x) = \sin x \cos 2x$

8. Verify that $\sin(2\pi x/p)$ and $\cos(2\pi x/p)$ are periodic with period p.

2   Arbitrary period and half-range expansions

In Section 1 we found a way to represent a periodic function of period 2π with a Fourier series. It is not necessary to restrict ourselves to this period. In fact, we may broaden the idea of Fourier series to include functions of any period by a simple rescaling of the variables. Let us suppose that a function f is periodic with period $2a$. (We use $2a$ in place of p for later convenience.) Then we may relate f to a series of the functions 1, $\sin(\pi x/a)$, $\cos(\pi x/a)$, $\sin(2\pi x/a)$, $\cos(2\pi x/a)$, $\ldots$, all having period $2a$, in the form

$$f(x) \sim a_0 + \sum_{n=1}^{\infty} a_n \cos(n\pi x/a) + b_n \sin(n\pi x/a).$$

The coefficients of this Fourier series may be determined either by scaling from the formulas of Section 1 or through the concept of orthogonality. In either case, the coefficients are

$$a_0 = \frac{1}{2a} \int_{-a}^{a} f(x)\, dx, \qquad a_n = \frac{1}{a} \int_{-a}^{a} f(x) \cos(n\pi x/a)\, dx$$

$$b_n = \frac{1}{a} \int_{-a}^{a} f(x) \sin(n\pi x/a)\, dx.$$

Example   Let $f(x) = |\sin \pi x|$, which is periodic with period 1. The Fourier coefficients of f are thus ($a = \frac{1}{2}$)

$$a_0 = \int_{-1/2}^{1/2} |\sin \pi x|\, dx = \frac{2}{\pi}$$

$$a_n = 2 \int_{-1/2}^{1/2} |\sin \pi x| \cos 2n\pi x\, dx = -\frac{4}{\pi} \frac{1}{4n^2 - 1}$$

$$b_n = 2 \int_{-1/2}^{1/2} |\sin \pi x| \sin 2n\pi x\, dx = 0$$

Consequently, the Fourier series of the function is

$$|\sin \pi x| \sim \frac{2}{\pi} - \frac{4}{\pi} \sum_{n=1}^{\infty} \frac{1}{4n^2 - 1} \cos 2n\pi x.$$

It is often necessary to represent by a Fourier series a function that has been defined only in a finite interval. We can justify such a representation by making the given function part of a periodic function. If the given function f is defined on the interval $-a < x < a$, we may construct $\bar{f}$, the *periodic extension* of period $2a$, by using the following definitions:

$$\bar{f}(x) = f(x), \qquad -a < x < a$$
$$\bar{f}(x) = f(x + 2a), \qquad -3a < x < -a$$
$$\bar{f}(x) = f(x - 2a), \qquad a < x < 3a$$

and so on, up and down the x-axis. Notice that the argument of f on the right-hand side always falls in the interval $-a < x < a$, where f was originally given. Graphically, this kind of extension amounts to making a template of the graph of f on $-a < x < a$ and then copying from the template in abutting intervals of length $2a$.

For the extended function with period $2a$, the formulas for the Fourier coefficients become

$$a_0 = \frac{1}{2a} \int_{-a}^{a} \bar{f}(x)\, dx, \qquad \begin{matrix} a_n \\ b_n \end{matrix} \Bigg\} = \frac{1}{a} \int_{-a}^{a} \bar{f}(x) \begin{Bmatrix} \cos(n\pi x/a) \\ \sin(n\pi x/a) \end{Bmatrix} dx.$$

If we are concerned with $f(x)$ only in the interval $-a < x < a$ where it was originally given, the process of periodic extension is strictly formal— the formulas for the coefficients involve f only on the original interval—and we may write

$$f(x) \sim a_0 + \sum a_n \cos(n\pi x/a) + b_n \sin(n\pi x/a), \qquad -a < x < a.$$

The inequality draws attention to the fact that f was defined only on the interval $-a$ to a.

Example Suppose $f(x) = x$ in the interval $-1 < x < 1$. The graph of its periodic extension (with period 2) is seen in Fig. 1.3, and the Fourier coefficients are

$$a_0 = 0, \qquad a_n = 0, \qquad b_n = \int_{-1}^{1} x \sin n\pi x\, dx = -\frac{2\cos n\pi}{n\pi} = \frac{2}{\pi}\frac{(-1)^{n+1}}{n}.$$

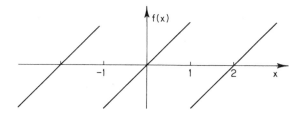

Figure 1.3 $f(x) = x$, $-1 < x < 1$, f periodic with period 2.

The sine and cosine functions which appear in a Fourier series have some special symmetry properties which are useful in evaluating the coefficients. The graph of the cosine function is symmetrical about the vertical axis, and that of the sine is antisymmetric. We formalize these properties with a definition.

Definition $g(x)$ is *even* if $g(-x) = g(x)$; $h(x)$ is *odd* if $h(-x) = -h(x)$.

Examples

> $\sin \alpha x$, x, x^3, and any odd power of x are all odd.
> $\cos \alpha x$, 1, x^2, and any even power of x are all even.

Most functions are neither even nor odd, but any function may be written as the sum of an even plus an odd function:

$$f(x) = \tfrac{1}{2}(f(x) + f(-x)) + \tfrac{1}{2}(f(x) - f(-x)).$$

It is a simple matter to show that the first term is an even function and the second is odd.

The important features about even and odd functions are:

1. The integral of an odd function over a symmetric interval is zero.

2. The integral of an even function over a symmetric interval is twice the integral over the right half of the interval. In symbols, we can write

$$\int_{-a}^{a} odd \; dx = 0$$

$$\int_{-a}^{a} even \; dx = 2 \int_{0}^{a} even \; dx.$$

The combinatorial properties of even and odd functions are

$$\text{even} \times \text{odd} = \text{odd}$$
$$\text{even} \times \text{even} = \text{even}$$
$$\text{odd} \times \text{odd} = \text{even}$$
$$\text{even} + \text{even} = \text{even}$$
$$\text{odd} + \text{odd} = \text{odd}.$$

Suppose now that g is an even function in the interval $-a < x < a$. Since the sine function is odd, and the product $g(x) \sin(n\pi x/a)$ is odd,

$$b_n = \frac{1}{a} \int_{-a}^{a} g(x) \sin(n\pi x/a) \, dx = 0.$$

That is, all the sine coefficients are zero. Also, since the cosine is even, so is $g(x) \cos(n\pi x/a)$, and then

$$a_n = \frac{1}{a} \int_{-a}^{a} g(x) \cos(n\pi x/a) \, dx = \frac{2}{a} \int_{0}^{a} g(x) \cos(n\pi x/a) \, dx.$$

Thus the cosine coefficients can be computed from an integral over the interval from 0 to a.

Parallel results hold for odd functions: the cosine coefficients are all zero and the sine coefficients can be simplified. We summarize the results:

Theorem If $g(x)$ is even on the interval $-a < x < a$ $(g(-x) = g(x))$, then

$$g(x) \sim a_0 + \sum a_n \cos(n\pi x/a), \qquad -a < x < a$$

$$\text{where} \quad a_0 = \frac{1}{a} \int_{0}^{a} g(x) \, dx, \qquad a_n = \frac{2}{a} \int_{0}^{a} g(x) \cos(n\pi x/a) \, dx.$$

If $h(x)$ is odd on the interval $-a < x < a$ $(h(-x) = -h(x))$, then

$$h(x) \sim \sum b_n \sin(n\pi x/a), \qquad -a < x < a$$

$$\text{where} \quad b_n = \frac{2}{a} \int_{0}^{a} h(x) \sin(n\pi x/a) \, dx.$$

Very frequently, a function given in an interval $0 < x < a$ must be represented in the form of a Fourier series. There are an infinite number of ways of doing this, but two ways are especially simple and useful: extending the given function to one defined on a symmetric interval $-a < x < a$ by making the extended function either odd or even.

Definition Let $f(x)$ be given for $0 < x < a$. The *odd extension* of f is defined by

$$f_o(x) = \begin{cases} f(x), & 0 < x < a \\ -f(-x), & -a < x < 0. \end{cases}$$

The *even extension* of f is defined by

$$f_e(x) = \begin{cases} f(x), & 0 < x < a \\ f(-x), & -a < x < 0 \end{cases}$$

Notice that if $-a < x < 0$, then $0 < -x < a$, so the functional values on the right are known from the given functions.

Graphically, the even extension is made by reflecting the graph in the vertical axis. The odd extension is made by reflecting first in the vertical then in the horizontal axis (see Fig. 1.4).

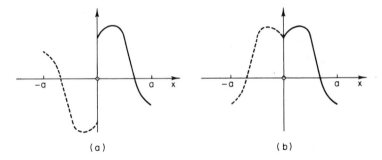

(a)   (b)

Figure 1.4 Odd and even extensions of a function originally given for $0 < x < a$.

Now the Fourier series of either extension may be calculated from the formulas above. Since f_e is even and f_o is odd, we have

$$f_e(x) \sim a_0 + \sum a_n \cos(n\pi x/a), \qquad -a < x < a$$
$$f_o(x) \sim \sum b_n \sin(n\pi x/a), \qquad -a < x < a.$$

If the series on the right converge, they actually represent periodic functions with period $2a$. The cosine series would represent the *even periodic* extension of f—the periodic extension of f_e; and the sine series would represent the *odd periodic* extension of f.

When the problem at hand is to represent the function $f(x)$ in the interval $0 < x < a$ where it was originally given, we may use either the Fourier sine series or the cosine series above, since both f_e and f_o coincide with f in that interval.

Thus we may summarize by saying: If $f(x)$ is given for $0 < x < a$, then

$$f(x) \sim a_0 + \sum a_n \cos(n\pi x/a), \qquad 0 < x < a$$

$$a_0 = \frac{1}{a} \int_0^a f(x)\,dx, \qquad a_n = \frac{2}{a} \int_0^a f(x) \cos(n\pi x/a)\,dx$$

and

$$f(x) \sim \sum b_n \sin(n\pi x/a), \qquad 0 < x < a$$

$$b_n = \frac{2}{a} \int_0^a f(x) \sin(n\pi x/a)\,dx.$$

These two representations are called *half-range expansions*. We shall need these, more than any other kind of Fourier series, in the applications we make later in this book.

Example Let us suppose that the function f has the formula

$$f(x) = x, \qquad 0 < x < 1.$$

Then the odd periodic extension of f is as shown in Figure 1.5 and the Fourier sine coefficients of f are

$$b_n = 2 \int_0^1 x \sin n\pi x\,dx = -\frac{2}{n\pi} \cos n\pi$$

The even periodic extension of f is shown in Figure 1.6. The Fourier cosine coefficients are

$$a_0 = \int_0^1 x\,dx = \frac{1}{2}$$

$$a_n = 2 \int_0^1 x \cos n\pi x\,dx = -\frac{2}{n^2 \pi^2}(1 - \cos n\pi).$$

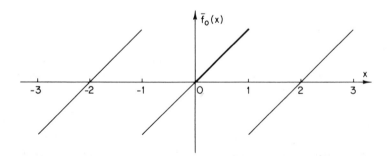

Figure 1.5 Odd periodic extension (period 2) of $f(x) = x$, $0 < x < 1$.

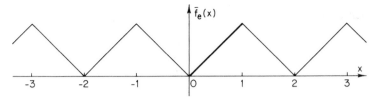

Figure 1.6   Even periodic extension (period 2) of $f(x) = x$, $0 < x < 1$.

The following six correspondences (we will later show them to be equalities) follow from the ideas of this section. Note that the inequalities showing the applicable range of x are crucial.

$$\sum_{n=1}^{\infty} \frac{-2 \cos n\pi}{n\pi} \sin n\pi x \sim \begin{cases} f(x) = x, & 0 < x < 1 \\ f_o(x) = x, & -1 < x < 1 \\ f_o(x), & -\infty < x < \infty \end{cases}$$

$$\frac{1}{2} - \sum_{n=1}^{\infty} \frac{2(1 - \cos n\pi)}{n^2 \pi^2} \cos n\pi x \sim \begin{cases} f(x) = x, & 0 < x < 1 \\ f_o(x) = |x|, & -1 < x < 1 \\ \bar{f_e}(x), & -\infty < x < \infty \end{cases}$$

Exercises

1. Find the Fourier series of each of the following functions. Sketch the graph of the periodic extension of f for at least two periods.

   a. $f(x) = |x|$, $-1 < x < 1$
   b. $f(x) = \begin{cases} -1, & -2 < x < 0 \\ 1, & 0 < x < 2 \end{cases}$
   c. $f(x) = x^2$, $-\tfrac{1}{2} < x < \tfrac{1}{2}$

2. Show that the functions $\cos(n\pi x/a)$ and $\sin(n\pi x/a)$ satisfy orthogonality relations similar to those given in Section 1.

3. Suppose a Fourier series is needed for a function defined in the interval $0 < x < 2a$. Show how to construct a periodic extension with period $2a$ and give formulas for the Fourier coefficients which use only integrals from 0 to $2a$. (Hint: see Exercise 5, Section 1.)

4. Show that the formula

$$e^x = \cosh x + \sinh x$$

   gives the decomposition of the function e^x into a sum of an even and an odd function.

5. Identify each of the following as being even, odd, or neither. Sketch.

    a. $f(x) = x$ **b.** $f(x) = |x|$
    c. $f(x) = |\cos x|$ **d.** $f(x) = \arc \sin x$
    e. $f(x) = x \cos x$ **f.** $f(x) = x + \cos(x + 1)$

6. If $f(x)$ is given in the interval $0 < x < a$, what other ways are there to extend it to a function on $-a < x < a$?

7. Find the Fourier series of the functions:

    a. $f(x) = x,$ $-1 < x < 1$ **b.** $f(x) = 1,$ $-2 < x < 2$
    c. $f(x) = \begin{cases} x, & -\frac{1}{2} < x < \frac{1}{2} \\ 1 - x, & \frac{1}{2} < x < \frac{3}{2}. \end{cases}$

8. Is it true that, if all the sine coefficients of a function f defined on $-a < x < a$ are zero, then f is even?

9. We know that if $f(x)$ is odd on the interval $-a < x < a$, its Fourier series is composed only of sines. What additional symmetry condition on f will make the sine coefficients with even indices be zero? Give an example.   symmetric about the origin.

10. Sketch both the even and odd extensions of the functions:

    a. $f(x) = 1,$ $0 < x < a$ **b.** $f(x) = x,$ $0 < x < a$
    c. $f(x) = \sin x,$ $0 < x < 1$ **d.** $f(x) = \sin x,$ $0 < x < \pi.$

11. Find the Fourier sine series and the cosine series for the functions given in Exercise 10. Sketch the even and odd periodic extensions for several periods.

12. Prove the orthogonality relations

$$\int_0^a \sin(n\pi x/a) \sin(m\pi x/a) \, dx = \begin{cases} 0, & n \neq m \\ a/2, & n = m \end{cases}$$

$$\int_0^a \cos(m\pi x/a) \cos(n\pi x/a) \, dx = \begin{cases} 0, & n \neq m \\ a/2, & n = m \neq 0 \\ a, & n = m = 0. \end{cases}$$

13. If $f(x)$ is continuous on the interval $0 < x < a$, is its even periodic extension continuous? What about the odd periodic extension? Check especially at $x = 0$ and $\pm a$.

3   Convergence of Fourier series

Now we are ready to take up the second question of Section 1: Does the Fourier series of a function actually represent that function? The word "represent" has many interpretations, but for most practical purposes, we really want to know the answer to the question:

If a value of x is chosen, the numbers $\cos(n\pi x/a)$ and $\sin(n\pi x/a)$ are computed for each n and inserted into the Fourier series of f, and the sum of the series is calculated, is that sum equal to the functional value $f(x)$?

In this section we shall state, without proof, some theorems which answer the question. But first we need a few definitions about limits and continuity.

The ordinary limit $\lim_{x \to x_0} f(x)$ can be rewritten as $\lim_{h \to 0} f(x_0 + h)$. Here, h may approach zero in any manner. But if h is required to be positive only, we get what is called the right-hand limit of f at x_0, defined by

$$f(x_0 +) = \lim_{\substack{h \to 0+}} f(x_0 + h) = \lim_{\substack{h \to 0 \\ h > 0}} f(x_0 + h).$$

The left-hand limit is defined similarly:

$$f(x_0 -) = \lim_{\substack{h \to 0-}} f(x_0 + h) = \lim_{\substack{h \to 0 \\ h < 0}} f(x_0 + h) = \lim_{\substack{h \to 0+}} f(x_0 - h).$$

If both left- and right-hand limits exist and are equal, the ordinary limit exists and is equal to the one-handed limits. It is quite possible that the left- and right-handed limits exist but are different. This happens, for instance, at $x = 0$ for the function

$$f(x) = \begin{cases} 1, & 0 < x < \pi \\ -1, & -\pi < x < 0. \end{cases}$$

In this case, the left-hand limit at $x_0 = 0$ is -1, while the right-hand limit is $+1$. A discontinuity at which the one-handed limits exist, but do not agree, is called a *jump* discontinuity.

It is also possible that at some point both limits exist and agree, but that the function is not defined at that point. In such a case a function is said to have a removable discontinuity. If the value of the function at the troublesome point is redefined to be equal to the limit, the function will become continuous. For example, the function $f(x) = (\sin x)/x$ has a removable discontinuity at $x = 0$. The discontinuity is eliminated by redefining $f(x) = (\sin x)/x \; (x \neq 0)$, $f(0) = 1$. Removable discontinuities are so simple that *we shall ignore them from now on.*

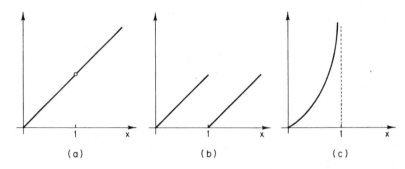

(a)   (b)   (c)

Figure 1.7 (a) Removable discontinuity $f(x) = (x - x^2)/(1 - x)$. (b) Jump discontinuity $f(x) = x \, (0 < x < 1) = x - 1 \, (1 < x)$. (c) "Bad" discontinuity $f(x) = -\ln(1 - x)$.

Other discontinuities are more serious. They occur if one or both of the one-handed limits fail to exist. Each of the functions $\sin(1/x)$, $e^{1/x}$, $1/x$ has a discontinuity at $x = 0$ which is neither removable nor a jump (see Fig. 1.7).

We shall say that a function is *sectionally continuous* on an interval $a \leq x \leq b$, if it is continuous, except possibly for a finite number of jumps and removable discontinuities. A function is sectionally continuous (without qualification) if it is sectionally continuous on every interval of finite length. For instance, if a periodic function is sectionally continuous on any interval of length one period, it is sectionally continuous (see Fig. 1.8).

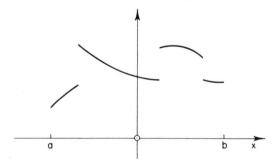

Figure 1.8 Typical sectionally continuous function made up of four continuous "sections."

Examples

1. The "square wave," defined by

$$f(x) = \begin{cases} 1, & 0 < x < a, \\ -1, & -a < x < 0, \end{cases} \qquad f(x + 2a) = f(x)$$

is sectionally continuous.

2.   The function $f(x) = 1/x$ cannot be sectionally continuous on any interval which contains 0 or even has 0 as an endpoint, because the function is not bounded at $x = 0$.

3.   If $f(x) = x$, $-1 < x < 1$, then f is continuous on that interval. Its periodic extension is *sectionally* continuous but not continuous.

The examples clarify a couple of facts about the meaning of sectional continuity. Most important is that a sectionally continuous function must not "blow up" at any point—even an endpoint—of an interval. Note also that a function need not be defined at every point in order to qualify as sectionally continuous. No value was given for the "square wave" function at $x = 0$, $\pm a$, but the function remains sectionally continuous, no matter what values are assigned for these points.   Sectionally smooth

A function is *sectionally smooth* in an interval $a < x < b$ if f is sectionally continuous; $f'(x)$ exists, except perhaps at a finite number of points; and $f'(x)$ is sectionally continuous. The graph of a sectionally smooth function, then, has a finite number of removable discontinuities, jumps, and corners. (The derivative will not exist at these points.) *Between* these points, the graph will be continuous, with a continuous derivative. No vertical tangents are allowed, for these indicate that the derivative is infinite.

Examples

1. $f(x) = |x|^{1/2}$ is continuous, but not sectionally smooth in any interval which contains 0, because $|f'(x)| \to \infty$ as $x \to 0$.

2.   The square wave is sectionally smooth, but not continuous.

Now we can state the main convergence theorem.

Theorem   If $f(x)$ is sectionally smooth and periodic with period $2a$, then at each point x the Fourier series corresponding to f converges and

$$a_0 + \sum \left[a_n \cos\left(\frac{n\pi x}{a}\right) + b_n \sin\left(\frac{n\pi x}{a}\right) \right] = \frac{f(x+) + f(x-)}{2}.$$

$f(x) = S(x)$

This theorem gives an answer to the question at the beginning of the section. Recall that a sectionally smooth function has only a finite number of jumps and no bad discontinuities, in every finite interval. Hence $f(x+)$ and $f(x-)$ are equal to each other and to $f(x)$, except perhaps at a finite number of points. For this reason, if f satisfies the hypotheses of the theorem, we write f *equal* to its Fourier series, even though the equality may fail at jumps.

In constructing the periodic extension of a function, we never defined the values of $f(x)$ at the endpoints. Since the Fourier coefficients are given by integrals, the value assigned to $f(x)$ at one point cannot influence them, so in that sense, the value of f at $x = \pm a$ is unimportant. But because of the averaging features of the Fourier series, it is reasonable to define

$$f(a) = f(-a) = \tfrac{1}{2}(f(a-) + f(-a+)).$$

That is, the value of f at the endpoints is the average of the one-handed limits at the endpoints, each limit taken from the interior. For instance, if $f(x) = 1 + x$, $0 < x < 1$, and $f(x) = 0$, $-1 < x < 0$, then $f(\pm 1)$ should be taken to be 1, and $f(0)$ should be $\tfrac{1}{2}$.

Examples

1. The square wave function

$$f(x) = \begin{cases} 1, & 0 < x < 1 \\ -1, & -1 < x < 0 \end{cases}$$

is sectionally smooth; therefore the corresponding Fourier series converges to

$$\begin{array}{ll} 1, & 0 < x < 1 \\ -1, & -1 < x < 0 \\ 0, & x = 0,\ 1,\ -1 \end{array}$$

and is periodic with period 2.

2. For the function $f(x) = |x|^{1/2}$, $-\pi < x < \pi$, $f(x + 2\pi) = f(x)$, the theorem above does not guarantee convergence of the Fourier series at any point, even though the function is continuous. Nevertheless, the series does converge at every point x! This shows that the conditions in the theorem above are perhaps too strong. (But they are useful.)

Exercises

1. In each part below, a function is equated to its Fourier series. By evaluating both sides of that equality at the appropriate value of x, derive the second equality.

a. $|x| = \dfrac{1}{2} - \dfrac{4}{\pi^2} \displaystyle\sum_{k=0}^{\infty} \dfrac{1}{(2k+1)^2} \cos(2k+1)\pi x$, $\qquad -1 < x < 1$

$\dfrac{\pi^2}{8} = 1 + \dfrac{1}{9} + \dfrac{1}{25} + \cdots$ *Evaluate at zero; trivially follows*

evaluate
_at_ $\frac{1}{2}$

b. $\dfrac{4}{\pi}\displaystyle\sum_{k=0}^{\infty}\dfrac{1}{2k+1}\sin(2k+1)\pi x = \begin{cases} 1, & 0 < x < 1 \\ -1, & -1 < x < 0 \end{cases}$

$\dfrac{\pi}{4} = 1 - \dfrac{1}{3} + \dfrac{1}{5} - \dfrac{1}{7} + \cdots$

_at_ $n\pi$

$2m^2\pi$

c. $|\sin x| = \dfrac{1}{\pi} - \dfrac{2}{\pi}\displaystyle\sum_{n=1}^{\infty}\dfrac{1}{4n^2 - 1}\cos 2nx$

$\dfrac{1}{2} = \dfrac{1}{3} + \dfrac{1}{15} + \dfrac{1}{35} + \cdots$

2. Check each function described below to see if it is sectionally smooth. If it is, state the value to which its Fourier series converges at each point x in the interval and at the end points. Sketch.

a. $f(x) = |x| + x, \quad -1 < x < 1$

$= \dfrac{x}{|x|} + 1$

b. $f(x) = x\cos x, \quad -\dfrac{\pi}{2} < x < \dfrac{\pi}{2}$

c. $f(x) = x\cos x, \quad -1 < x < 1$

d. $f(x) = \begin{cases} 0, & 1 < x < 3 \\ 1, & -1 < x < 1 \\ x, & -3 < x < -1 \end{cases}$

3. To what value does the Fourier series of f converge if f is a *continuous*, sectionally smooth periodic function?

4. State convergence theorems for the Fourier sine and cosine series which arise from half-range expansions.

4   Uniform convergence

The theorem of the preceding section treats convergence at individual points of an interval. A stronger kind of convergence is uniform convergence in an interval. Let

$$S_N(x) = a_0 + \sum_{n=1}^{N}(a_n\cos(n\pi x/a) + b_n\sin(n\pi x/a))$$

be the partial sum of the Fourier series of a function f. The maximum deviation between the graphs of $S_N(x)$ and $f(x)$ is

$$\delta_N = \max|f(x) - S_N(x)|, \qquad -a \le x \le a$$

where the maximum is taken over all x in the interval, including the end points. If the maximum deviation tends to zero as N increases, we say that the series *converges uniformly* in the interval $-a \le x \le a$.

Roughly speaking, if a Fourier series converges uniformly, then a finite number N of terms give a good approximation—to within $\pm\delta_N$—of the value of $f(x)$ at *any* and *every* point of the interval. Furthermore, by taking a large enough N, one can make the error as small as necessary.

There are two important facts about uniform convergence. If a Fourier series converges uniformly in a period–interval, then (1) it must converge to a continuous function, and (2) it must converge to the (continuous) function which generates the series. Thus, a function that has a nonremovable discontinuity *cannot* have a uniformly convergent Fourier series. (And not all continuous functions have uniformly convergent Fourier series.)

In Fig. 1.9 are graphs of some partial sums of a square-wave function. It is easy to see that for every N there are points near $x = 0$ and $x = \pm\pi$ where $|f(x) - S_N(x)|$ is nearly equal to 1, so convergence is *not* uniform. (Incidentally, the graphs in Fig. 1.9 also show the partial sums of $f(x)$ overshooting their mark near $x = 0$. This feature of Fourier series is called Gibbs' phenomenon and always occurs near a jump.) On the other hand, Fig. 1.10 shows graphs of a "sawtooth" function and the partial sums of its Fourier series. The maximum deviation always occurs at $x = 0$, and the convergence is uniform.

One of the ways of proving uniform convergence is by examining the coefficients.

Theorem   If the series $\sum(|a_n| + |b_n|)$ converges, then the Fourier series

$$a_0 + \sum a_n \cos(n\pi x/a) + b_n \sin(n\pi x/a)$$

converges uniformly in the interval $-a \le x \le a$, and, in fact, on the whole real line.

Example   For the function

$$f(x) = |x|, \qquad -1 < x < 1$$

the Fourier coefficients are

$$a_0 = \frac{1}{2}, \qquad a_n = \frac{2}{\pi^2}\frac{-1 + (-1)^n}{n^2}, \qquad b_n = 0.$$

Since the series $\sum 1/n^2$ converges, the series of absolute values of the coefficients converges, and so the Fourier series converges uniformly on the interval $-1 \le x \le 1$ to $|x|$. The Fourier series converges uniformly to the periodic extension of $f(x)$ on the whole real line.

Another way of proving uniform convergence of a Fourier series is by examining the function f which generates it.

Theorem If $f(x)$ is periodic, continuous, and has a sectionally continuous derivative, then the Fourier series corresponding to f converges uniformly to $f(x)$ on the entire real axis.

While this theorem is stated for a periodic function, it may be adapted to a function $f(x)$ given on the interval $-a < x < a$. If the *periodic extension* of f satisfies the conditions of the theorem, then the Fourier series of f converges uniformly on the interval $-a \leq x \leq a$.

Example Consider the function

$$f(x) = x, \qquad -1 < x < 1.$$

Although $f(x)$ is continuous and has a continuous derivative in the interval $-1 < x < 1$, the periodic extension of f is *not* continuous. The Fourier series cannot converge uniformly in any interval containing $x = \pm 1$.

On the other hand, the function $f(x) = |\sin x|$, periodic with period 2π, is continuous and has a sectionally continuous derivative. Therefore, its Fourier series converges uniformly to $f(x)$ everywhere.

A restatement of the theorem is:

Theorem If $f(x)$ is given on $-a < x < a$, if f is continuous and bounded and has a sectionally continuous derivative, and if $f(-a+) = f(a-)$, then the Fourier series of f converges uniformly to f on the interval $-a \leq x \leq a$. (The series converges to $f(a-) = f(-a+)$ at $x = \pm a$).

In this statement, the condition on the limits at the end points replaces the condition of continuity of the periodic extension of f.

If an odd periodic function is to be continuous, it must have value 0 at $x = 0$ and at the endpoints of the symmetric period–interval. Thus, the odd periodic extension of a function given in $0 < x < a$ may have jump discontinuities even though it is continuous where originally given. The even periodic extension causes no such difficulty, however.

Theorem If $f(x)$ is given on $0 < x < a$, if f is continuous and bounded and has a sectionally continuous derivative, and if $f(0+) = f(a-) = 0$, then the Fourier sine series of f converges uniformly to f in the interval $0 \leq x \leq a$. (The series converges to 0 at $x = 0$ and $x = a$.)

Theorem If $f(x)$ is given on $0 < x < a$, if f is continuous and bounded and has a sectionally continuous derivative, then the Fourier cosine series of f converges uniformly to f in the interval $0 \leq x \leq a$. (The series converges to $f(0+)$ at $x = 0$ and to $f(a-)$ at $x = a$).

$$f(x)= \begin{cases} 1, & 0 < x < \pi \\ -1, & -\pi < x < 0 \end{cases}$$

(a)

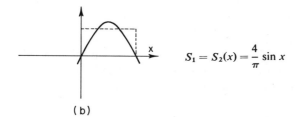

$$S_1 = S_2(x) = \frac{4}{\pi} \sin x$$

(b)

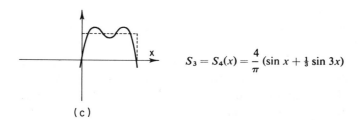

$$S_3 = S_4(x) = \frac{4}{\pi} (\sin x + \tfrac{1}{3} \sin 3x)$$

(c)

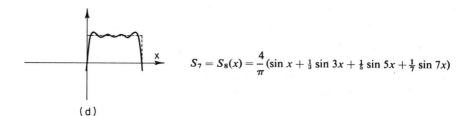

$$S_7 = S_8(x) = \frac{4}{\pi} (\sin x + \tfrac{1}{3} \sin 3x + \tfrac{1}{5} \sin 5x + \tfrac{1}{7} \sin 7x)$$

(d)

$$S_{11} = S_{12}(x) = \frac{4}{\pi} (\sin x + \cdots + \tfrac{1}{11} \sin 11x).$$

(e)

Figure 1.9 Partial sums of the square wave function. Convergence is *not* uniform.

$$f(x) = |x|, \qquad -\pi < x < \pi$$
$$f(x + 2\pi) = f(x)$$

(a)

$$S_0(x) = \frac{\pi}{2}$$

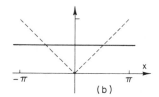

(b)

$$S_2(x) = S_1(x) = \frac{\pi}{2} - \frac{4}{\pi} \cos x$$

(c)

$$S_3(x) = S_4(x) = \frac{\pi}{2} - \frac{4}{\pi} (\cos x + \tfrac{1}{9} \cos 3x)$$

(d)

$$S_5(x) = S_6(x) = \frac{\pi}{2} - \frac{4}{\pi} (\cos x + \tfrac{1}{9} \cos 3x + \tfrac{1}{25} \cos 5x)$$

(e)

Figure 1.10 Partial sums of a saw-tooth function. Convergence *is* uniform.

Exercises

1. Determine whether the series of the following functions converge uniformly or not. Sketch each function.

a. $f(x) = e^x$, $\quad -1 < x < 1$ $\qquad$ b. $f(x) = \sinh x$, $\qquad -\pi < x < \pi$
c. $f(x) = \sin x$, $\quad -\pi < x < \pi$
d. $f(x) = \sin x + |\sin x|$, $\quad -\pi < x < \pi$
e. $f(x) = x + |x|$, $\quad -\pi < x < \pi$ $\quad$ f. $f(x) = x(x^2 - 1)$, $\quad -1 < x < 1$
g. $f(x) = 1 + 2x - 2x^3$, $\quad -1 < x < 1$

2. The Fourier series of the function

$$f(x) = (\sin x)/x, \qquad -\pi < x < \pi$$

converges at every point. To what value does the series converge at $x = 0$? at $x = \pi$? The convergence is uniform. Why?

3. Determine whether the sine and cosine series of the following functions converge uniformly. Sketch.

a. $f(x) = \sinh x$, $\quad 0 < x < \pi$ $\qquad$ b. $f(x) = \sin x$, $\quad 0 < x < \pi$
c. $f(x) = \sin \pi x$, $\quad 0 < x < \frac{1}{2}$ $\qquad$ d. $f(x) = 1/(1 + x)$, $\quad 0 < x < 1$
e. $f(x) = 1/(1 + x^2)$, $\quad 0 < x < 2$

4. If a_n and b_n tend to zero as n tends to infinity, show that the series

$$a_0 + \sum e^{-\alpha n}(a_n \cos nx + b_n \sin nx)$$

converges uniformly ($\alpha > 0$).

5. Determine which of the series of Exercise 1 of the preceding section converge uniformly.

5 Operations on Fourier series

In the course of this book we shall have to perform certain operations on Fourier series. The purpose of this section is to find conditions under which they are legitimate. Two things must be noted, however. First, the theorems stated here are not the best possible: there are theorems with weaker hypotheses and the same conclusions. Second, in applying mathematics, we often carry out operations formally, legitimate or not. The results must then be checked for correctness.

Throughout this section we shall state results about functions and Fourier series with period 2π, for typographic convenience. The results remain true when the period is $2a$ instead. For functions defined only on a finite interval, the periodic extension must fulfill the hypotheses. We shall refer to a function $f(x)$ with the series shown.

$$f(x) \sim a_0 + \sum_{n=1}^{\infty} a_n \cos nx + b_n \sin nx \qquad (1)$$

Theorem 1   The Fourier series of the function $cf(x)$ has coefficients ca_0, ca_n, and cb_n (c a constant).

This theorem is a simple consequence of the fact that a constant passes through an integral. The fact that the integral of a sum is the sum of the integrals leads to the following.

Theorem 2   The Fourier coefficients of the sum $f(x) + g(x)$ are the sums of the corresponding coefficients of $f(x)$ and $g(x)$.

These two theorems are so natural that the reader has probably used them already without thinking about it. The theorems that follow are much more difficult to prove, but extremely important.

Theorem 3   If $f(x)$ is periodic and sectionally continuous, then the Fourier series of f may be integrated term by term:

$$\int_a^b f(x)\, dx = \int_a^b a_0\, dx + \sum_{n=1}^{\infty} \int_a^b (a_n \cos nx + b_n \sin nx)\, dx. \qquad (2)$$

Theorem 4   If $f(x)$ is periodic and sectionally continuous, and if $g(x)$ is sectionally continuous for $a \le x \le b$, then

$$\int_a^b f(x)g(x)\, dx = \int_a^b a_0 g(x)\, dx + \sum_{n=1}^{\infty} \int_a^b (a_n \cos nx + b_n \sin nx)g(x)\, dx. \qquad (3)$$

In Theorems 3 and 4, the function $f(x)$ is only required to be sectionally continuous. It is not necessary that the Fourier series of $f(x)$ converge at all. Nevertheless, the theorems guarantee that the series on the right converges and equals the integral on the left in Eqs. (2) and (3).

One important application of Theorem 4 was the derivation of the formulas for the Fourier coefficients in Section 1. An application of Theorems 3 and 4 is given below.

Example   The periodic function $g(x)$ whose formula in the interval $0 < x < 2\pi$ is

$$g(x) = x, \qquad 0 < x < 2\pi$$

has the Fourier series

$$g(x) \sim \pi - 2 \sum_{n=1}^{\infty} \frac{\sin nx}{n}.$$

By applying Theorems 1 and 2, we find that the function $f(x)$ defined by $f(x) = [\pi - g(x)]/2$ has the series

$$f(x) \sim \sum_{n=1}^{\infty} \frac{\sin nx}{n}.$$

This manipulation would be simple algebra if the correspondence $\sim$ were an equality.

The function $f(x)$ satisfies the hypotheses of Theorem 3. Thus we may integrate the series above from 0 to b to get

$$\int_0^b f(x)\, dx = \sum_{n=1}^{\infty} \frac{1 - \cos nb}{n^2}.$$

Theorem 3 guarantees that this equality holds for any b. In the interval from 0 to 2π we have $g(x) = x$ and $f(x) = (\pi - x)/2$. Hence

$$\int_0^b f(x)\, dx = \frac{\pi b}{2} - \frac{b^2}{4} = \sum_{n=1}^{\infty} \frac{1 - \cos nb}{n^2}, \qquad 0 \le b \le 2\pi.$$

Now, replacing b by x, we have

$$\frac{x(2\pi - x)}{4} = \sum_{n=1}^{\infty} \frac{1}{n^2} - \sum_{n=1}^{\infty} \frac{\cos nx}{n^2}, \qquad 0 \le x \le 2\pi. \tag{4}$$

Outside the indicated interval, the periodic extension of the function on the left equals the series on the right.

It is worthwhile to mention that the series on the right of Eq. (4) is the Fourier series of the function on the left. That is to say,

$$\frac{1}{2\pi} \int_0^{2\pi} \frac{x(2\pi - x)}{4}\, dx = \sum_{n=1}^{\infty} \frac{1}{n^2} \tag{5}$$

$$\frac{1}{\pi} \int_0^{2\pi} \frac{x(2\pi - x)}{4} \cos nx = \frac{-1}{n^2} \tag{6}$$

$$\frac{1}{\pi} \int_0^{2\pi} \frac{x(2\pi - x)}{4} \sin nx = 0. \tag{7}$$

Equations (6) and (7) can be verified directly, of course, but Theorem 4, together with the orthogonality relations of Section 1, also guarantees them. In addition, Eq. (5) gives us a way to evaluate the series on the right.

The last operation to be discussed is differentiation, one that plays a principal role in applications.

Theorem 5   If $f(x)$ is periodic, continuous, and sectionally smooth, then the differentiated Fourier series of $f(x)$ converges to $f'(x)$ at every point x where $f''(x)$ exists:

$$f'(x) = \sum_{n=1}^{\infty} (-na_n \sin nx + nb_n \cos nx). \qquad (8)$$

The hypotheses on $f(x)$ itself imply (see Section 4) that the Fourier series of $f(x)$ converges uniformly. If $f(x)$ (or its periodic extension) fails to be continuous, it is certain that the differentiated series of $f(x)$ will fail to converge, at some points at least.

As an example, take the function which is periodic with period 2π and has the formula

$$f(x) = |x|, \qquad -\pi < x < \pi.$$

This function is indeed continuous and sectionally smooth and is equal to its Fourier series,

$$f(x) = \frac{\pi}{2} - \frac{4}{\pi}\left(\cos x + \frac{\cos 3x}{9} + \frac{\cos 5x}{25} + \cdots\right).$$

According to Theorem 5, the differentiated series

$$\frac{4}{\pi}\left(\sin x + \frac{\sin 3x}{3} + \frac{\sin 5x}{5} + \cdots\right)$$

converges to $f'(x)$ at any point x where $f''(x)$ exists. Now, the derivative of the sawtooth function $f(x)$ (see Fig. 1.10) is the square-wave function

$$f'(x) = \begin{cases} 1, & 0 < x < \pi \\ -1, & -\pi < x < 0 \end{cases} \qquad (9)$$

(see Fig. 1.9). Moreover, we know that the sine series above is the Fourier series of the square wave $f'(x)$ and that it converges to the values given by Eq. (9), except at the points where $f'(x)$ has a jump. These are precisely the points where $f''(x)$ does not exist.

Although the uniqueness property stated in this theorem is so very natural that we tend to assume it is true without checking, it really is a consequence of Theorem 4.

Theorem 6   If $f(x)$ is periodic and sectionally continuous, its Fourier series is unique.

That is to say, only one series can correspond to $f(x)$. We often make use of uniqueness in this way: if two Fourier series are equal (or correspond to the same function), then the coefficients of like terms must match.

Exercises

1. Evaluate the sum of the series $\sum 1/n^2$ by performing the integration indicated in Eq. (5).

2. Sketch the graphs of the periodic extension of the function

$$f(x) = \frac{\pi - x}{2}, \qquad 0 < x < 2\pi$$

and of its derivative $f'(x)$ and of

$$F(x) = \int_0^x f(t)\, dt.$$

3. Suppose that a function has the formula $f(x) = x$, $0 < x < \pi$. What is its derivative? Can the Fourier sine series be differentiated term by term? What about the cosine series?

4. Verify Eqs. (6) and (7) by integration.

5. Suppose that a function $f(x)$ is continuous and sectionally smooth in the interval $0 < x < a$. What additional conditions must $f(x)$ satisfy in order to guarantee that its sine series can be differentiated term by term? The cosine series?

6. Is the derivative of a periodic function periodic? Is the integral of a periodic function periodic?

7. It is known that the equality

$$\ln\left| 2 \cos \frac{x}{2} \right| = \sum_{n=1}^{\infty} \frac{(-1)^{n+1}}{n} \cos nx$$

is valid except when x is an integer multiple of π. Can the Fourier series be differentiated term by term?

8. Use the series below, together with integration and differentiation to find a Fourier series for the function $p(x) = x(\pi - x)$, $0 < x < \pi$.

$$x = 2 \sum_{n=1}^{\infty} \frac{(-1)^{n+1}}{n} \sin nx, \qquad 0 < x < \pi$$

6 Mean error and convergence in mean

While we can study the behavior of infinite series, we must almost always use finite series in practice. Fortunately, Fourier series have some properties which make them very useful in this setting. Before going on to these properties, we shall develop a useful formula.

Suppose f is a function defined in the interval $-a < x < a$, for which

$$\int_{-a}^{a} (f(x))^2 \, dx$$

is a finite number. Let

$$f(x) \sim a_0 + \sum a_n \cos(n\pi x/a) + b_n \sin(n\pi x/a)$$

and let $g(x)$ have a finite Fourier series

$$g(x) = A_0 + \sum_{1}^{N} A_n \cos(n\pi x/a) + B_n \sin(n\pi x/a).$$

Then we may perform the following operations:

$$\int_{-a}^{a} f(x)g(x) \, dx = \int_{-a}^{a} f(x)\left[A_0 + \sum_{1}^{N} A_n \cos(n\pi x/a) + B_n \sin(n\pi x/a) \right] dx$$

$$= A_0 \int_{-a}^{a} f(x) \, dx + \sum_{1}^{N} A_n \int_{-a}^{a} f(x) \cos(n\pi x/a) \, dx$$

$$2a\,a_0 = \qquad + \sum_{1}^{N} B_n \int_{-a}^{a} f(x) \sin(n\pi x/a) \, dx.$$

We recognize the integrals as multiples of the Fourier coefficients of f, and rewrite

$$\frac{1}{a} \int_{-a}^{a} f(x)g(x) \, dx = 2a_0 A_0 + \sum_{1}^{N} (a_n A_n + b_n B_n). \tag{1}$$

Now suppose we wish to approximate $f(x)$ by a *finite* Fourier series. The difficulty here is deciding what "approximate" means. Of the many ways we can measure approximation, the one that is easiest to use is the following:

$$\left\| f(x) - g(x) \right\|_2^2 = E_N = \int_{-a}^{a} (f(x) - g(x))^2 \, dx. \quad \geq 0 \tag{2}$$

(Here g is the function with a Fourier series containing terms up to and including $\cos(N\pi x/a)$. Clearly, E_N can never be negative, and if f and g are "close," then E_N will be small. Thus our problem is to choose the coefficients of g so as to minimize E_N. (We assume N fixed.)

To compute E_N, we first expand the integrand:

$$E_N = \int_{-a}^{a} f^2(x)\, dx - 2 \int_{-a}^{a} f(x)g(x)\, dx + \int_{-a}^{a} g^2(x)\, dx. \tag{3}$$

The first integral has nothing to do with g; the other two integrals clearly depend on the choice of g and can be manipulated so as to minimize E_N. We already have an expression for the middle integral. The last one can be found by replacing f with g in Eq. (1)

$$\int_{-a}^{a} g^2(x)\, dx = a\left[2A_0^2 + \sum_{1}^{N} A_n^2 + B_n^2\right]. \tag{4}$$

Now we have a formula for E_N in terms of the variables A_0, A_n, B_n:

$$E_N = \int_{-a}^{a} f^2(x)\, dx - 2a\left[2A_0 a_0 + \sum_{1}^{N} A_n a_n + B_n b_n\right]$$

$$+ a\left[2A_0^2 + \sum_{1}^{N} A_n^2 + B_n^2\right]. \tag{5}$$

The error E_N takes its minimum value when all the partial derivatives with respect to the variables are zero. We must then solve the equations

$$\frac{\partial E_N}{\partial A_0} = -4aa_0 + 4aA_0 = 0$$

$$\frac{\partial E_N}{\partial A_n} = -2aa_n + 2aA_n = 0$$

$$\frac{\partial E_N}{\partial B_n} = -2ab_n + 2aB_n = 0.$$

These equations require that $A_0 = a_0$, $A_n = a_n$, $B_n = b_n$. Thus g should be chosen to be the truncated Fourier series of f,

$$g(x) = a_0 + \sum_{1}^{N} a_n \cos(n\pi x/a) + b_n \sin(n\pi x/a)$$

in order to minimize E_N.

Now that we know which choice of As and Bs minimizes E_N, we can compute that minimum value. After some algebra, we see that

$$\text{min } E_N = \int_{-a}^{a} f^2(x)\, dx - a\left[2a_0^2 + \sum_{1}^{N} a_n^2 + b_n^2\right]. \tag{6}$$

Even this minimum error must be greater than or equal to zero, and thus we have the *Bessel inequality*

$$\frac{1}{a}\int_{-a}^{a} f^2(x)\, dx \geq 2a_0^2 + \sum_{1}^{N} a_n^2 + b_n^2. \tag{7}$$

This inequality is valid for any N and therefore is also valid in the limit as N tends to infinity. The actual fact is that, in the limit, the inequality becomes *Parseval's equality*:

$$\frac{1}{a}\int_{-a}^{a} f^2(x)\, dx = 2a_0^2 + \sum_{1}^{\infty} [a_n^2 + b_n^2]. \tag{8}$$

Another very important consequence of Bessel's inequality is that the two series Σa_n^2 and Σb_n^2 must converge if the left-hand side of Eqs. (7) and (8) is finite. Thus, the numbers a_n and b_n must tend to 0 as n tends to infinity.

By comparing Eqs. (6) and (8), we get a different expression for the minimum error

$$\text{min } E_N = a \sum_{N+1}^{\infty} [a_n^2 + b_n^2].$$

This quantity decreases steadily to zero as N increases. Since min E_N is, according to Eq. (2), a mean deviation between f and the truncated Fourier series of f, we often say "the Fourier series of f converges to f in the mean." (Another kind of convergence!)

Summary   If $f(x)$ has been defined in the interval $-a < x < a$, and if

$$\int_{-a}^{a} f^2(x)\, dx$$

is finite, then:

   a.   Among all finite series of the form

$$g(x) = A_0 + \sum_{1}^{N} [A_n \cos(n\pi x/a) + B_n \sin(n\pi x/a)]$$

the one which best approximates f in the sense of the error described by Eq. (2) is the truncated Fourier series of f:

$$a_0 + \sum_{1}^{N} [a_n \cos(n\pi x/a) + b_n \sin(n\pi x/a)].$$

Parseval's Equality

b. $\quad \dfrac{1}{a} \displaystyle\int_{-a}^{a} f^2(x)\, dx = 2a_0^2 + \sum_{1}^{\infty} [a_n^2 + b_n^2].$

c. $\quad a_n = \dfrac{1}{a} \displaystyle\int_{-a}^{a} f(x) \cos(n\pi x/a)\, dx \to 0$

$\qquad b_n = \dfrac{1}{a} \displaystyle\int_{-a}^{a} f(x) \sin(n\pi x/a)\, dx \to 0, \qquad$ as $\quad n \to \infty.$

d. The Fourier series of f converges to f in the sense of the mean.

Properties b and c are very useful for checking computed values of Fourier coefficients.

Exercises

1. Evaluate the definite integral

$$\frac{1}{\pi} \int_{-\pi}^{\pi} \left(\ln \left| 2 \cos \frac{x}{2} \right| \right)^2 dx.$$

(Hint: see Section 9, Eq. (4) and Section 5, Eq. (5).)

2. Verify Parseval's equality for the functions below:

    a. $f(x) = x, \quad -1 < x < 1$ **b.** $f(x) = \sin x, \quad -\pi < x < \pi.$

3. What can be said about the behavior of the Fourier coefficients of the following functions as $n \to \infty$?

    a. $f(x) = |x|^{1/2}, \quad -1 < x < 1$ **b.** $f(x) = |x|^{-1/2}, \quad -1 < x < 1$

4. How do we know that E_N has a minimum and not a maximum?

5. If a function f defined on the interval $-a < x < a$ has Fourier coefficients

$$a_n = 0, \qquad b_n = 1/\sqrt{n}$$

what can you say about

$$\int_{-a}^{a} f^2(x)\, dx?$$

6. Show that, as $n \to \infty$, the Fourier sine coefficients of the function

$$f(x) = 1/x, \qquad -\pi < x < \pi$$

tend to a constant. (Since this is an odd function, we can take the cosine coefficients to be zero, although strictly speaking they do not exist.) Use the fact that

$$\int_0^\infty \frac{\sin t}{t}\, dt = \frac{\pi}{2}.$$

7 Numerical determination of Fourier coefficients

There are many functions whose Fourier coefficients cannot be determined analytically because the integrals involved are not known in terms of easily evaluated functions. Also, it may happen that a function is not known explicitly, but its value can be found at some points. In either case, if a Fourier series is to be found for the function, some numerical technique must be employed to approximate the integrals which give the Fourier coefficients. It turns out that one of the crudest numerical integration techniques is the best.

We shall assume that the function whose coefficients are to be approximated is continuous, periodic, and sectionally smooth. If this is not the case, a sectionally smooth function can be modified to make it fit this description by following the procedure illustrated in Fig. 1.11.

Suppose then that $f(x)$ is continuous, sectionally smooth, and periodic with period $2a$. Let the interval $-a \le x \le a$ be divided into r equal subintervals whose endpoints are $-a = x_0, x_1, \ldots, x_r = a$. Then the approximate Fourier coefficients of f are (carets indicate approximate values):

$$\hat{a}_0 = \frac{f(x_1) + f(x_2) + \cdots + f(x_r)}{r}$$

$$\hat{a}_n = \frac{(f(x_1)\cos(n\pi/a)x_1 + \cdots + f(x_r)\cos(n\pi/a)x_r)\,\Delta x}{(\cos^2(n\pi/a)x_1 + \cdots + \cos^2(n\pi/a)x_r)\,\Delta x}$$

$$\hat{b}_n = \frac{(f(x_1)\sin(n\pi/a)x_1 + \cdots + f(x_r)\cos(n\pi/a)x_r)\,\Delta x}{(\sin^2(n\pi/a)x_1 + \cdots + \sin^2(n\pi/a)x_r)\,\Delta x}.$$

The coefficient a_0 represents an approximate mean value of f. In the other formulas $\Delta x = 2a/r$.

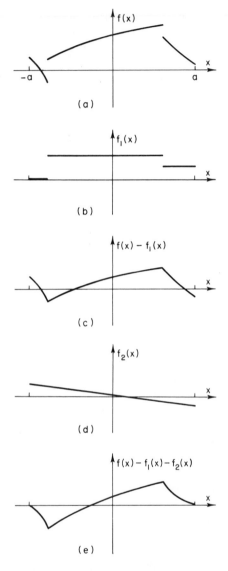

Figure 1.11 Preparation of a function for numerical integration of Fourier coefficients. (a) Graph of sectionally smooth function $f(x)$ given on $-a < x < a$. (b) Graph of $f_1(x)$, which has jumps of the same magnitude and position as $f(x)$. Coefficients can be found analytically. (c) Graph of $f(x) - f_1(x)$. This function has no jumps in $-a < x < a$. (d) Graph of $f_2(x)$. The periodic extension of $f_2(x)$ and of $f(x) - f_1(x)$ have jumps of the same magnitude at $x = \pm a$, and so forth. The coefficients of f_2 can be found analytically. (e) Graph of $f_3(x) = f(x) - f_1(x) - f_2(x)$. The Fourier series of $f_3(x)$ converges uniformly (the coefficients tend to zero rapidly).

The denominators of the expressions above can be found by some tedious calculations:

$$\cos^2 \frac{n\pi}{a} x_1 + \cdots + \cos^2 \frac{n\pi}{a} x_r = \begin{cases} r, & \text{if } n = 0, \dfrac{r}{2}, r, \ldots \\ \dfrac{r}{2}, & \text{otherwise} \end{cases}$$

$$\sin^2 \frac{n\pi}{a} x_1 + \cdots + \sin^2 \frac{n\pi}{a} x_r = \begin{cases} 0, & \text{if } n = 0, \dfrac{r}{2}, r, \ldots \\ \dfrac{r}{2}, & \text{otherwise.} \end{cases}$$

Of course, n can equal $r/2$ only if r is even.

Now we can simplify our formulas. Taking into account the results above, we have

$$\hat{a}_0 = \frac{1}{r} (f(x_1) + \cdots + f(x_r))$$

$$\hat{a}_n = \frac{2}{r} \left(f(x_1) \cos \frac{n\pi}{a} x_1 + \cdots + f(x_r) \cos \frac{n\pi}{a} x_r \right)$$

$$\hat{b}_n = \frac{2}{r} \left(f(x_1) \sin \frac{n\pi}{a} x_1 + \cdots + f(x_r) \sin \frac{n\pi}{a} x_r \right).$$

These are valid for $n < r/2$. If r is even ($r = 2q$) then the sine coefficients $\hat{b}_q$ is undefined, and the cosine coefficient is

$$\hat{a}_q = \frac{1}{r} \left(f(x_1) \cos \frac{q\pi}{a} x_1 + \cdots + f(x_r) \cos \frac{q\pi}{a} x_r \right).$$

How many coefficients calculated numerically are reasonably accurate? Since the values of the function at $x_1, \ldots, x_r$ represent r pieces of information, we may expect r coefficients to be reasonably accurate; and this is indeed the case. Thus, if r is an even number, say $r = 2q$, we can find

$$\hat{a}_0, \quad \hat{a}_1, \quad \hat{b}_1, \ldots, \hat{a}_q$$

but not $\hat{b}_q$, which is undefined. (Remember that $\hat{a}_q$ is found from the special formula.) If r is an odd number, say $r = 2q + 1$, then we can find

$$\hat{a}_0, \quad \hat{a}_1, \quad \hat{b}_1, \ldots, \hat{a}_q, \quad \hat{b}_q.$$

When $f(x)$ is given in the interval $0 \leq x \leq a$, and the sine or cosine coefficients are to be determined, the formulas may be derived from those above.

Let the interval be divided into s equal subintervals with endpoints $0 = x_0$, $x_1, \ldots, x_s = a$. Then we find the approximate Fourier coefficients to be

$$\hat{a}_0 = \frac{1}{s}\left(\frac{1}{2}f(x_0) + f(x_1) + \cdots + f(x_{s-1}) + \frac{1}{2}f(x_s)\right)$$

$$\hat{a}_n = \frac{2}{s}\left(\frac{1}{2}f(x_0) + f(x_1)\cos\frac{n\pi}{a}x_1 + \cdots + \frac{1}{2}f(x_s)\cos\frac{n\pi}{a}x_s\right), \quad n = 1, \ldots, s-1$$

$$\hat{a}_s = \frac{1}{s}\left(\frac{1}{2}f(x_0) + f(x_1)\cos\frac{s\pi}{a}x_1 + \cdots + \frac{1}{2}f(x_s)\cos\frac{s\pi}{a}x_s\right)$$

or

$$b_n = \frac{2}{s}\left(f(x_1)\sin\frac{n\pi}{a}x_1 + \cdots + f(x_{s-1})\sin\frac{n\pi}{a}x_{s-1}\right), \quad n = 1, 2, \ldots, s.$$

Notice that in the formulas for $\hat{a}_0$ and $\hat{a}_n$, the endpoints have a factor of $\frac{1}{2}$ in the sum. In the formula for the b_n, the endpoints are omitted, since the sine function would be zero there.

An important feature of the approximate Fourier coefficients is this: If

$$F(x) = \hat{a}_0 + \hat{a}_1\cos\frac{\pi x}{a} + \hat{b}_1\sin\frac{\pi x}{a} + \cdots$$

is a finite Fourier series using a total of r $\hat{a}$s and $\hat{b}$s calculated from the formulas above, then

$$F(x_i) = f(x_i), \quad i = 1, 2, \ldots, r.$$

Thus the graph of $F(x)$ cuts the graph of $f(x)$ at the points x_i.

Example Calculate the approximate Fourier coefficients of $f(x) = (\sin x)/x$ in $-\pi < x < \pi$.

Since f is even, it will have a cosine series. We simplify computation by using the half-range formulas and making s even. We take $s = 6$, $x_0 = 0$, $x_1 = \pi/6, \ldots, x_5 = 5\pi/6$, $x_6 = \pi$. The numerical information is in Table 1.1.

The results of the calculation are given in Table 1.2. On the left are the approximate coefficients calculated from the table. On the right are the correct values (to five decimals), obtained with the aid of a table of the sine integral (see Exercise 2).

For hand calculation, choosing s to be a muliple of 4 makes many of the cosines "easy" numbers such as 1 or 0.5. When the calculation is done by digital computer, there are no "easy" numbers, and it seems that there is an advantage to using a prime number for s. (Why?)

Table 1.1

i	x_i	$\cos x_i$	$\cos 2x_i$	$\cos 3x_i$	$(\sin x_i)/x_i$
0	0	1.0	1.0	1.0	1.0
1	$\dfrac{\pi}{6}$	0.86603	0.5	0	0.95493
2	$\dfrac{\pi}{3}$	0.5	-0.5	-1.0	0.82699
3	$\dfrac{\pi}{2}$	0	-1.0	0	0.63662
4	$\dfrac{2\pi}{3}$	-0.5	-0.5	1.0	0.41350
5	$\dfrac{5\pi}{6}$	-0.86603	0.5	0	0.19099
6	π	-1.0	1.0	-1.0	0.0

Table 1.2

n	$\hat{a}_n$	a_n	error
0	0.58717	0.58949	0.00232
1	0.45611	0.45141	0.00470
2	-0.06130	-0.05640	0.00490
3	0.02884	0.02356	0.00528

Exercises

1. Since the table in the example gives $(\sin x/x)$ for seven points, seven cosine coefficients can be calculated. Find $\hat{a}_6$.

2. Express the Fourier cosine coefficients of the example in terms of integrals of the form

$$\mathrm{Si}((n+1)\pi) = \int_0^{(n+1)\pi} \frac{\sin t}{t}\, dt.$$

This is the sine integral function and is tabulated in many books, especially *Handbook of Mathematical Functions*, Abramowitz and Stegun [1].

3. Each entry in the table below represents the depth of the water in Lake Ontario (minus the low-water datum of 242.8 feet) on the first of the corresponding month. Assuming that the water level is a periodic function of period one year, and that the observations are taken at equal intervals, compute the Fourier coefficients $\hat{a}_0, \hat{a}_1, \hat{b}_1, \hat{a}_2, \hat{b}_2, \ldots$, thus identifying the mean level, and fluctuations of period 12 months, 6 months, 4 months, and so forth.

Jan.	0.75	Jul.	2.35
Feb.	0.60	Aug.	2.15
Mar.	0.65	Sept.	1.75
Apr.	1.15	Oct.	1.05
May	1.80	Nov.	1.00
Jun.	2.25	Dec.	0.90

8 Fourier integral

In Sections 1 and 2 of this chapter we developed the representation of a periodic function in terms of sines and cosines with the same period. Then, by means of periodic extension, we obtained series representations for functions defined only on a finite interval. Now we must deal with nonperiodic functions defined for x between $-\infty$ and ∞. Can such functions also be represented in terms of sines and cosines?

That such a representation may exist for some nonperiodic functions is evident from some simple integral formulas. For instance, it is known that the following definite integral depends on the parameter x as shown.

$$\int_0^\infty \frac{\sin \lambda x}{\lambda} \, d\lambda = \begin{cases} \dfrac{\pi}{2}, & x > 0 \\ 0, & x = 0 \\ -\dfrac{\pi}{2}, & x < 0 \end{cases} \tag{1}$$

From this point it is not difficult to determine that the integral

$$K(x, h) = \frac{2}{\pi} \int_0^\infty \frac{\sin \lambda h}{\lambda} \cos \lambda x \, d\lambda \tag{2}$$

depends on the two parameters x and h, in this way (assuming that $h > 0$):

$$K(x, h) = \begin{cases} 1, & |x| < h \\ 0, & |x| > h. \end{cases} \tag{3}$$

$$\frac{1}{2}, \quad |x| = h$$

The fact that this function (whose graph as a function of x is a rectangular pulse of width $2h$ and height 1) can be represented in terms of $\cos \lambda x$ is the key to representing other functions.

Suppose that a function $f(x)$ is defined for all x and is sectionally continuous in every finite interval. Then the mean value theorem guarantees that the equality

$$f(x) = \lim_{h \to 0+} \frac{1}{2h} \int_{x-h}^{x+h} f(x')\, dx' \tag{4}$$

is valid at every point x where f is continuous. We incorporate sines and cosines into this equality by rewriting the right-hand side of Eq. (4) in this way:

$$f(x) = \lim_{h \to 0+} \frac{1}{2h} \int_{-\infty}^{\infty} f(x') K(x - x', h)\, dx'. \tag{5}$$

Figure 1.12   Graph of $K(x - x', h)$.

This is exactly the same as Eq. (4), because $K(x - x', h)$ is 1 for x' between $x - h$ and $x + h$ and is 0 on the rest of the interval (see Fig. 1.12). Thus we have

$$f(x) = \lim_{h \to 0+} \frac{1}{2h} \int_{-\infty}^{\infty} f(x') \frac{2}{\pi} \int_{0}^{\infty} \frac{\sin \lambda h}{\lambda} \cos \lambda(x - x')\, d\lambda\, dx'. \tag{6}$$

If $f(x)$ is such that the order of integration may be reversed, we may progress to the form

$$f(x) = \lim_{h \to 0+} \int_{0}^{\infty} \frac{\sin \lambda h}{\lambda h} \frac{1}{\pi} \int_{-\infty}^{\infty} f(x') \cos \lambda(x - x')\, dx'\, d\lambda. \tag{7}$$

In the limit as h approaches 0, $(\sin \lambda h)/\lambda h$ approaches 1. Thus we expect to obtain the representation

$$f(x) = \int_{0}^{\infty} \frac{1}{\pi} \int_{-\infty}^{\infty} f(x') \cos \lambda(x - x')\, dx'\, d\lambda \tag{8}$$

for appropriate functions $f(x)$. When $\cos \lambda(x - x')$ is expanded by the formula for the cosine of a difference, Eq. (8) can be written in the following form, which we shall use almost exclusively from now on:

$$f(x) = \int_0^\infty (A(\lambda) \cos \lambda x + B(\lambda)\sin \lambda x) \, d\lambda,$$

$$A(\lambda) = \frac{1}{\pi} \int_{-\infty}^\infty f(x) \cos \lambda x \, dx, \qquad B(\lambda) = \frac{1}{\pi} \int_{-\infty}^\infty f(x) \sin \lambda x \, dx.$$

This is called the *Fourier integral representation* of the function f; $A(\lambda)$ and $B(\lambda)$ are called the Fourier integral *coefficient functions* of f. The following theorem gives some conditions under which the representation is valid.

Theorem Let $f(x)$ be sectionally smooth on every finite interval, and let $\int_{-\infty}^\infty |f(x)| \, dx$ be finite. Then at every point x,

$$\int_0^\infty (A(\lambda) \cos \lambda x + B(\lambda) \sin \lambda x) \, d\lambda = \frac{1}{2}(f(x+) + f(x-))$$

where

$$\left.\begin{matrix} A(\lambda) \\ B(\lambda) \end{matrix}\right\} = \frac{1}{\pi} \int_{-\infty}^\infty f(x) \begin{Bmatrix} \cos \lambda x \\ \sin \lambda x \end{Bmatrix} dx.$$

We have by no means proved the theorem in our derivation, but we have some indication of the roles played by the various hypotheses. Incidentally, the hypotheses of the theorem eliminate periodic functions from consideration; no periodic function, other than 0, can meet the condition that

$$\int_{-\infty}^\infty |f(x)| \, dx$$

be finite.

Examples

1. The function

$$f(x) = \begin{cases} 1, & |x| < 1 \\ 0, & |x| > 1, \end{cases}$$

which is also $K(x, 1)$, has the Fourier integral coefficient functions

$$A(\lambda) = \frac{1}{\pi} \int_{-\infty}^\infty f(x) \cos \lambda x \, dx = \frac{1}{\pi} \int_{-1}^1 \cos \lambda x \, dx = \frac{2 \sin \lambda}{\pi \lambda}$$

$$B(\lambda) = 0.$$

Since $f(x)$ is sectionally smooth, the Fourier integral representation is legitimate, and we write

$$f(x) = \int_0^\infty \frac{2 \sin \lambda}{\pi \lambda} \cos \lambda x \, d\lambda .$$

(Actually, the integral above equals $\frac{1}{2}$ at $x = \pm 1$, so equality fails at these two points.)

2. Let $f(x) = \exp(-|x|)$. Then direct integration gives

$$A(\lambda) = \frac{1}{\pi} \int_{-\infty}^\infty \exp(-|x|) \cos \lambda x \, dx$$

$$= \frac{2}{\pi} \int_0^\infty e^{-x} \cos \lambda x \, dx$$

$$= \frac{2}{\pi} \frac{e^{-x}(-\cos \lambda x + \lambda \sin \lambda x)}{1 + \lambda^2} \bigg|_0^\infty = \frac{2}{\pi} \frac{1}{1 + \lambda^2}.$$

$B(\lambda) = 0$, since $\exp(-|x|)$ is even. Since $\exp(-|x|)$ is continuous and sectionally smooth, we may write

$$\exp(-|x|) = \frac{2}{\pi} \int_0^\infty \frac{\cos \lambda x}{1 + \lambda^2} \, d\lambda .$$

These two examples illustrate the fact that, in general, one cannot evaluate the integral in the Fourier integral representation. It is the theorem stated above that allows us to write the equality between a suitable function and its Fourier integral.

If $f(x)$ is defined only in the interval $0 < x < \infty$, one can construct an even or odd extension whose Fourier integral contains only $\cos \lambda x$ or $\sin \lambda x$. These are called the Fourier *cosine* and *sine integral representations* of f, respectively. Thus we may write

$$f(x) = \int_0^\infty A(\lambda) \cos \lambda x \, d\lambda, \qquad 0 < x < \infty$$

$$A(\lambda) = \frac{2}{\pi} \int_0^\infty f(x) \cos \lambda x \, dx$$

or

$$f(x) = \int_0^\infty B(\lambda) \sin \lambda x \, d\lambda, \qquad 0 < x < \infty$$

$$B(\lambda) = \frac{2}{\pi} \int_0^\infty f(x) \sin \lambda x \, dx .$$

Exercises

1. Sketch the even and odd extensions of each of the following functions, and find the Fourier cosine and sine integrals for f. Each function is given in the interval $0 < x < \infty$.

   a. $f(x) = e^{-x}$

   b. $f(x) = \begin{cases} 1, & 0 < x < 1 \\ 0, & 1 < x \end{cases}$

   c. $f(x) = \begin{cases} \pi - x, & 0 < x < \pi \\ 0, & \pi < x \end{cases}$

2. Change the variable of integration in the formulas for A and B and justify each step of the following string of equalities. (Do not worry about changing order of integration.)

   $$f(x) = \frac{1}{\pi} \int_0^\infty \int_{-\infty}^\infty f(t)(\cos \lambda t \cos \lambda x + \sin \lambda t \sin \lambda x) \, dt \, d\lambda$$

   $$= \frac{1}{\pi} \int_{-\infty}^\infty f(t) \int_0^\infty \cos \lambda(t - x) \, d\lambda \, dt$$

   $$= \frac{1}{\pi} \int_{-\infty}^\infty f(t) \left[\lim_{\omega \to \infty} \frac{\sin \omega(t - x)}{t - x} \right] dt$$

   $$= \lim_{\omega \to \infty} \frac{1}{\pi} \int_{-\infty}^\infty f(t) \frac{\sin \omega(t - x)}{t - x} \, dt$$

   The last integral is called Fourier's single integral. Sketch the function $(\sin \omega v)/v$ as a function of v for several values of ω. What happens near $v = 0$? Sometimes notation is compressed and, instead of the last line, we write

   $$f(x) = \int_{-\infty}^\infty f(t) \, \delta(t - x) \, dt.$$

   Although δ is not, strictly speaking, a function, it is called "Dirac's delta function."

3. Find the Fourier integral representation of each of the following functions.

   a. $f(x) = 1/(1 + x^2)$ **b.** $f(x) = (\sin x)/x$

   (Hint: To evaluate the Fourier integral coefficient functions, consult the Fourier integral representations found in the Examples.)

4. In Exercise 3b, the integral $\int_{-\infty}^{\infty} |f(x)|\ dx$ is not finite. Nevertheless, $A(\lambda)$ and $B(\lambda)$ do exist $(B(\lambda) = 0)$. Find a rationale in the convergence theorem for saying that this function can be represented by its Fourier integral. (Hint: See Example 1.)

5. Suppose that $f(x)$ is a continuous, differentiable function with Fourier integral coefficient functions $A(\lambda)$ and $B(\lambda)$. How are the Fourier integral coefficient functions of the derivative $f'(x)$ related to A and B?

6. Show that if k and K are positive, then

a. $\displaystyle\int_0^\infty e^{-kx} \sin x\ dx = \frac{1}{1+k^2}$

b. $\displaystyle\int_0^\infty \frac{1 - e^{-Kx}}{x} \sin x\ dx = \tan^{-1} K$

c. $\displaystyle\int_0^\infty \frac{\sin x}{x}\ dx = \frac{\pi}{2}.$

(Part (a) by direct integration, (b) by integration of (a) with respect to k, (c) by limit of (b) as $K \to \infty$.)

7. Starting from Exercise 6c, derive Eq. (1).

8. Use trigonometric identities for the product $\sin \lambda h \cos \lambda x$ to derive Eq. (3) from Eq. (1).

9. Find the Fourier integral representation of each of these functions:

a. $f(x) = \begin{cases} \sin x, & -\pi < x < \pi \\ 0, & |x| > \pi \end{cases}$

b. $f(x) = \begin{cases} \sin x, & 0 < x < \pi \\ 0, & \text{otherwise} \end{cases}$

c. $f(x) = \begin{cases} |\sin x|, & -\pi < x < \pi \\ 0, & \text{otherwise.} \end{cases}$

9   Complex methods

Suppose that a function $f(x)$ corresponds to the Fourier series

$$f(x) \sim a_0 + \sum_1^\infty a_n \cos nx + b_n \sin nx.$$

(We use period 2π for simplicity only.) A famous formula of Euler states that

$$e^{i\theta} = \cos \theta + i \sin \theta, \qquad \text{where} \quad i^2 = -1.$$

Some simple algebra then gives the exponential definitions of the sine and cosine:

$$\cos \theta = \frac{1}{2}(e^{i\theta} + e^{-i\theta}), \qquad \sin \theta = \frac{1}{2i}(e^{i\theta} - e^{-i\theta}).$$

By substituting the exponential forms into the Fourier series of f we arrive at the alternate form

$$f(x) \sim a_0 + \frac{1}{2}\sum_1^\infty a_n(e^{inx} + e^{-inx}) - ib_n(e^{inx} - e^{-inx})$$

$$\sim a_0 + \frac{1}{2}\sum_1^\infty (a_n - ib_n)e^{inx} + (a_n + ib_n)e^{-inx}.$$

We are now led to define *complex Fourier coefficients* for f:

$$c_0 = a_0, \qquad c_n = \tfrac{1}{2}(a_n - ib_n), \qquad c_{-n} = \tfrac{1}{2}(a_n + ib_n), \qquad n = 1, 2, 3, \ldots .$$

In terms of these new coefficients, we have

$$f(x) \sim c_0 + \sum_1^\infty c_n e^{inx} + c_{-n} e^{-inx} \sim \sum_{-\infty}^\infty c_n e^{inx}. \tag{1}$$

This is the complex form of the Fourier series for f. It is easy to derive the universal formula

wrong

$$c_n = \frac{1}{2\pi}\int_{-\pi}^\pi f(x)e^{-inx}\, dx \tag{2}$$

which is valid for all integers n, positive, negative, or zero. The complex form is used especially in physics and electrical engineering.

Frequently the function corresponding to a Fourier series can be recognized by use of the complex form. For instance, the series

$$\sum_1^\infty \frac{(-1)^{n+1}}{n} \cos nx$$

may be considered the real part of

$$\sum_1^\infty \frac{(-1)^{n+1}}{n} e^{inx} = \sum_1^\infty \frac{(-1)^{n+1}}{n} (e^{ix})^n \tag{3}$$

because the real part of $e^{i\theta}$ is $\cos \theta$. The series on the right in Eq. (3) is recognized as a Taylor series

$$\sum_1^\infty \frac{(-1)^{n+1}}{n} (e^{ix})^n = \ln(1 + e^{ix}).$$

Some manipulations yield

$$1 + e^{ix} = e^{ix/2}(e^{ix/2} + e^{-ix/2}) = 2e^{ix/2} \cos \frac{x}{2}$$

$$\ln(1 + e^{ix}) = \frac{ix}{2} + \ln\left(2 \cos \frac{x}{2}\right).$$

The real part of $\ln(1 + e^{ix})$ is $\ln [2 \cos (x/2)]$ when $-\pi < x < \pi$. Thus, we derive the relation

$$\ln\left|2 \cos \frac{x}{2}\right| \sim \sum_{1}^{\infty} \frac{(-1)^{n+1}}{n} \cos nx. \tag{4}$$

(The series actually converges except at $x = \pm\pi, \pm3\pi, \dots$.)

The Fourier integral of a function $f(x)$ defined in the entire interval $-\infty < x < \infty$ can also be cast in complex form:

$$f(x) = \int_{-\infty}^{\infty} C(\lambda)e^{i\lambda x} \, d\lambda. \tag{5}$$

The complex Fourier integral coefficient function is given by

$$C(\lambda) = \frac{1}{2\pi} \int_{-\infty}^{\infty} f(x)e^{-i\lambda x} \, dx. \tag{6}$$

It is simple to show that

$$C(\lambda) = A(\lambda) + iB(\lambda) \tag{7}$$

where A and B are the usual Fourier integral coefficients. The complex Fourier integral coefficient is often called the Fourier transform of the function $f(x)$.

Exercises

1. Relate the functions and series below by using complex form and Taylor series.

    a. $1 + \sum_{1}^{\infty} r^n \cos nx = \dfrac{1 - r \cos x}{1 - 2r \cos x + r^2},\qquad 0 \le r < 1$

    b. $\sum_{1}^{\infty} \dfrac{\sin nx}{n!} = e^{\cos x} \sin(\sin x)$

2. Show by integrating that

$$\int_{-\pi}^{\pi} e^{inx} e^{-imx} \, dx = \begin{cases} 0, & n \neq m \\ 2\pi, & n = m \end{cases}$$

and develop the formula for the complex Fourier coefficients using this idea of orthogonality.

3. Use the complex form

$$a_n - ib_n = \frac{1}{\pi} \int_{-\pi}^{\pi} f(x) \, e^{-inx} \, dx, \qquad n \neq 0$$

to find the Fourier series of the function

$$f(x) = e^{ax}, \qquad -\pi < x < \pi.$$

4. Develop formula (2) for the complex Fourier coefficients c_n from the formulas for a_n and b_n.

5. Find the complex Fourier integral representation of the following functions:

a. $f(x) = \begin{cases} e^{-x}, & x > 0 \\ 0, & x < 0 \end{cases}$

b. $f(x) = \begin{cases} \sin x, & 0 < x < \pi \\ 0, & \text{elsewhere.} \end{cases}$

10 Applications of Fourier series and integrals

Fourier series and integrals are among the most basic tools of applied mathematics. We give below just a few applications which do not fall within the scope of the rest of this book.

A Nonhomogeneous differential equation

Many mechanical and electrical systems may be described by the differential equation

$$\ddot{y} + \alpha \dot{y} + \beta y = f(t).$$

The function $f(t)$ is called the "forcing function," βy the "restoring term," and $\alpha \dot{y}$ the "damping term."
    If $f(t)$ is periodic with period 2π, let its Fourier series be

$$f(t) = a_0 + \sum_{1}^{\infty} a_n \cos nt + b_n \sin nt.$$

If $y(t)$ is periodic with period 2π, it and its derivatives have Fourier series:

$$y(t) = A_0 + \sum A_n \cos nt + B_n \sin nt$$
$$\dot{y}(t) = \sum -nA_n \sin nt + n\,B_n \cos nt$$
$$\ddot{y}(t) = \sum -n^2 A_n \cos nt - n^2 B_n \sin nt.$$

Then the differential equation can be written in the form:

$$\beta A_0 + \sum (-n^2 A_n + \alpha n B_n + \beta A_n)\cos nt + \sum (-n^2 B_n - \alpha n A_n + \beta B_n)\sin nt$$
$$= a_0 + \sum a_n \cos nt + b_n \sin nt.$$

The As and Bs are now determined by matching coefficients

$$\beta A_0 = a_0$$
$$(\beta - n^2)A_n + \alpha n B_n = a_n$$
$$-\alpha n A_n + (\beta - n^2)B_n = b_n.$$

When these equations are solved for the As and Bs, we find

$$A_n = \frac{(\beta - n^2)a_n - \alpha n b_n}{\Delta}, \qquad B_n = \frac{(\beta - n^2)b_n + \alpha n a_n}{\Delta}$$

where

$$\Delta = (\beta - n^2)^2 + \alpha^2 n^2.$$

Now, given the function f, the as and bs can be determined, thus giving the As and Bs. The function $y(t)$ represented by the series found is the periodic part of the response. Depending on the initial conditions, there may also be a transient response which dies out as t increases.

B   Boundary value problems

By way of introduction to the next chapter, we apply the idea of Fourier series to the solution of the boundary value problem

$$\frac{d^2 u}{dx^2} + pu = f(x), \qquad 0 < x < a$$

$$u(0) = 0, \qquad u(a) = 0.$$

First, we will assume that $f(x)$ is equal to its Fourier sine series,

$$f(x) = \sum b_n \sin(n\pi x/a), \qquad 0 < x < a.$$

And second, we will assume that the solution $u(x)$, which we are seeking, equals its Fourier sine series,

$$u(x) = \sum B_n \sin(n\pi x/a), \qquad 0 < x < a$$

and that this series may be differentiated twice to give

$$\frac{d^2u}{dx^2} = \sum - \left(\frac{n^2\pi^2}{a^2} B_n\right) \sin(n\pi x/a), \qquad 0 < x < a.$$

When we insert the series forms for u, u'', and $f(x)$ into the differential equation, we find that

$$\sum \left(-\frac{n^2\pi^2}{a^2} B_n + pB_n\right) \sin(n\pi z/\pi) = \sum b_n(n\pi x/a), \qquad 0 < x < a.$$

Since the coefficients of like terms in the two series must match, we may conclude that

$$\left(p - \frac{n^2\pi^2}{a^2}\right) B_n = b_n, \qquad n = 1, 2, 3, \ldots.$$

If it should happen that $p = m^2\pi^2/a^2$ for some positive integer m, there is no value of B_m which satisfies

$$\left(p - \frac{m^2\pi^2}{a^2}\right) B_m = b_m$$

unless that $b_m = 0$ also, in which case *any* value of B_m is satisfactory. In summary, we may say that

$$B_n = b_n \bigg/ \left(p - \frac{n^2\pi^2}{a^2}\right)$$

and

$$u(x) = \sum \frac{a^2 b_n}{a^2 p - n^2\pi^2} \sin(n\pi x/a)$$

with the agreement that a zero denominator must be handled separately.

As an example, consider the boundary value problem

$$\frac{d^2u}{dx^2} - u = -x, \qquad 0 < x < 1$$

$$u(0) = 0, \qquad u(1) = 0.$$

It may readily be verified that

$$-x = \sum_{n=1}^{\infty} \frac{2(-1)^n}{\pi n} \sin n\pi x, \qquad 0 < x < 1.$$

Thus, by the preceding development, the solution must be

$$u(x) = \sum_{n=1}^{\infty} \frac{2}{\pi} \frac{(-1)^{n+1}}{n(n^2\pi^2 + 1)} \sin n\pi x, \qquad 0 < x < 1.$$

Although this particular series belongs to a known function, one would not, in general, know any formula for the solution $u(x)$ other than its Fourier sine series.

C   The sampling theorem

One of the most important results of information theory is the sampling theorem, which is based on a combination of the Fourier series and the Fourier integral in their complex forms. What the electrical engineer calls a signal is just a function $f(t)$ defined for all t. If the function is integrable, there is a Fourier integral representation for it:

$$f(t) = \int_{-\infty}^{\infty} C(\omega) \exp(i\omega t)\, d\omega$$

$$C(\omega) = \frac{1}{2\pi} \int_{-\infty}^{\infty} f(t) \exp(-i\omega t)\, dt.$$

A signal is called *band limited* if its Fourier transform is zero except in a finite interval; that is, if

$$C(\omega) = 0, \qquad \text{for} \quad |\omega| > \Omega.$$

Then Ω is called the cutoff frequency. If f is band limited, we can write it in the form

$$f(t) = \int_{-\Omega}^{\Omega} C(\omega) \exp(i\omega t)\, d\omega \tag{1}$$

since $C(\omega)$ is zero outside the interval $-\Omega < \omega < \Omega$. We focus our attention on this interval by writing $C(\omega)$ as a Fourier series:

$$C(\omega) = \sum_{-\infty}^{\infty} c_n \exp\left(\frac{in\pi\omega}{\Omega}\right), \qquad -\Omega < \omega < \Omega. \tag{2}$$

The (complex) coefficients are

$$c_n = \frac{1}{2\Omega} \int_{-\Omega}^{\Omega} C(\omega) \exp\left(\frac{-in\pi\omega}{\Omega}\right) d\omega.$$

The point of the sampling theorem is to observe that the integral for c_n actually is a value of $f(t)$ at a particular time. In fact, from the integral Eq. (1), we see that

$$c_n = \frac{1}{2\Omega} f\left(\frac{-n\pi}{\Omega}\right).$$

Thus there is an easy way of finding the Fourier transform of a band-limited function. We have

$$C(\omega) = \frac{1}{2\Omega} \sum_{-\infty}^{\infty} f\left(\frac{-n\pi}{\Omega}\right) \exp\left(\frac{in\pi\omega}{\Omega}\right)$$

$$= \frac{1}{2\Omega} \sum_{-\infty}^{\infty} f\left(\frac{n\pi}{\Omega}\right) \exp\left(\frac{-in\pi\omega}{\Omega}\right), \qquad -\Omega < \omega < \Omega.$$

By utilizing Eq. (1) again, we can reconstruct $f(t)$:

$$f(t) = \int_{-\Omega}^{\Omega} C(\omega) \exp(i\omega t) \, d\omega$$

$$= \frac{1}{2\Omega} \sum_{-\infty}^{\infty} f\left(\frac{n\pi}{\Omega}\right) \int_{-\Omega}^{\Omega} \exp\left(\frac{-in\pi\omega}{\Omega}\right) \exp(i\omega t) \, d\omega.$$

Carrying out the integration and using the identity

$$\sin \theta = \frac{(e^{i\theta} - e^{-i\theta})}{2i},$$

we find

$$f(t) = \sum_{-\infty}^{\infty} f\left(\frac{n\pi}{\Omega}\right) \frac{\sin(\Omega t - n\pi)}{\Omega t - n\pi}.$$

This is the main result of the sampling theorem. It says that the band-limited function $f(t)$ may be reconstructed from the samples of f at $t = 0, \pm \pi/\Omega, \ldots$. The sampling theorem can be realized physically. Modern telephone equipment employs sampling to send many conversations over one wire. In fact it seems that the sampling is audible on some transoceanic cable calls.

More information about the sampling theorem may be found in *An Introduction to Information Theory*, Reza [26].

11   Comments and references

The first use of trigonometric series occurred in the middle of the 18th century. Euler seems to have originated the use of orthogonality for the determination of coefficients. In the early 19th century Fourier made extensive use of trigonometric series in studying problems of heat conduction (see Chapter 2). His claim, that an arbitrary function could be represented as a trigonometric series, led to an extensive reexamination of the foundations of calculus. The concept of function was sharpened first. Dirichlet (about 1830) established sufficient conditions for the convergence of Fourier series. Later, Riemann was led to redefine the integral in an attempt to discover necessary and sufficient conditions for the convergence of a Fourier series. (His attempt failed—the problem has not been resolved.) Many other great mathematicians have founded important theories (the theory of sets, for one) in the course of studying Fourier series. They continue to be a topic of active research: a long-standing question was settled as recently as 1966.

Historical interest aside, Fourier series and integrals are extremely important in applied mathematics, physics, and engineering, and they merit further study. A superbly written and ogranized book is Tolstov's *Fourier Series* [30]; mathematical prerequisites are not too high. *Fourier Series and Boundary Value Problems* by Churchill [7] is a standard text for engineering applications. *Numerical Analysis for Scientists and Engineers* by Hamming [16] has information on numerical aspects of Fourier series.

Miscellaneous exercises

1. Find the Fourier sine series of the trapezoidal function given for $0 < x < \pi$ by

$$f(x) = \begin{cases} x/\alpha, & 0 < x < \alpha \\ 1, & \alpha < x < \pi - \alpha \\ (\pi - x)/\alpha, & \pi - \alpha < x < \pi. \end{cases}$$

2. Show that the series found in Exercise 1 converges uniformly.

3. When α approaches 0, the function of Exercise 1 approaches a square wave. Do the sine coefficients found in Exercise 1 approach those of a square wave?

4. Find the Fourier cosine series of the function

$$F(x) = \int_0^x f(t)\, dt,$$

where f denotes the function in Exercise 1. Sketch.

5. Find the Fourier sine series of the function given in the interval $0 < x < a$ by the formula (α is a parameter between 0 and 1)

$$f(x) = \begin{cases} hx/\alpha a, & 0 < x < \alpha a \\ h(a - x)/(1 - \alpha)a, & \alpha a < x < a. \end{cases}$$

6. Sketch the function of Exercise 5. To what does its Fourier sine series converge at $x = 0$? at $x = \alpha a$? at $x = a$?

7. Suppose that $f(x) = 1, 0 < x < a$. Sketch and find the Fourier series of the following extensions of $f(x)$:

   a. even extension
   b. odd extension
   c. periodic extension (period a)
   d. even periodic extension
   e. odd periodic extension
   f. the one corresponding to $f(x) = 0, \ -a < x < 0$.

8. Same task as in Exercise 7, but $f(x) = x, 0 < x < a$.

9. Find the Fourier series of the function given by

$$f(x) = \begin{cases} 0, & -a < x < 0 \\ 2x, & 0 < x < a. \end{cases}$$

   Sketch the graph of $f(x)$ and its periodic extension. To what values does the series converge at $x = -a$, $x = -a/2$, $x = 0$, $x = a$, and $x = 2a$?

10. Sketch the odd periodic extension and find the Fourier sine series of the function given by

$$f(x) = \begin{cases} 1, & 0 < x < \pi/2 \\ \tfrac{1}{2}, & \pi/2 < x < \pi. \end{cases}$$

   To what values does the series converge at $x = 0$, $x = \pi/2$, $x = \pi$, $x = 3\pi/2$, and $x = 2\pi$?

11. Sketch the even periodic extension of the function given in Exercise 10. Find its Fourier cosine series. To what values does the series converge at $x = 0$, $x = \pi/2$, $x = \pi$, $x = 3\pi/2$, and $x = 2\pi$?

12. Find the Fourier cosine series of the function

$$g(x) = \begin{cases} 1 - x, & 0 < x < 1 \\ 0, & 1 < x < 2. \end{cases}$$

   Sketch the graph of the sum of the cosine series.

13. Find the Fourier sine series of the function defined by $f(x) = 1 - 2x$, $0 < x < 1$. Sketch the graph of the odd periodic extension of $f(x)$, and determine the sum of the sine series at points where the graph has a jump.

14. Same requirements as Exercise 13, but use the cosine series and the even periodic extension.

15. Find the Fourier series of the function given by

$$f(x) = \begin{cases} 0, & -\pi < x < -\pi/2 \\ \sin 2x, & -\pi/2 < x < \pi/2 \\ 0, & \pi/2 < x < \pi. \end{cases}$$

Sketch the graph of the function.

16. Show that the function given by the formula $f(x) = (\pi - x)/2, 0 < x < 2\pi$, has the Fourier series

$$f(x) = \sum_{1}^{\infty} \frac{\sin nx}{n}, \qquad 0 < x < 2\pi.$$

Sketch $f(x)$ and its periodic extension.

17. Use complex methods and a finite geometric series to show that

$$\sum_{n=1}^{N} \cos nx = \frac{\sin(N + \frac{1}{2})x - \sin \frac{1}{2}x}{2 \sin \frac{1}{2}x}.$$

Then use trigonometric identities to identify

$$\sum_{n=1}^{N} \cos nx = \frac{\sin \frac{1}{2}Nx \cos \frac{1}{2}(N + 1)x}{\sin \frac{1}{2}x}.$$

18. Identify the partial sums of the Fourier series in Exercise 16 as

$$S_N(x) = \sum_{n=1}^{N} \frac{\sin nx}{n}.$$

The series of Exercise 17 is $S_N'(x)$. Use this information to locate the maxima and minima of $S_N(x)$ in the interval $0 \le x \le \pi$. Find the value of $S_N(x)$ at the first point in the interval $0 < x < \pi$ where $S_N'(x) = 0$ for $N = 5$. Compare to $(\pi - x)/2$ at that point.

19. Find the Fourier sine series of the function given by

$$f(x) = \begin{cases} \sin(\pi x/a), & 0 < x < a \\ 0, & a < x < \pi \end{cases}$$

assuming that $0 < a < \pi$.

20. Find the Fourier cosine series of the function given in Exercise 19.

21. Find the Fourier integral representation of the function given by

$$f(x) = \begin{cases} 1, & 0 < x < a \\ 0, & x < 0 \quad \text{or} \quad x > a. \end{cases}$$

22. Find the Fourier sine and cosine integral representations of the function given by

$$f(x) = \begin{cases} (a - x)/a, & 0 < x < a \\ 0, & a < x. \end{cases}$$

23. Find the Fourier sine integral representation of the function

$$f(x) = \begin{cases} \sin x, & 0 < x < \pi \\ 0, & \pi < x. \end{cases}$$

24. Find the Fourier integral representation of the function

$$f(x) = \begin{cases} 1/\varepsilon, & \alpha < x < \alpha + \varepsilon \\ 0, & \text{elsewhere.} \end{cases}$$

25. Use integration by parts to establish the equality

$$\int_0^\infty e^{-\lambda} \cos \lambda x \, d\lambda = \frac{1}{1 + x^2}.$$

26. The equation in Exercise 25 is valid for all x. Explain why its validity implies that

$$\frac{2}{\pi} \int_0^\infty \frac{\cos \lambda x}{1 + x^2} \, dx = e^{-\lambda}, \qquad \lambda > 0.$$

27. Integrate both sides of the equality in Exercise 25 from 0 to t to derive the equality

$$\int_0^\infty \frac{e^{-\lambda} \sin \lambda t}{\lambda} \, d\lambda = \tan^{-1} t.$$

28. Does the equality in Exercise 27 imply that

$$\frac{2}{\pi} \int_0^\infty \tan^{-1} t \sin \lambda t \, dt = \frac{e^{-\lambda}}{\lambda}?$$

29. From Exercise 27 derive the equality

$$\int_0^\infty \frac{1 - e^{-\lambda}}{\lambda} \sin \lambda x \, d\lambda = \frac{\pi}{2} - \tan^{-1} x, \qquad x > 0.$$

30. Without using integration, obtain the Fourier series (period 2π) of each of the following functions:

a. $2 + 4 \sin 50x - 12 \cos 41x$ b. $\sin^2 5x$
c. $\sin(4x + 2)$ d. $\sin 3x \cos 5x$
e. $\cos^3 x$ f. $\cos(2x + \frac{1}{3}\pi)$.

31. Let the function $f(x)$ be given in the interval $0 < x < 1$ by the formula

$$f(x) = 1 - x.$$

Find (a) a sine series (b) a cosine series (c) a sine integral (d) a cosine integral that equals the given function for $0 < x < 1$. In each case, sketch the function to which the series or integral converges in the interval $-2 < x < 2$.

32. Verify the Fourier integral

$$\int_0^\infty \cos \lambda q \exp(-\lambda^2 t)\, d\lambda = \sqrt{\frac{\pi}{4t}} \exp\left(-\frac{q^2}{4t}\right) \qquad (t > 0)$$

by transforming the left-hand side according to these steps: (a) convert to an integral from $-\infty$ to ∞ by using the evenness of the integrand; (b) replace $\cos \lambda q$ by $\exp(i\lambda q)$ (justify this step); (c) complete the square in the exponent; (d) change the variable of integration; (e) use the equality

$$\int_{-\infty}^\infty \exp(-u^2)\, du = \sqrt{\pi}.$$

33. Approximate the first seven cosine coefficients ($\hat{a}_0$, $\hat{a}_1$, ..., $\hat{a}_6$) of the function

$$f(x) = \frac{1}{1 + x^2}, \qquad 0 < x < 1.$$

34. Use Fourier sine series representations of $u(x)$ and of the function $f(x) = x$, $0 < x < a$, to solve the boundary value problem

$$\frac{d^2 u}{dx^2} - \gamma^2 u = -x, \qquad 0 < x < a,$$

$$u(0) = 0, \qquad u(a) = 0.$$

35–43. For each of these exercises,

    a. find the Fourier cosine series of the function;

    b. determine the value to which the series converges at the given values of x;

    c. sketch the even periodic extension of the given function for at least two periods.

44–52. For each of these exercises,

  a. find the Fourier sine series of the function;

  b. determine the value to which the series converges at the given values of x;

  c. sketch the odd periodic extension of the given function for at least two periods.

35 & 44.

$$f(x) = \begin{cases} 0, & 0 < x < a/3 \\ x - a/3, & a/3 < x < 2a/3 \\ a/3, & 2a/3 < x < a \end{cases}$$

$x = 0, a/3, a, -a/2$

36 & 45. $f(x) = \begin{cases} \frac{1}{2}, & 0 < x < a/2 \\ 1, & a/2 < x < a \end{cases}$

$x = a/2, 2a, 0, -a$

37 & 46. $f(x) = \begin{cases} 2x/a, & 0 < x < a/2 \\ (3a - 2x)/2a, & a/2 < x < a \end{cases}$

$x = 0, a/2, a, 3a/2$

38 & 47. $f(x) = \begin{cases} x, & 0 < x < a/2 \\ a/2, & a/2 < x < a \end{cases}$

$x = 0, a, -a/2$

39 & 48. $f(x) = (a - x)/a, \qquad 0 < x < a$

$x = 0, a, -a/2$

40 & 49. $f(x) = \begin{cases} 0, & 0 < x < a/4 \\ 1, & a/4 < x < 3a/4 \\ 0, & 3a/4 < x < a \end{cases}$

$x = 0, a/4, a/2, a, -3a/4$

41 & 50. $f(x) = x(a - x), \qquad 0 < x < a$

$x = 0, -a, -a/2$

42 & 51. $f(x) = e^{kx}, \qquad 0 < x < a$

$x = 0, a/2, a, -a$

43 & 52. $f(x) = \begin{cases} 0, & 0 < x < a/2 \\ 1, & a/2 < x < a \end{cases}$

$x = -a, a/2, a$

53–58. For each of these exercises,

    a. find the Fourier cosine integral representation of the function;
    b. sketch the even extension of the function.

59–64. For each of these exercises,

    a. find the Fourier sine integral representation of the function;
    b. sketch the odd extension of the function.

53 & 59. $f(x) = e^{-x}, \quad 0 < x$

54 & 60. $f(x) = \begin{cases} e^{-x}, & 0 < x < a \\ 0, & a < x \end{cases}$

55 & 61. $f(x) = \begin{cases} 1, & 0 < x < b \\ 0, & b < x \end{cases}$

56 & 62. $f(x) = \begin{cases} \cos x, & 0 < x < \pi \\ 0, & \pi < x \end{cases}$

57 & 63. $f(x) = \begin{cases} 1 - x, & 0 < x < 1 \\ 0, & 1 < x \end{cases}$

58 & 64. $f(x) = \begin{cases} 1, & 0 < x < 1 \\ 2 - x, & 1 < x < 2 \\ 0, & 2 < x \end{cases}$

chapter 2
The
Heat
Equation

1   Derivation and boundary conditions

As the first example of the derivation of a partial differential equation, we consider the problem of describing the temperature in a rod of conducting material. In order to simplify the problem as much as possible, we shall assume that the rod has a uniform cross section (like an extrusion) and that the temperature does not vary from point to point on a section. Thus, if we use a coordinate system as suggested in Fig. 2.1, we may say that the temperature depends only on position x and time t.

The basic idea in developing the partial differential equation is to apply the laws of physics to a small piece of the rod. Specifically, we apply the law of conservation of energy to a slice of the rod which lies between x and $x + \Delta x$ (Fig. 2.2).

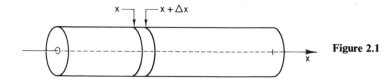

Figure 2.1

Figure 2.2

$q(x,t)$ $q(x + \Delta x, t)$

The law of conservation of energy states that the amount of heat which enters a region plus what is generated inside is equal to the amount of heat which leaves plus the amount stored. The law is equally valid in terms of rates per unit time instead of amounts.

Now let $q(x, t)$ be the rate of heat flow at point x and time t. The dimensions of q are* $[q] = H/tL^2$, and q is taken to be positive when heat flows to the right. The rate at which heat enters the slice through the surface at x is $Aq(x, t)$, where A is the area of a cross section. The rate at which heat leaves the slice through the surface at $x + \Delta x$ is $Aq(x + \Delta x, t)$.

The rate of heat storage in the slice of material is proportional to the rate of change of temperature. Thus, if ρ is the density and c is the heat capacity per unit mass ($[c] = H/mT$), we may approximate the rate of heat storage in the slice by

$$\rho c A \, \Delta x \, \frac{\partial u}{\partial t} \, (x, t)$$

where $u(x, t)$ is the temperature.

In some cases heat may be generated in the material of the rod—for instance, by resistance to an electrical current, or by chemical or nuclear reaction. If the rate of heat generation per unit volume is r, $[r] = H/tL^3$, then the rate at which heat is generated in the slice is $A \, \Delta x \, r$. (Note that r may depend on x, t, and even u.)

We have now quantified the law of conservation of energy for the slice of rod in the form

$$Aq(x, t) + A \, \Delta x \, r = Aq(x + \Delta x, t) + A \, \Delta x \, \rho c \, \frac{\partial u}{\partial t}.$$

After some algebraic manipulation, we have

$$\frac{q(x, t) - q(x + \Delta x, t)}{\Delta x} + r = \rho c \, \frac{\partial u}{\partial t}.$$

The ratio

$$\frac{q(x + \Delta x, t) - q(x, t)}{\Delta x}$$

---

* Square brackets are used to symbolize "dimension of." $H =$ heat, $t =$ time, $T =$ temperature, $L =$ length, $m =$ mass, and so forth.

should be recognized as a difference quotient. If we allow Δx to decrease, this quotient becomes, in the limit,

$$\lim_{\Delta x \to 0} \frac{q(x + \Delta x, t) - q(x, t)}{\Delta x} = \frac{\partial q}{\partial x}.$$

The limit process thus leaves the law of conservation of energy in the form

$$-\frac{\partial q}{\partial x} + r = \rho c \frac{\partial u}{\partial t}.$$

We are not finished, since there are two dependent variables, q and u, in this equation. We need another equation relating q and u. This relation is Fourier's law of heat conduction, which in one dimension may be written:

$$q = -\kappa \frac{\partial u}{\partial x}.$$

In words, heat flows downhill (q is positive when $\partial u/\partial x$ is negative) at a rate proportional to the gradient of the temperature. The "constant" of proportionality κ may depend on x if the rod is not uniform, and also on temperature.

Substituting Fourier's law in the heat balance equation yields

$$\frac{\partial}{\partial x}\left(\kappa \frac{\partial u}{\partial x}\right) + r = \rho c \frac{\partial u}{\partial t}.$$

Note that κ, ρ, and c may all be functions. If, however, they are independent of x, t, and u, we may write

$$\frac{\partial^2 u}{\partial x^2} + \frac{r}{\kappa} = \frac{\rho c}{\kappa}\frac{\partial u}{\partial t}. \tag{1}$$

The equation is applicable in the region $0 < x < a$ and for $t > 0$. The quantity $\kappa/\rho c$ is often written as k, and is called the diffusivity. For some time we will be working with the heat equation without generation:

$$\frac{\partial^2 u}{\partial x^2} = \frac{1}{k}\frac{\partial u}{\partial t}, \qquad 0 < x < a, \quad 0 < t \tag{2}$$

which, to review, is supposed to describe the temperature u in a rod of length a with uniform properties and cross section, in which no heat is generated, and whose cylindrical surface is insulated.

This equation alone is not enough information to completely specify the temperature, however. Each of the functions

$$u(x, t) = x^2 + 2kt$$

$$u(x, t) = e^{-kt} \sin x$$

satisfies the partial differential equation and so do their sum and difference.

Clearly this is not a satisfactory situation either from the mathematical or physical viewpoint; we would like the temperature to be uniquely determined. More conditions must be placed on the function u. The appropriate additional conditions are those that describe:

1. the initial temperature distribution in the rod, and

2. what is happening at the ends of the rod.

The *initial condition* is described mathematically as

$$u(x, 0) = f(x), \qquad 0 < x < a$$

where $f(x)$ is a given function of x alone.

The *boundary conditions* may take a variety of forms. First, the temperature at either end may be held constant, for instance by exposing the end to an ice-water bath or to condensing steam. We can describe these conditions by the equations

$$u(0, t) = T_0, \qquad u(a, t) = T_1, \qquad t > 0$$

where T_0 and T_1 may be the same or different. More generally, the temperature at the boundary may be controlled in some way, without being held constant. If x_0 symbolizes an endpoint, the condition is

$$u(x_0, t) = \alpha(t) \qquad\qquad (I)$$

where α is a function of time. Of course, the case of a constant function is included here. This type of boundary condition is called a "Dirichlet condition" or "condition of the first kind."

Another possibility is that the heat flow rate is controlled. Since Fourier's law associates the heat flow rate and the gradient of the temperature, we can write

$$\frac{\partial u}{\partial x}(x_0, t) = \beta(t) \qquad\qquad (II)$$

where β is a function of time. This is called a "Neumann condition" or "condition of the second kind." We most frequently take $\beta(t)$ identically zero. Then the condition

$$\frac{\partial u}{\partial x}(x_0, t) = 0$$

corresponds to an *insulated* surface, for this equation says that the heat flow is zero.

Still another possible boundary condition is

$$c_1 u(x_0, t) + c_2 \frac{\partial u}{\partial x}(x_0, t) = \gamma(t) \qquad\qquad \text{(III)}$$

called "third kind" or "Robin's condition." This kind of condition can also be realized physically. If the surface at $x = a$ is exposed to air or other fluid, then the heat conducted up to that surface from inside the rod is carried away by convection. Newton's law of cooling says that the rate at which heat is transferred from the body to the air is proportional to the difference in temperature between the body and the air. In symbols, we have

$$q(a, t) = h(u(a, t) - T(t))$$

where $T(t)$ is the air temperature. After application of Fourier's law, this becomes

$$-\kappa \frac{\partial u}{\partial x}(a, t) = hu(a, t) - hT(t). \qquad\qquad \text{(3)}$$

This equation can be put into form III. (Note: h is called the convection coefficient. $[h] = H/L^2 tT$.)

All of the boundary conditions above involve the function u and/or its derivative at one point. If more than one point is involved, the boundary condition is called "mixed." For example, if a uniform rod is bent into a ring and the ends $x = 0$ and $x = a$ are joined, appropriate boundary conditions would be

$$u(0, t) = u(a, t), \qquad t > 0$$

$$\frac{\partial u}{\partial x}(0, t) = \frac{\partial u}{\partial x}(a, t), \qquad t > 0$$

(IV)

both of mixed type.

Many other kinds of boundary conditions exist and are even realizable, but the four kinds mentioned above are the most commonly encountered. An important feature common to all four types is that they involve a *linear* operation on the function u.

The heat equation, an initial condition and a boundary condition for each end form what is called an *initial value–boundary value problem*. For instance, one possible problem would be

$$\frac{\partial^2 u}{\partial x^2} = \frac{1}{k}\frac{\partial u}{\partial t}, \qquad 0 < x < a, \quad 0 < t \tag{4}$$

$$u(0, t) = T_0, \qquad 0 < t \tag{5}$$

$$-\kappa \frac{\partial u}{\partial x}(a, t) = h(u(a, t) - T_1), \qquad 0 < t \tag{6}$$

$$u(x, 0) = f(x), \qquad 0 < x < a. \tag{7}$$

Notice that the boundary conditions may be of different kinds at different ends.

Although we shall not prove it, it is true that there is one, and only one, solution to a complete initial value–boundary value problem.

Exercises

1. Put Eq. (3) into form III. Notice that the signs still indicate that heat flows in the direction of lower temperature. That is, if $u(a, t) > T(t)$, then $q(a, t)$ is positive and the gradient of u is negative. Show that, if the surface at $x = 0$ (left end) is exposed to convection, the boundary condition would read

$$\kappa \frac{\partial u}{\partial x}(0, t) = hu(0, t) - hT(t).$$

Explain the signs.

2. Suppose the surface at $x = a$ is exposed to radiation. The Stefan–Boltzmann law of radiation says that the rate of radiation heat transfer is proportional to the difference of the fourth powers of the *absolute* temperatures of the bodies. Show that a radiation boundary condition at $x = a$ with a transfer to a body at absolute temperature T would be

$$q(a, t) = \sigma(u^4(a, t) - T^4).$$

3. Restate this condition in terms of the gradient or derivative of u at a.

4. The difference above may be written

$$u^4 - T^4 = (u - T)(u^3 + u^2 T + u T^2 + T^3).$$

Under what conditions might the second factor on the right be taken approximately constant? If the factor were constant the boundary condition would be linear.

5. Give a physical interpretation for the problem in Eqs. (4)–(7).

6. Suppose that the end of the rod at $x = 0$ is immersed in an insulated container of water or other fluid; that the temperature of the fluid is the same as the temperature of the end of the rod; that the heat capacity of the fluid is C units of heat per degree. Show that this situation is represented mathematically by the equation

$$C \frac{\partial u}{\partial t} (0, t) = \kappa A \frac{\partial u}{\partial x} (0, t)$$

where A is the cross-sectional area of the rod.

2   Steady-state temperatures

Before tackling a complete heat conduction problem, we shall solve a simplified version called the steady-state or equilibrium problem. We begin with this example:

$$\frac{\partial^2 u}{\partial x^2} = \frac{1}{k} \frac{\partial u}{\partial t}, \qquad 0 < x < a, \quad 0 < t \tag{1}$$

$$u(0, t) = T_0, \qquad 0 < t \tag{2}$$

$$u(a, t) = T_1, \qquad 0 < t \tag{3}$$

$$u(x, 0) = f(x), \qquad 0 < x < a. \tag{4}$$

We may think of $u(x, t)$ as the temperature in a cylindrical rod, with insulated lateral surface, whose ends are held at temperatures T_0 and T_1.

Experience indicates that after a long time under the same conditions, the variation of temperature with time dies away. In terms of the function $u(x, t)$ that represents temperature, we thus expect that the limit of $u(x, t)$, as t tends to infinity, exists and depends only on x:

$$\lim_{t \to \infty} u(x, t) = v(x)$$

and also that

$$\lim_{t \to \infty} \frac{\partial u}{\partial t} = 0.$$

The function $v(x)$, called the *steady state* temperature distribution, must still satisfy the boundary conditions and the heat equation, which are valid for all $t > 0$. Therefore $v(x)$ should be the solution to the problem

$$\frac{d^2v}{dx^2} = 0, \qquad 0 < x < a \tag{5}$$

$$v(0) = T_0, \qquad v(a) = T_1. \tag{6}$$

On integrating the differential equation twice, we find

$$\frac{dv}{dx} = A, \qquad v(x) = Ax + B.$$

The constants A and B are to be chosen so that $v(x)$ satisfies the boundary conditions:

$$v(0) = B = T_0, \qquad v(a) = Aa + B = T_1.$$

When these two equations are solved for A and B, the steady-state distribution becomes

$$v(x) = T_0 + (T_1 - T_0)\frac{x}{a}. \tag{7}$$

Of course, Eqs. (5) and (6), which together form the steady-state problem corresponding to Eqs. (1)–(4), could have been derived from scratch as was done in Chapter 0, Section 3. Here, however, we see it as part of a more comprehensive problem.

We can establish this rule for setting up the steady-state problem corresponding to a given heat conduction problem: take limits in all equations valid for large t (the partial differential equation and the boundary conditions), replacing u and its derivatives with respect to x by v and its derivatives, and replacing $\partial u/\partial t$ by 0.

Now we take as an example Eqs. (4)–(7) of Section 1, which were

$$\frac{\partial^2 u}{\partial x^2} = \frac{1}{k}\frac{\partial u}{\partial t}, \qquad 0 < x < a, \;\; 0 < t \tag{8}$$

$$u(0, t) = T_0, \qquad 0 < t \tag{9}$$

$$-\kappa\frac{\partial u}{\partial x}(a, t) = h(u(a, t) - T_1), \qquad 0 < t \tag{10}$$

$$u(x, 0) = f(x), \qquad 0 < x < a. \tag{11}$$

The rule above, when applied to this problem, leads us to the following equations:

$$\frac{d^2v}{dx^2} = 0, \qquad 0 < x < a$$

$$v(0) = T_0, \qquad -\kappa v'(a) = h(v(a) - T_1).$$

The solution of the differential equation is $v(x) = A + Bx$. The boundary conditions require that A and B satisfy

$$v(0) = A = T_0$$

$$-\kappa v'(a) = -\kappa B = h(A + Ba - T_1).$$

Solving simultaneously, we find

$$A = T_0, \qquad B = \frac{h(T_1 - T_0)}{\kappa + ha}.$$

Thus the steady-state solution of Eqs. (8)–(11) is (see Fig. 2.6)

$$v(x) = T_0 + \frac{xh(T_1 - T_0)}{\kappa + ha}. \tag{12}$$

In both of these examples, the steady-state temperature distribution has been uniquely determined by the differential equation and boundary conditions. This is usually the case, but not always. For the problem

$$\frac{\partial^2 u}{\partial x^2} = \frac{1}{k}\frac{\partial u}{\partial t}, \qquad 0 < x < a, \quad 0 < t \tag{13}$$

$$\frac{\partial u}{\partial x}(0, t) = 0, \qquad 0 < t \tag{14}$$

$$\frac{\partial u}{\partial x}(a, t) = 0, \qquad 0 < t \tag{15}$$

$$u(x, 0) = f(x), \qquad 0 < x < a, \tag{16}$$

which describes the temperature in an insulated rod having insulated ends also, the corresponding steady-state problem for $v(x) = \lim_{t \to \infty} u(x, t)$ is

$$\frac{d^2v}{dx^2} = 0, \qquad 0 < x < a$$

$$\frac{dv}{dx}(0) = 0, \qquad \frac{dv}{dx}(a) = 0.$$

It is easy to see that $v(x) = T$ (any constant) is a solution to this problem.

The solution is not unique. However, there are physical considerations by which T can be determined. Since the lateral surface of the rod is insulated, and since the ends are also insulated in this problem, the rod does not exchange heat with the rest of the universe. Thus the heat present at $t = 0$ is still present for any other time.

If ρ, c, and A have the same meaning as before, we may say that the heat content of a slice of the rod between x and $x + \Delta x$ is given approximately by $\rho c A \, \Delta x u(x, t)$, and so the total heat content of the rod at time t is

$$\int_0^a \rho c A u(x, t) \, dx.$$

At time $t = 0$, $u(x, 0) = f(x)$, while in the limit $u(x, t) \to v(x)$. The heat content at $t = 0$ and in the limit must be the same:

$$\int_0^a \rho c A f(x) \, dx = \int_0^a \rho c A v(x) \, dx = \int_0^a \rho c A T \, dx = \rho c A a T.$$

The coefficients ρ, c, and A are independent of x and may be cancelled. Thus we find

$$T = \frac{1}{a} \int_0^a f(x) \, dx$$

which says that T, the final temperature in the rod, is the average of the temperature at $t = 0$.

It should not be supposed that every steady-state temperature distribution has a straight-line graph. This is certainly not the case in the problem of Exercise 1.

While the steady-state solution gives us some valuable information about the solution of an initial value–boundary value problem, it also has an important position as the first step in finding the complete solution. We now isolate the "rest" of the unknown temperature $u(x, t)$ by defining the *transient* temperature distribution,

$$w(x, t) = u(x, t) - v(x).$$

The name "transient" is appropriate because, according to our assumptions about the behavior of u for large values of t, we expect $w(x, t)$ to tend to zero as t tends to infinity.

In general, the transient also satisfies an initial value–boundary value problem that is similar to the original one but is distinguished by having a homogeneous partial differential equation and boundary conditions. To illustrate this point, we shall treat the problem stated in Eqs. (1)–(4) whose steady-state solution is given by Eq. (7).

By using the equality $u(x, t) = w(x, t) + v(x)$, and what we know about v —that is, Eqs. (5) and (6)—we make the original problem into a new problem for $w(x, t)$. We have the following relations:

$$\frac{\partial^2 u}{\partial x^2} = \frac{\partial^2 w}{\partial x^2} + \frac{d^2 v}{dx^2} = \frac{\partial^2 w}{\partial x^2}$$

$$\frac{\partial u}{\partial t} = \frac{\partial w}{\partial t} + \frac{dv}{dt} = \frac{\partial w}{\partial t}$$

$$u(0, t) = T_0 = w(0, t) + v(0) = w(0, t) + T_0$$

$$u(a, t) = T_1 = w(a, t) + v(a) = w(a, t) + T_1$$

$$u(x, 0) = w(x, 0) + v(x).$$

By substituting into Eqs. (1)–(4), we get the following initial value–boundary value problem for w:

$$\frac{\partial^2 w}{\partial x^2} = \frac{1}{k} \frac{\partial w}{\partial t}, \qquad 0 < x < a, \quad 0 < t \qquad (17)$$

$$w(0, t) = 0, \qquad 0 < t \qquad (18)$$

$$w(a, t) = 0, \qquad 0 < t \qquad (19)$$

$$w(x, 0) = f(x) - \left[T_0 + (T_1 - T_0)\frac{x}{a} \right] \qquad (20)$$

$$= g(x), \qquad 0 < x < a.$$

It is evident from Eqs. (17)–(19) that the homogeneous partial differential equation and boundary conditions predicted for w have been established. We shall see in the next section how the transient temperature can be found.

Exercises

1. State and solve the steady-state problem corresponding to

$$\frac{\partial^2 u}{\partial x^2} - \gamma^2(u - T) = \frac{1}{k} \frac{\partial u}{\partial t}, \qquad 0 < x < a, \quad 0 < t$$

$$u(0, t) = T_0, \, u(a, t) = T_1, \qquad 0 < t$$

$$u(x, 0) = 0, \qquad 0 < x < a.$$

2. State the problem satisfied by the transient temperature distribution corresponding to the problem in Exercise 1.

3. Obtain the steady-state solution of the problem

$$\frac{\partial^2 u}{\partial x^2} + \gamma^2(u - T) = \frac{1}{k}\frac{\partial u}{\partial t}, \qquad 0 < x < a, \quad 0 < t$$

$$u(0, t) = T, u(a, t) = T, \qquad 0 < t$$

$$u(x, 0) = T_1 x/a, \qquad 0 < x < a.$$

Give a physical interpretation of this problem. What happens if $\gamma = \pi/a$?

4. State the initial value–boundary value problem satisfied by the transient temperature distribution corresponding to Eqs. (8)–(11).

5. Find the steady-state solution of the problem

$$\frac{\partial}{\partial x}\left(\kappa \frac{\partial u}{\partial x}\right) = c\rho \frac{\partial u}{\partial t}, \qquad 0 < x < a, \quad 0 < t$$

$$u(0, t) = T_0, u(a, t) = T_1, \qquad 0 < t$$

if

$$\kappa(x) = b + dx, \qquad b, d \quad \text{constants.}$$

6. Find and sketch the steady-state solution of the heat conduction problem in the interval $0 < x < a$ for the given boundary conditions.

a. $\dfrac{\partial u}{\partial x}(0, t) = 0, \quad u(a, t) = T_0$

b. $u(0, t) - \dfrac{\partial u}{\partial x}(0, t) = T_0, \quad \dfrac{\partial u}{\partial x}(a, t) = 0$

c. $u(0, t) - \dfrac{\partial u}{\partial x}(0, t) = T_0, \quad u(a, t) + \dfrac{\partial u}{\partial x}(a, t) = T_1$

7. Find the steady-state solution of the inhomogeneous problems:

$$\frac{\partial^2 u}{\partial x^2} + r = \frac{1}{k}\frac{\partial u}{\partial t}, \qquad 0 < x < a, \quad 0 < t$$

a. $u(0, t) = T_0, \quad \dfrac{\partial u}{\partial x}(a, t) = 0, \quad r = \text{const.}$

b. $u(0, t) = T_0, \quad u(a, t) = T_1, \quad r = \alpha - \beta^2 u.$

3 Example: Fixed end temperatures

In Section 1 we saw that the temperature $u(x, t)$ in a uniform rod with insulated lateral surface would be determined by the problem

$$\frac{\partial^2 u}{\partial x^2} = \frac{1}{k} \frac{\partial u}{\partial t}, \qquad 0 < x < a, \quad 0 < t \tag{1}$$

$$u(0, t) = T_0, \qquad 0 < t \tag{2}$$

$$u(a, t) = T_1, \qquad 0 < t \tag{3}$$

$$u(x, 0) = f(x), \qquad 0 < x < a \tag{4}$$

if the ends of the rod are held at fixed temperatures and if the initial temperature distribution is $f(x)$. In Section 2 we found that the steady-state temperature distribution,

$$v(x) = \lim_{t \to \infty} u(x, t),$$

satisfied the boundary value problem

$$\frac{d^2 v}{dx^2} = 0, \qquad 0 < x < a \tag{5}$$

$$v(0) = T_0, \qquad v(a) = T_1. \tag{6}$$

In fact, we were able to find $v(x)$ explicitly:

$$v(x) = T_0 + (T_1 - T_0) \frac{x}{a}. \tag{7}$$

We also defined the transient temperature distribution as

$$w(x, t) = u(x, t) - v(x)$$

and determined that w satisfies the boundary value–initial value problem

$$\frac{\partial^2 w}{\partial x^2} = \frac{1}{k} \frac{\partial w}{\partial t}, \qquad 0 < x < a, \quad 0 < t \tag{8}$$

$$w(0, t) = 0, \qquad 0 < t \tag{9}$$

$$w(a, t) = 0, \qquad 0 < t \tag{10}$$

$$w(x, 0) = f(x) - v(x) = g(x), \qquad 0 < x < a. \tag{11}$$

Our objective now is to determine the transient temperature distribution, $w(x, t)$, and—since $v(x)$ is already known—the unknown temperature

$$u(x, t) = v(x) + w(x, t). \tag{12}$$

The problem in w can be attacked by a method called product method, separation of variables, or Fourier's method. For this method to work, it is essential to have *homogeneous boundary conditions*. Thus, the method may be applied to the transient distribution w but *not* to the original function u. Of course, since both the partial differential equation and the boundary conditions satisfied by $w(x, t)$ are homogeneous, the function $w \equiv 0$ satisfies them. Because this solution is obvious and is of no help in satisfying the initial condition, it is called the *trivial* solution. We are seeking the unobvious, nontrivial solutions, so we shall avoid the trivial solution at every turn.

The general idea of the method is to assume that the solution of the partial differential equation has the form of a product: $w(x, t) = \phi(x)T(t)$. Since each of the factors depends on only one variable, we have

$$\frac{\partial^2 w}{\partial x^2} = \phi''(x)T(t), \qquad \frac{\partial w}{\partial t} = \phi(x)T'(t).$$

The partial differential equation becomes

$$\phi''T = \frac{1}{k}\phi T'$$

and on dividing through by ϕT, we find

$$\frac{\phi''(x)}{\phi(x)} = \frac{T'(t)}{kT(t)}.$$

On the left is a function of x and on the right a function of t. If this equation is to hold for $0 < x < a$ and $0 < t$, the mutual value of these two functions must be constant:

$$\frac{\phi''}{\phi} = p, \qquad \frac{T'}{kT} = p.$$

Now we have two ordinary differential equations for the two factor functions:

$$\phi'' - p\phi = 0, \qquad T' - pkT = 0. \tag{13}$$

The two boundary conditions on w may also be stated in the product form:

$$w(0, t) = \phi(0)T(t) = 0, \qquad w(a, t) = \phi(a)T(t) = 0.$$

There are two ways that these equations can be satisfied for all $t > 0$. Either the function $T(t) = 0$ for all t, or the other factors must be zero. But in the first case, $w = \phi(x)T(t)$ is also identically zero, leading us back to the trivial solution. Therefore we take the other alternative and choose

$$\phi(0) = 0, \qquad \phi(a) = 0. \tag{14}$$

Our job now is to solve the equations (13) and satisfy the boundary conditions (14) while avoiding the trivial solution. The solutions of Eq. (13) are (assuming $p > 0$)

$$\phi(x) = a' \cosh \sqrt{p}\, x + b' \sinh \sqrt{p}\, x, \qquad T(t) = c' e^{pkt}.$$

Applying the boundary conditions (14) to $\phi(x)$ leads us to conclude first that $a' = 0$ and then that $b' = 0$, so that $\phi(x) \equiv 0$. But this gives the trivial solution, $w(x, t) = 0$, which is unsatisfactory. The same conclusion is reached if we try $p = 0$.

We now try a negative constant. Replacing p by $-\lambda^2$ in Eq. (13) gives us the two equations

$$\phi'' + \lambda^2 \phi = 0, \qquad T' + \lambda^2 k T = 0$$

whose solutions are

$$\phi(x) = a' \cos \lambda x + b' \sin \lambda x, \qquad T(t) = c' \exp(-\lambda^2 kt).$$

If ϕ has the form given above, the boundary conditions require that $\phi(0) = a' = 0$, leaving $\phi(x) = b' \sin \lambda x$. Then $\phi(a) = b' \sin \lambda a = 0$.

We now have two choices: either $b' = 0$, making $\phi(x) = 0$ for all values of x; or $\sin \lambda a = 0$. We reject the first possibility, for it leads to the trivial solution $w(x, t) = 0$. In order for the second possibility to hold, we must have $\lambda = n\pi/a$, where $n = \pm 1, \pm 2, \pm 3, \ldots$. The negative values of n do not give any new functions, since $\sin(-\theta) = -\sin\theta$. Hence we allow $n = 1, 2, 3, \ldots$ only. We shall set $\lambda_n = n\pi/a$.

Incidentally, since the differential equation (13) and the boundary conditions (14) for $\phi(x)$ are homogeneous, any constant multiple of a solution is still a solution. We shall therefore remember this fact and drop the constant b' in $\phi(x)$. Likewise, we delete the c' in $T(t)$.

To review our position, we have for each $n = 1, 2, 3, \ldots$, a function $\phi_n(x) = \sin \lambda_n x$, and an associated function $T_n = \exp(-\lambda_n^2 kt)$. The product $w_n(x, t) = \sin \lambda_n x \exp(-\lambda_n^2 kt)$ has the properties:

1. $w_n(x, t)$ satisfies the heat equation;
2. $w_n(0, t) = 0$;
3. $w_n(a, t) = 0$.

We now apply the *principle of superposition* and form a linear combination of the ws

$$w(x, t) = \sum_{n=1}^{\infty} b_n \sin \lambda_n x \exp(-\lambda_n^2 kt). \qquad (15)$$

As each term satisfies the heat equation and since that is a linear, homogeneous equation, the sum of the series should satisfy the heat equation.

(There is a mathematical question about convergence which we shall ignore.) Also, since each term is 0 at $x = 0$ and at $x = a$, the sum of the series must also be 0 at those two points, for any choice of constants b_n. Therefore, the function $w(x, t)$ satisfies the differential equation and the two boundary conditions of the boundary value–initial value problem. Of the four parts of the original problem, only the initial condition has not yet been satisfied. At $t = 0$, the exponentials in Eq. (15) are all unity. Thus the initial condition takes the form

$$w(x, 0) = \sum_{n=1}^{\infty} b_n \sin(n\pi x/a) = g(x), \qquad 0 < x < a.$$

We immediately recognize a problem in Fourier series, which is solved by choosing the constants b_n according to the formula

$$b_n = \frac{2}{a} \int_0^a g(x) \sin\left(\frac{n\pi x}{a}\right) dx.$$

If the function $g(x)$ is continuous and sectionally smooth, we know that the Fourier series actually converges to g in the interval $0 < x < a$, so the solution which we have found for $w(x, t)$ actually satisfies all requirements set on w. Even if g does not satisfy these conditions, it can be shown that the solution we have arrived at is the best one can do.

Once the transient temperature has been determined, we find the original unknown $u(x, t)$ as the sum of the transient and the steady state solutions,

$$u(x, t) = v(x) + w(x, t).$$

Example   Suppose the original problem to be

$$\frac{\partial^2 u}{\partial x^2} = \frac{1}{k} \frac{\partial u}{\partial t}, \qquad 0 < x < a, \quad 0 < t$$

$$u(0, t) = T_0, \qquad 0 < t$$

$$u(a, t) = T_1, \qquad 0 < t$$

$$u(x, 0) = 0, \qquad 0 < x < a.$$

The steady-state solution is

$$v(x) = T_0 + (T_1 - T_0)\frac{x}{a},$$

$w(x, t) = u(x, t) - v(x)$ satisfies

$$\frac{\partial^2 w}{\partial x^2} = \frac{1}{k}\frac{\partial w}{\partial t}, \qquad\qquad 0 < x < a, \quad 0 < t$$

$$w(0, t) = 0, \qquad\qquad 0 < t$$

$$w(a, t) = 0, \qquad\qquad 0 < t$$

$$w(x, 0) = -T_0 - (T_1 - T_0)\frac{x}{a} = g(x), \qquad 0 < x < a.$$

According to our calculations above, w has the form

$$w(x, t) = \sum b_n \sin \lambda_n x \exp(-\lambda_n^2 kt) \qquad (16)$$

and the initial condition is

$$w(x, 0) = \sum b_n \sin\left(\frac{n\pi x}{a}\right) = g(x), \qquad 0 < x < a.$$

The coefficients b_n are given by

$$b_n = \frac{2}{a}\int_0^a \left[-T_0 - (T_1 - T_0)\frac{x}{a}\right]\sin\left(\frac{n\pi x}{a}\right)dx$$

$$= -\frac{2T_0}{a}\left.\frac{\cos(n\pi x/a)}{(n\pi/a)}\right|_0^a$$

$$\quad -\frac{2}{a^2}(T_1 - T_0)\left.\frac{\sin(n\pi x/a) - (n\pi x/a)\cos(n\pi x/a)}{(n\pi/a)^2}\right|_0^a$$

$$= -\frac{2T_0}{n\pi}(1 - (-1)^n) + \frac{2(T_1 - T_0)}{n\pi}(-1)^n$$

$$b_n = \frac{-2}{n\pi}(T_0 - T_1(-1)^n).$$

Now the complete solution (see Fig. 2.3) is

$$u(x, t) = w(x, t) + T_0 + (T_1 - T_0)\frac{x}{a}$$

$$w(x, t) = -\frac{2}{\pi}\sum_1^\infty \frac{T_0 - T_1(-1)^n}{n}\sin \lambda_n x \exp(-\lambda_n^2 kt). \qquad (17)$$

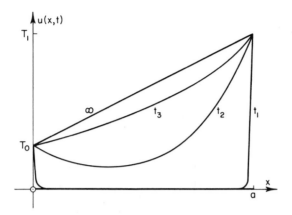

Figure 2.3   The solution $u(x, t)$, corresponding to $u(x, 0) = 0$, graphed at $t = 0 <$ $t_1 < t_2 < t_3 < \infty$.

We can discover certain features of $u(x, t)$ by examining the solution. First, $u(x, 0)$ really is zero $(0 < x < a)$ because the Fourier series converges. Second, when t is positive but very small, the series for $w(x, t)$ will almost equal $-T_0 - (T_1 - T_0)x/a$. But at $x = 0$ and $x = a$, the series adds up to zero (and $w(x, t)$ is a continuous function of x); thus $u(x, t)$ satisfies the boundary conditions. Third, when t is large, $\exp(-\lambda_1^2 kt)$ is small, and the other exponentials are still smaller. Then $w(x, t)$ may be well approximated by the first term (or first few terms) of the series. Finally, as $t \to \infty$, $w(x, t)$ disappears completely.

Exercises

1.   Write out the first few terms of the series for $w(x, t)$ in Eq. (17).

 2.   If $k = 1 \text{ cm}^2/\text{sec}$, $a = 1$ cm, show that after $t = 0.5$ sec the other terms of the series for w are negligible compared with the first term. Sketch $u(x, t)$ for $t = 0$, $t = 0.5$, $t = 1.0$, and $t = \infty$.

3.   Reformulate the heat conduction problem in terms of the dimensionless distance x/a and time kt/a^2. Is there an appropriate dimensionless temperature?

4.   Sketch the functions ϕ_1, ϕ_2, and ϕ_3, and verify that they satisfy the boundary conditions $\phi(0) = 0$, $\phi(a) = 0$.

5. Find a function $w(x, t)$ that satisfies Eqs. (8)–(11) for each of these initial conditions:

    a. $g(x) = T_0$

    b. $g(x) = \beta x$

    c. $g(x) = \beta(a - x)$.

4 Example: Insulated bar

We shall consider again the uniform bar which was discussed in Section 1. Let us suppose now that the ends of the bar at $x = 0$ and $x = a$ are insulated instead of being held at constant temperatures. The boundary value–initial value problem which describes the temperature in this rod is:

$$\frac{\partial^2 u}{\partial x^2} = \frac{1}{k}\frac{\partial u}{\partial t}, \qquad\qquad 0 < x < a, \quad 0 < t \qquad (1)$$

$$\frac{\partial u}{\partial x}(0, t) = 0, \qquad \frac{\partial u}{\partial x}(a, t) = 0, \qquad 0 < t \qquad (2)$$

$$u(x, 0) = f(x), \qquad\qquad 0 < x < a \qquad (3)$$

where $f(x)$ is supposed to be a given function.

We saw in Section 2 that the solution of the steady-state problem is not unique; by invoking a physical argument we determined that the steady-state temperature should be

$$\lim_{t \to \infty} u(x, t) = \frac{1}{a}\int_0^a f(x)\,dx.$$

Since the boundary conditions and the partial differential equation are homogeneous in this example, it is not necessary for us to establish the transient problem. We may look for $u(x, t)$ directly.

Assume that u has the product form $u(x, t) = \phi(x)T(t)$. The heat equation becomes

$$\phi''T = \frac{1}{k}\phi T'$$

and the variables are separated by dividing through by ϕT, leaving

$$\frac{\phi''(x)}{\phi(x)} = \frac{T'(t)}{kT(t)}, \qquad 0 < x < a, \quad 0 < t.$$

In order that a function of x equal a function of t, their mutual value must be a constant. If that constant were positive, T would be an increasing exponential function of time, which would be unacceptable. It is also easy to show that if the constant were positive, ϕ could not satisfy the boundary conditions without being identically zero.

Assuming then a negative constant, we can write

$$\frac{\phi''}{\phi} = -\lambda^2 = \frac{T'}{kT}$$

and separate these equalities into two ordinary differential equations linked by the common parameter λ^2:

$$\phi'' + \lambda^2 \phi = 0, \qquad 0 < x < a \qquad (4)$$
$$T' + \lambda^2 kT = 0, \qquad 0 < t. \qquad (5)$$

The boundary conditions on u can be translated into conditions on ϕ, because they are homogeneous conditions. The boundary conditions in product form are

$$\frac{\partial u}{\partial x}(0, t) = \phi'(0)T(t) = 0, \qquad 0 < t$$

$$\frac{\partial u}{\partial x}(a, t) = \phi'(a)T(t) = 0, \qquad 0 < t.$$

To satisfy these equations, we must have the function $T(t)$ always zero (which would make $u(x, t) \equiv 0$), or else

$$\phi'(0) = 0, \qquad \phi'(a) = 0.$$

The second alternative avoids the trivial solution.

We now have a homogeneous differential equation for ϕ together with homogeneous boundary conditions:

$$\phi'' + \lambda^2 \phi = 0, \qquad 0 < x < a \qquad (6)$$
$$\phi'(0) = 0, \qquad \phi'(a) = 0. \qquad (7)$$

A problem of this kind is called an *eigenvalue problem*. We are looking for those values of the parameter λ^2 for which nonzero solutions of Eqs. (6) and (7) may exist. Those values are called *eigenvalues*, and the corresponding solutions are called eigenfunctions. Note that the significant parameter is λ^2, not λ. The square is used only for convenience.

The general solution of the differential equation in Eq. (6) is

$$\phi(x) = A \cos \lambda x + B \sin \lambda x.$$

Applying the boundary condition at $x = 0$, we see that $\phi'(0) = B\lambda = 0$, giving $B = 0$ or $\lambda = 0$. We put aside the case $\lambda = 0$ and assume $B = 0$, so $\phi(x) = A \cos \lambda x$. Then the second boundary condition requires that $\phi'(a) = -A\lambda \sin \lambda a = 0$. Once again, we may have $A = 0$ or $\sin \lambda a = 0$. But $A = 0$ makes $\phi(x) \equiv 0$, and therefore $u(x, t) \equiv 0$; this is a trivial solution. We choose therefore to make $\sin \lambda a = 0$ by restricting λ to the values π/a, $2\pi/a$, $3\pi/a$, ... We label the eigenvalues with a subscript:

$$\lambda_n{}^2 = \left(\frac{n\pi}{a}\right)^2, \qquad \phi_n(x) = \cos \lambda_n(x), \qquad n = 1, 2, \dots .$$

Returning to the case $\lambda = 0$, we see that Eqs. (6) and (7) become

$$\phi'' = 0, \qquad 0 < x < a,$$

$$\phi'(0) = 0, \qquad \phi'(a) = 0.$$

Any constant function certainly satisfies these conditions. Therefore, we designate

$$\lambda_0{}^2 = 0, \qquad \phi_0(x) = 1.$$

Now that the numbers $\lambda_n{}^2$ are known, we can solve Eq. (5) for $T(t)$, finding

$$T_0 = 1, \qquad T_n = \exp(-\lambda_n{}^2 kt).$$

The products $\phi_n T_n$ give solutions of the partial differential equation (1) which satisfy the boundary conditions, Eq. (2):

$$u_n(x, t) = \phi_n(x)T_n(t) = \cos \lambda_n x \exp(-\lambda_n{}^2 kt).$$

Because the partial differential equation and the boundary conditions are all linear and homogeneous, any linear combination of solutions is also a solution. The solution $u(x, t)$ of the whole system may therefore have the form

$$u(x, t) = a_0 + \sum_1^\infty a_n \cos \lambda_n x \exp(-\lambda_n{}^2 kt).$$

There is only one condition of the original set remaining to be satisfied, the initial condition Eq. (3). For $u(x, t)$ in the form above, the initial condition is

$$u(x, 0) = a_0 + \sum_1^\infty a_n \cos \lambda_n x = f(x), \qquad 0 < x < a.$$

Here we recognize a problem in Fourier series and can immediately cite formulas for the coefficients:

$$a_0 = \frac{1}{a} \int_0^a f(x)\, dx, \qquad a_n = \frac{2}{a} \int_0^a f(x) \cos\left(\frac{n\pi x}{a}\right) dx.$$

When these coefficients are computed and substituted in the formulas for
$u(x, t)$, that function becomes the solution to the initial value-boundary value
problem Eqs. (1)–(3). Notice that the first term of the series solution a_0 is
exactly the function we found for a steady-state solution before; and, in fact,
when $t \to \infty$ all other terms in $u(x, t)$ *do* disappear leaving

$$\lim_{t \to \infty} u(x, t) = a_0 = \frac{1}{a} \int_0^a f(x)\, dx.$$

In Fig. 2.4 is a graph of the solution to Eqs. (1)–(3) for initial condition
$f(x) = T_0 + (T_1 - T_0)x/a$.

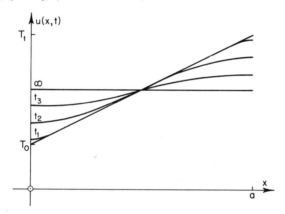

Figure 2.4 The solution $u(x, t)$, corresponding to $u(x, 0) = T_0 + (T_1 - T_0)x/a$,
graphed at $t = 0 < t_1 < t_2 < t_3 < \infty$.

Exercises

1. If f is sectionally continuous, the coefficients $a_n \to 0$ as $n \to \infty$. For
   $t = t_1 > 0$, fixed, the solution is

$$u(x, t_1) = a_0 + \sum_1^\infty a_n \exp(-\lambda_n^2 k t_1) \cos \lambda_n x$$

and the coefficients of this cosine series are

$$A_n(t_1) = a_n \exp(-\lambda_n^2 k t_1).$$

Show that $A_n(t_1) \to 0$ so rapidly as $n \to \infty$ that the series above converges
uniformly $0 \le x \le a$. Show the same for the series which represents

$$\frac{\partial^2 u}{\partial x^2}(x, t_1).$$

2. Sketch the functions ϕ_1, ϕ_2, and ϕ_3 and verify graphically that they satisfy the boundary conditions of Eq. (7).

3. Using the initial condition

$$u(x, 0) = T_1 \frac{x}{a}, \qquad 0 < x < a$$

find the solution $u(x, t)$ of this example problem. Sketch $u(x, 0)$, $u(x, t)$ for some $t > 0$ (using the first three terms of the series), and the steady-state solution.

4. Verify that $u_n(x, t)$ satisfies the partial differential equation (1) and the boundary conditions, Eq. (2).

5. Repeat Exercise 3 using the initial condition

$$u(x, 0) = T_0 + T_1 \left(\frac{x}{a}\right)^2, \qquad 0 < x < a.$$

6. Show that if the "separation constant" were positive, then ϕ could not satisfy the boundary conditions without being identically zero.

5 Example: Different boundary conditions

In many cases, boundary conditions at the two endpoints will be of different kinds. In this section we shall solve the problem of finding the temperature in a rod having one end insulated and the other held at a constant temperature. The boundary value–initial value problem satisfied by the temperature in the rod is

$$\frac{\partial^2 u}{\partial x^2} = \frac{1}{k} \frac{\partial u}{\partial t}, \qquad 0 < x < a, \quad 0 < t \tag{1}$$

$$u(0, t) = T_0, \qquad 0 < t \tag{2}$$

$$\frac{\partial u}{\partial x} (a, t) = 0, \qquad 0 < t \tag{3}$$

$$u(x, 0) = f(x), \qquad 0 < x < a. \tag{4}$$

It is easy to verify that the steady-state solution of this problem is

$v(x) = T_0$. Using this information, we can find the boundary value–initial value problem satisfied by the transient temperature $w(x, t) = u(x, t) - T_0$:

$$\frac{\partial^2 w}{\partial x^2} = \frac{1}{k}\frac{\partial w}{\partial t}, \qquad\qquad 0 < x < a, \quad 0 < t \qquad\qquad (5)$$

$$w(0, t) = 0, \qquad\qquad 0 < t \qquad\qquad (6)$$

$$\frac{\partial w}{\partial x}(a, t) = 0, \qquad\qquad 0 < t \qquad\qquad (7)$$

$$w(x, 0) = f(x) - T_0 = g(x), \qquad 0 < x < a. \qquad\qquad (8)$$

Since this problem is homogeneous, we can attack it by the method of separation of variables. The assumption that $w(x, t)$ has the form of a product, $\phi(x)T(t)$, and insertion of w in that form into the partial differential equation (5) lead to the separated equations

$$\phi''(x) + \lambda^2\phi = 0, \qquad 0 < x < a \qquad\qquad (9)$$

$$T' + \lambda^2 kT = 0, \qquad 0 < t. \qquad\qquad (10)$$

Moreover, the boundary conditions take the form

$$\phi(0)T(t) = 0, \qquad 0 < t \qquad\qquad (11)$$

$$\phi'(a)T(t) = 0, \qquad 0 < t. \qquad\qquad (12)$$

As before, we conclude that $\phi(0)$ and $\phi'(a)$ should both be zero:

$$\phi(0) = 0, \qquad \phi'(a) = 0. \qquad\qquad (13)$$

Now, the general solution of the differential equation (9) is

$$\phi(x) = a' \cos \lambda x + b' \sin \lambda x.$$

The boundary condition, $\phi(0) = 0$, requires that $a' = 0$, leaving

$$\phi(x) = b' \sin \lambda x.$$

The boundary condition at $x = a$ now takes the form

$$\phi'(a) = b'\lambda \cos \lambda a = 0.$$

The three choices are: $b' = 0$, which gives the trivial solution; $\lambda = 0$, which should be investigated separately (Exercise 2), and $\cos \lambda a = 0$. The third alternative—the only acceptable one—requires that λa be an odd multiple of $\pi/2$, which we may express as

$$\lambda_n = (2n - 1)\pi/2a, \qquad n = 1, 2, \dots . \qquad\qquad (14)$$

The eigenfunctions are given by the formula

$$\phi_n(x) = \sin \lambda_n x. \tag{15}$$

With the eigenfunctions and eigenvalues now in hand, we return to the differential equation (10), whose solution is

$$T_n(t) = \exp(-\lambda_n^2 kt).$$

As in previous cases, we assemble the general solution of the homogeneous problem expressed in equations (5), (6), and (7) by forming a general linear combination of our product solutions,

$$w(x, t) = \sum_1^\infty b_n \sin \lambda_n x \exp(-\lambda_n^2 kt). \tag{16}$$

The choice of the coefficients, b_n, must be made so as to satisfy the initial condition, Eq. (8). Using the form of w given by equation (16), we find that the initial condition is

$$w(x, 0) = \sum_1^\infty b_n \sin \frac{(2n-1)\pi x}{2a} = g(x), \qquad 0 < x < a \tag{17}$$

A routine Fourier sine series for the interval $0 < x < a$ would involve the functions $\sin(n\pi x/a)$, rather than the functions we have. By one of several means (Exercises 3–8), it may be shown that the series in Eq. (17) represents the function $g(x)$ provided that g is sectionally smooth and that we choose the coefficients by the formula

$$b_n = \frac{2}{a} \int_0^a g(x) \sin \frac{(2n-1)\pi x}{2a} \, dx. \tag{18}$$

Now the original problem is completely solved. The solution is

$$u(x, t) = T_0 + \sum_1^\infty b_n \sin \lambda_n x \exp(-\lambda_n^2 kt). \tag{19}$$

It is good practice to verify that all the requirements of Eqs. (1)–(4) are indeed satisfied.

By way of example, suppose that the initial condition (4) is

$$u(x, 0) = T_1, \qquad 0 < x < a.$$

Then $g(x) = T_1 - T_0$, $0 < x < a$, and the coefficients as determined by Eq. (18) are

$$b_n = (T_1 - T_0) \frac{4}{\pi(2n-1)}.$$

Therefore, the complete solution of the boundary value–initial value problem with initial condition $u(x, 0) = T_1$ would be

$$u(x, t) = T_0 + (T_1 - T_0)\frac{4}{\pi}\sum_1^\infty \frac{1}{2n-1}\sin\lambda_n x\,\exp(-\lambda_n^2 kt). \qquad (20)$$

Now that we have been through three examples, we can outline the method we have been using to solve linear boundary value–initial value problems. Up to this moment we have seen only homogeneous partial differential equations, but a nonhomogeneity which is independent of t can be treated by the same technique.

1. If the partial differential equation or a boundary condition or both are not homogeneous, first find a function $v(x)$, independent of t, which satisfies the partial differential equation and the boundary conditions. Since $v(x)$ does not depend on t, the partial differential equation applied to $v(x)$ becomes an ordinary differential equation. Finding $v(x)$ is just a matter of solving a two-point boundary value problem.

2. Determine the initial value–boundary value problem satisfied by the "transient solution" $w(x, t) = u(x, t) - v(x)$. This must be a *homogeneous* problem. That is, the partial differential equation and the boundary conditions (but not usually the initial condition) are satisfied by the constant function 0.

3. Assuming that $w(x, t) = \phi(x)T(t)$, separate the partial differential equation into two ordinary differential equations, one for $\phi(x)$ and one for $T(t)$, linked by the separation constant, $-\lambda^2$. Also reduce the boundary conditions to conditions on ϕ alone.

4. Solve the eigenvalue problem for ϕ. That is, find the values of λ^2 for which the eigenvalue problem has nonzero solutions. Label the eigenfunctions and eigenvalues $\phi_n(x)$ and λ_n^2.

5. Solve the ordinary differential equation for the time factors, $T_n(t)$.

6. Form the general solution of the homogeneous problem as a sum of constant multiples of the product solutions:

$$w(x, t) = \sum c_n \phi_n(x)T_n(t).$$

7. Choose the c_n so that the initial condition is satisfied. This may or may not be a routine Fourier series problem.

8. Form the solution of the original problem

$$u(x, t) = v(x) + w(x, t)$$

and check that all conditions are satisfied.

Exercises

1. Find the steady-state solution of the problem stated in Eqs. (1)–(4).

2. Determine whether 0 is an eigenvalue of the eigenvalue problem stated in Eqs. (9) and (13).

3. To justify the expansion of Eq. (17), for an arbitrary sectionally smooth $g(x)$,

$$\sum b_n \sin \frac{(2n-1)\pi x}{2a} = g(x), \qquad 0 < x < a,$$

construct the function $G(x)$ with these properties:

$$G(x) = g(x), \qquad 0 < x < a$$
$$G(x) = g(2a - x), \qquad a < x < 2a.$$

Show that $G(x)$ corresponds to the series

$$G(x) \sim \sum B_N \sin \frac{N\pi x}{2a}, \qquad 0 < x < 2a.$$

4. Show that the B_N of the series above satisfy

$$B_N = 0 \quad (N \text{ even}), \qquad B_N = \frac{2}{a} \int_0^a g(x) \sin \frac{N\pi x}{2a} \, dx \quad (N \text{ odd}).$$

5. Sketch $g(x)$ and $G(x)$, and find the corresponding series, if $g(x)$ is given by

   a. $g(x) = x, \quad 0 < x < a$

   b. $g(x) = T, \quad 0 < x < a.$

6. Show that the eigenfunctions found in this section are orthogonal. That is, prove that

$$\int_0^a \sin \lambda_n x \sin \lambda_m x \, dx = \begin{cases} 0 & (m \neq n) \\ \dfrac{a}{2} & (m = n) \end{cases}$$

when $\lambda_n = (2n-1)\pi/2a.$

7. Solve the problem stated in Eqs. (1)–(4), taking $f(x) = Tx/a.$

8. Use the orthogonality relation in Exercise 6 to justify the formula in Eq. (18).

9. Solve the nonhomogeneous problem

$$\frac{\partial^2 u}{\partial x^2} = \frac{1}{k}\frac{\partial u}{\partial t} + \frac{T}{a^2}, \qquad 0 < x < a, \quad 0 < t$$

$$u(0, t) = T_0, \quad \frac{\partial u}{\partial x}(a, t) = 0, \qquad 0 < t$$

$$u(x, 0) = T_0, \qquad 0 < x < a.$$

10. Compare the solution of Exercise 9 with Eq. (20). Can one be turned into the other?

11. Solve the problem

$$\frac{\partial^2 u}{\partial x^2} = \frac{1}{k}\frac{\partial u}{\partial t}, \qquad 0 < x < a, \quad 0 < t$$

$$\frac{\partial u}{\partial x}(0, t) = 0, \, u(a, t) = T_0, \qquad 0 < t$$

$$u(x, 0) = T_1, \qquad 0 < x < a.$$

6 Example: Convection

We have seen three examples in which boundary conditions specified either u or $\partial u/\partial x$. Now we shall study a case where a condition of the third kind is involved. The physical model is conduction of heat in a rod with insulated lateral surface whose left end is held at constant temperature and whose right end is exposed to convective heat transfer. The boundary value–initial value problem satisfied by the temperature in the rod is

$$\frac{\partial^2 u}{\partial x^2} = \frac{1}{k}\frac{\partial u}{\partial t}, \qquad 0 < x < a, \quad 0 < t \qquad (1)$$

$$u(0, t) = T_0, \qquad 0 < t \qquad (2)$$

$$-\kappa \frac{\partial u}{\partial x}(a, t) = h(u(a, t) - T_1), \qquad 0 < t \qquad (3)$$

$$u(x, 0) = f(x), \qquad 0 < x < a. \qquad (4)$$

We found in Section 2 that the steady-state solution of this problem is

$$v(x) = T_0 + \frac{xh(T_1 - T_0)}{\kappa + ha}. \qquad (5)$$

Now, since the original boundary conditions were nonhomogeneous, we form the problem for the transient solution $w(x, t) = u(x, t) - v(x)$. By direct substitution it is found that

$$\frac{\partial^2 w}{\partial x^2} = \frac{1}{k}\frac{\partial w}{\partial t}, \qquad\qquad\qquad 0 < x < a, \quad 0 < t \qquad (6)$$

$$w(0, t) = 0, \qquad hw(a, t) + \kappa\frac{\partial w}{\partial x}(a, t) = 0, \qquad 0 < t \qquad (7)$$

$$w(x, 0) = f(x) - v(x) = g(x), \qquad\qquad 0 < x < a. \qquad (8)$$

The solution for $w(x, t)$ can now be found by the product method. On the assumption that w has the form of a product $\phi(x)T(t)$, the variables can be separated exactly as before, giving two ordinary differential equation linked by a common parameter λ^2:

$$\phi'' + \lambda^2\phi = 0, \qquad 0 < x < a$$
$$T' + \lambda^2 kT = 0, \qquad 0 < t.$$

Also, since the boundary conditions are linear and homogeneous, they can be translated directly into conditions on ϕ:

$$w(0, t) = \phi(0)T(t) = 0$$

$$\kappa\frac{\partial w}{\partial x}(a, t) + hw(a, t) = [\kappa\phi'(a) + h\phi(a)]T(t) = 0.$$

Either $T(t)$ is identically zero (which would make $w(x, t)$ identically zero), or

$$\phi(0) = 0, \qquad \kappa\phi'(a) + h\phi(a) = 0.$$

Combining the differential equation and boundary conditions on ϕ, we get the eigenvalue problem

$$\phi'' + \lambda^2\phi = 0, \qquad 0 < x < a \qquad\qquad (9)$$

$$\phi(0) = 0, \qquad \kappa\phi'(a) + h\phi(a) = 0. \qquad\qquad (10)$$

The general solution of the differential equation is

$$\phi(x) = a'\cos \lambda x + b'\sin \lambda x.$$

The boundary condition at zero requires that $\phi(0) = a' = 0$, leaving $\phi(x) = b'\sin \lambda x$. Now, at the other boundary,

$$\kappa\phi'(a) + h\phi(a) = b'(\kappa\lambda\cos \lambda a + h\sin \lambda a) = 0.$$

Discarding the possibilities $b' = 0$ and $\lambda = 0$, which both lead to the trivial solution, we are left with the equation

$$\kappa\lambda \cos \lambda a + h \sin \lambda a = 0, \qquad \text{or} \qquad \tan \lambda a = -\frac{\kappa}{h}\lambda.$$

From sketches of the graphs of $\tan \lambda a$ and $-\kappa\lambda/h$ (Fig. 2.5), we see that there are an infinite number of solutions, $\lambda_1, \lambda_2, \lambda_3, \ldots$, and that, for very large n, λ_n is given *approximately* by

$$\lambda_n \simeq \frac{2n-1}{2}\frac{\pi}{a}.$$

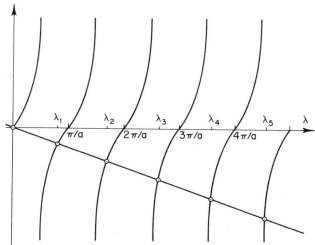

Figure 2.5   Graphs of $\tan \lambda a$ and $-\kappa\lambda/h$.

(Note: Solutions are tabulated in *Handbook of Mathematical Functions* by Abramowitz and Stegun [1].)

Thus we have for each $n = 1, 2, \ldots$, an eigenvalue λ_n^2 and an eigenfunction $\phi_n(x)$ which satisfies the eigenvalue problem Eqs. (9) and (10). Accompanying $\phi_n(x)$ is the function

$$T_n(t) = \exp(-\lambda_n^2 k t)$$

that makes $w_n(x, t) = \phi_n(x)T_n(t)$ a solution of the partial differential equation (6) and the boundary conditions Eq. (7). Since Eq. (6) and the conditions Eq. (7) are linear and homogeneous, any linear combination of solutions is a solution also. Therefore, the transient solution will have the form

$$w(x, t) = \sum_1^\infty b_n \sin \lambda_n x \exp(-\lambda_n^2 k t)$$

and the remaining condition to be satisfied, the initial condition Eq. (8), is

$$w(x, 0) = \sum_1^\infty b_n \sin \lambda_n x = g(x), \qquad 0 < x < a. \tag{11}$$

This is to say that the constants b_n are to be chosen so as to make the infinite series equal $g(x)$.

Although Eq. (11) looks like a Fourier series problem, it is not, because λ_2, λ_3, and so forth, are not all integer multiples of λ_1. If we attempt to use the idea of orthogonality, we can still find a way to select the b_n, for it may be shown by direct computation that

$$\int_0^a \sin \lambda_n x \sin \lambda_m x \, dx = 0, \qquad \text{if} \quad n \neq m.$$

Then if we multiply both sides of the proposed Eq. (11) by $\sin \lambda_m x$ (where m is fixed) and integrate from 0 to a, all the terms of the series disappear, except the one in which $n = m$, yielding an equation for b_m:

$$b_m = \frac{\int_0^a g(x) \sin \lambda_m x \, dx}{\int_0^a \sin^2 \lambda_m x \, dx}. \tag{12}$$

By this formula, the b_m may be calculated and inserted into the formula for $w(x, t)$. Then we may put together the solution $u(x, t)$ of the original problem Eqs. (1)–(4):

$$u(x, t) = v(x) + w(x, t)$$

$$= T_0 + \frac{xh(T_1 - T_0)}{(\kappa + ha)} + \sum b_n \sin \lambda_n x \exp(-\lambda_n^2 k t).$$

In Figure 2.6 is a graph of $w(x, t)$ corresponding to $f(x) = 0$.

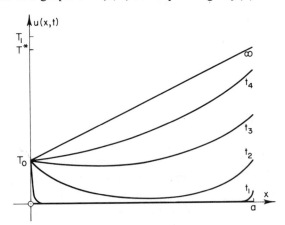

Figure 2.6 The solution $u(x, t)$, corresponding to $u(x, 0) = 0$, graphed at $t = 0 < t_1 < t_2 < t_3 < t_4 < \infty$.

Exercises

1. Sketch $v(x)$ as given in Eq. (5) assuming

    a. $T_1 > T_0$ **b.** $T_1 = T_0$ **c.** $T_1 < T_0$.

2. Verify the orthogonality integral by direct integration. It will be necessary to use the equation that defines the λ_n :

$$\kappa \lambda_n \cos \lambda_n a + h \sin \lambda_n a = 0.$$

3. Why have we ignored the negative solutions of the equation

$$\tan \lambda a = \frac{-\kappa\lambda}{h}?$$

4. Derive the formula Eq. (12) for the coefficients b_m .

5. Sketch the first two eigenfunctions of this example taking $\kappa/h = 0.5$. ($\lambda_1 = 2.29/a$, $\lambda_2 = 5.09/a$).

6. Verify that

$$\int_0^a \sin^2 \lambda_m x \; dx = \frac{a}{2} + \frac{\kappa \cos^2 \lambda_m a}{h} \frac{1}{2}.$$

7. Find the coefficients b_m corresponding to

$$g(x) = 1, \qquad 0 < x < a.$$

8. Using the solution of Exercise 7, write out the first few terms of the solution of Eqs. (6)–(8), where $g(x) = T$, $0 < x < a$.

9. Same as Exercise 7 for

$$g(x) = x, \qquad 0 < x < a.$$

7   Sturm–Liouville problems

At the end of the preceding section, we saw that ordinary Fourier series are not quite adequate for all the problems we can solve. We can make some generalizations, however, which do cover most cases that arise from separation of variables. In simple problems, we often find eigenvalue problems of the form

$$\phi'' + \lambda^2 \phi = 0, \qquad l < x < r \tag{1}$$

$$\alpha_1 \phi(l) - \alpha_2 \phi'(l) = 0 \tag{2}$$

$$\beta_1 \phi(r) + \beta_2 \phi'(r) = 0. \tag{3}$$

It is not difficult to determine the eigenvalues of this problem and to show the eigenfunctions orthogonal by direct calculation, but an indirect calculation is still easier.

Suppose that ϕ_n and ϕ_m are eigenfunctions corresponding to different eigenvalues λ_n^2 and λ_m^2. That is,

$$\phi_n'' + \lambda_n^2 \phi_n = 0, \qquad \phi_m'' + \lambda_m^2 \phi_m = 0,$$

and both functions satisfy the boundary conditions. Let us multiply the first differential equation by ϕ_m, the second by ϕ_n, subtract the two, and move the terms containing $\phi_n \phi_m$ to the other side:

$$\phi_n'' \phi_m - \phi_m'' \phi_n = (\lambda_m^2 - \lambda_n^2) \phi_n \phi_m .$$

The right-hand side is a constant (nonzero) multiple of the integrand in the orthogonality relation

$$\int_l^r \phi_n(x)\phi_m(x)\, dx = 0, \qquad n \neq m$$

which is proved true if the left-hand side is zero:

$$\int_l^r (\phi_n'' \phi_m - \phi_m'' \phi_n)\, dx = 0.$$

This integral is integrable by parts

$$\int_l^r (\phi_n'' \phi_m - \phi_m'' \phi_n)\, dx = [\phi_n'(x)\phi_m(x) - \phi_m'(x)\phi_n(x)]\Big|_l^r - \int_l^r (\phi_n' \phi_m' - \phi_m' \phi_n')\, dx.$$

The last integral is obviously zero, so we have

$$(\lambda_m^2 - \lambda_n^2) \int_l^r \phi_n(x)\phi_m(x)\, dx = [\phi_n'(x)\phi_m(x) - \phi_m'(x)\phi_n(x)]\Big|_l^r .$$

Both ϕ_n and ϕ_m satisfy the boundary condition at $x = r$,

$$\beta_1 \phi_m(r) + \beta_2 \phi_m'(r) = 0$$
$$\beta_1 \phi_n(r) + \beta_2 \phi_n'(r) = 0.$$

These two equations may be considered simultaneous equations in β_1 and β_2. At least one of the numbers β_1 and β_2 is different from zero; otherwise there would be no boundary condition. Hence the determinant of the equations must be zero:

$$\phi_m(r)\phi_n'(r) - \phi_n(r)\phi_m'(r) = 0.$$

A similar result holds at $x = l$. Thus

$$[\phi_n{}'(x)\phi_m(x) - \phi_m{}'(x)\phi_n(x)]\Big|_l^r = 0$$

and therefore we have proved the orthogonality relation

$$\int_l^r \phi_n(x)\phi_m(x)\,dx = 0, \qquad n \neq m$$

for eigenfunctions of Eqs. (1)–(3).

We may make a much broader generalization about orthogonality of eigenfunctions with very little trouble. Consider the model eigenvalue problem below, which might arise from separation of variables in a heat conduction problem (see Section 9).

$$[s(x)\phi'(x)]' - q(x)\phi(x) + \lambda^2 p(x)\phi(x) = 0, \qquad l < x < r$$
$$\alpha_1\phi(l) - \alpha_2\phi'(l) = 0$$
$$\beta_1\phi(r) + \beta_2\phi'(r) = 0.$$

Let us carry out the procedure used above with this problem. The eigenfunctions satisfy the differential equations

$$(s\phi_n{}')' - q\phi_n + \lambda_n{}^2 p\phi_n = 0$$
$$(s\phi_m{}')' - q\phi_m + \lambda_m{}^2 p\phi_m = 0.$$

Multiply the first by ϕ_m, the second by ϕ_n, subtract (the terms containing $q(x)$ cancel), and move the term containing $p\phi_n\phi_m$ to the other side:

$$(s\phi_n{}')'\phi_m - (s\phi_m{}')'\phi_n = (\lambda_m{}^2 - \lambda_n{}^2)p\phi_n\phi_m. \tag{4}$$

Integrate both sides from l to r, and apply integration by parts to the left-hand side:

$$\int_l^r [(s\phi_n{}')'\phi_m - (s\phi_m{}')'\phi_n]\,dx = [s\phi_n{}'\phi_m - s\phi_m{}'\phi_n]\Big|_l^r - \int_l^r (s\phi_n{}'\phi_m{}' - s\phi_n{}'\phi_m{}')\,dx.$$

The second integral is zero. From the boundary conditions we find that

$$\phi_n{}'(r)\phi_m(r) - \phi_m{}'(r)\phi_n(r) = 0$$
$$\phi_n{}'(l)\phi_m(l) - \phi_m{}'(l)\phi_n(l) = 0.$$

by the same reasoning as before. Hence, we discover the orthogonality relation

$$\int_l^r p(x)\phi_n(x)\phi_m(x)\,dx = 0, \qquad \lambda_n{}^2 \neq \lambda_m{}^2$$

for the eigenfunctions of the problem stated.

During these operations, we have made some tacit assumptions about integrability of functions after Eq. (4). In individual cases, where the coefficient functions s, q, and p and the eigenfunctions themselves are known, one can easily check the validity of the steps taken. In general, however, we would like to guarantee the existence of eigenfunctions and the legitimacy of computations after Eq. (4). To do so, we need the following:

Definition The problem

$$(s\phi')' - q\phi + \lambda^2 p\phi = 0, \qquad l < x < r \tag{5}$$

$$\alpha_1\phi(l) - \alpha_2\phi'(l) = 0 \tag{6}$$

$$\beta_1\phi(r) + \beta_2\phi'(r) = 0 \tag{7}$$

is called a *regular Sturm–Liouville problem* if the following conditions are fulfilled:

a. $s(x)$, $s'(x)$, $q(x)$, and $p(x)$ are continuous for $l \leq x \leq r$;
b. $s(x) > 0$ and $p(x) > 0$ for $l \leq x \leq r$;
c. $\alpha_1^2 + \alpha_2^2 > 0$, $\beta_1^2 + \beta_2^2 > 0$.
d. The parameter λ occurs only where shown.

Condition (a) and the first condition (b) together are imposed to be sure that the differential equation has continuous solutions. Notice that $s(l)$ and $s(r)$ must both be positive (not zero). Condition (c) just says that there are two boundary conditions: $\alpha_1^2 + \alpha_2^2 = 0$ only if $\alpha_1 = \alpha_2 = 0$, which would be no condition. The other requirements contribute to the desired properties in ways that are not obvious.

We are now ready to state the theorems that contain necessary information about eigenfunctions.

Theorem 1 The regular Sturm–Liouville problem has an infinite number of eigenfunctions ϕ_1, ϕ_2, ..., each corresponding to a different eigenvalue λ_1^2, λ_2^2, ... If $n \neq m$, the eigenfunctions ϕ_n and ϕ_m are orthogonal with *weight function* $p(x)$:

$$\int_l^r \phi_n(x)\phi_m(x)p(x)\, dx = 0, \qquad n \neq m.$$

The theorem is already proved, for the continuity of coefficients and eigenfunctions makes our previous calculations legitimate. It should be noted that any multiple of an eigenfunction is also an eigenfunction, but aside from a constant multiplier, the eigenfunctions of a Sturm–Liouville problem are unique.

A number of other properties of the Sturm–Liouville problem are known. We summarize a few below.

Theorem 2 (a) The regular Sturm–Liouville problem has an infinite number of eigenvalues, and $\lambda_n^2 \to \infty$ as $n \to \infty$.

(b) If the eigenvalues are numbered in order, $\lambda_1^2 < \lambda_2^2 < \dots$, then the eigenfunction corresponding to λ_n^2 has exactly $n - 1$ zeros in the interval $l < x < r$ (endpoints excluded).

(c) If $q(x) \ge 0$, and α_1, α_2, β_1, β_2, are all greater than or equal to zero, then all the eigenvalues are nonnegative.

Examples

1. We note that the eigenvalue problems in Sections 3–6 of this chapter are all regular Sturm–Liouville problems, as is (1)–(3) of this section. In particular, the problem

$$\phi'' + \lambda^2 \phi = 0, \qquad 0 < x < a$$
$$\phi(0) = 0, \qquad h\phi(a) + \kappa\phi'(a) = 0$$

is a regular Sturm–Liouville problem, in which

$$s(x) = p(x) = 1, \qquad q(x) = 0, \qquad \alpha_1 = 1, \qquad \alpha_2 = 0, \qquad \beta_1 = h, \qquad \beta_2 = \kappa.$$

All the conditions are met.

2. A less trivial example is

$$(x\phi')' + \lambda^2\left(\frac{1}{x}\right)\phi = 0, \qquad 1 < x < 2$$

$$\phi(1) = 0, \qquad \phi(2) = 0.$$

We identify $s(x) = x$, $p(x) = 1/x$, $q(x) = 0$. This is a regular Sturm–Liouville problem. The orthogonality relation is

$$\int_1^2 \phi_n(x)\phi_m(x)\frac{1}{x}\,dx = 0, \qquad n \ne m.$$

Exercises

1. The general solution of the differential equation in Example 2 is

$$\phi(x) = A\,\cos(\lambda \ln x) + B\,\sin(\lambda \ln x).$$

Find the eigenvalues and eigenfunctions and verify the orthogonality relation directly by integration.

2. Check the results of Theorem 2 for the problem consisting of

$$\phi'' + \lambda^2\phi = 0, \qquad 0 < x < a,$$

with boundary conditions

a. $\phi(0) = 0$, $\phi(a) = 0$ **b.** $\phi'(0) = 0$, $\phi'(a) = 0$.

In case (b), $\lambda_1^2 = 0$.

3. Find the eigenvalues and eigenfunctions and sketch the first few eigenfunctions of the problem

$$\phi'' + \lambda^2\phi = 0, \qquad 0 < x < a$$

with boundary conditions

a. $\phi(0) = 0$, $\phi'(a) = 0$
b. $\phi'(0) = 0$, $\phi(a) = 0$
c. $\phi(0) = 0$, $\phi(a) + \phi'(a) = 0$
d. $\phi(0) - \phi'(0) = 0$, $\phi'(a) = 0$
e. $\phi(0) - \phi'(0) = 0$, $\phi(a) + \phi'(a) = 0$.

4. In Eqs. (1)–(3), take $l = 0$, $r = a$, and show that

a. The eigenfunctions are $\phi_n(x) = \alpha_2\lambda_n \cos \lambda_n x + \alpha_1 \sin \lambda_n x$
b. The eigenvalues must be solutions of the equation

$$-\tan \lambda a = \frac{\lambda(\alpha_1\beta_2 + \alpha_2\beta_1)}{\alpha_1\beta_1 - \alpha_2\beta_2\lambda^2}.$$

5. Show that the eigenfunctions of each of the following problems are orthogonal, and state the orthogonality relation

a. $\phi'' + \lambda^2(1 + x)\phi = 0$, $\phi(0) = 0$, $\phi'(a) = 0$
b. $(e^x\phi')' + \lambda^2 e^x\phi = 0$, $\phi(0) - \phi'(0) = 0$, $\phi(a) = 0$
c. $\phi'' + (\lambda^2/x^2)\phi = 0$, $\phi(1) = 0$, $\phi'(2) = 0$
d. $\phi'' - \sin x\ \phi + e^x\lambda^2\phi = 0$, $\phi'(0) = 0$, $\phi'(a) = 0$.

6. Consider the problem

$$(s\phi')' - q\phi + \lambda^2 p\phi = 0, \qquad l < x < r$$
$$\phi(r) = 0$$

in which $s(l) = 0$, $s(x) > 0$ for $l < x \le r$, but p and q satisfy the conditions of a regular Sturm–Liouville problem. Require also that both $\phi(x)$ and $\phi'(x)$ have finite limits as $x \to l+$. Show that the eigenfunctions (if they exist) are orthogonal.

7. The problem below is not a regular Sturm–Liouville problem. Why? Show that the eigenfunctions are not orthogonal.

$$\phi'' + \lambda^2 \phi = 0, \qquad 0 < x < a$$
$$\phi(0) = 0, \qquad \phi'(a) - \lambda^2 \phi(a) = 0$$

8. Show that 0 is an eigenvalue of the problem

$$(s\phi')' + \lambda^2 p\phi = 0, \qquad l < x < r$$
$$\phi'(l) = 0, \qquad \phi'(r) = 0$$

where s and p satisfy the conditions of a regular Sturm–Liouville problem.

8 Expansion in series of eigenfunctions

We have seen that the eigenfunctions which arise from a regular Sturm–Liouville problem

$$(s\phi')' - q\phi + \lambda^2 p\phi = 0, \qquad l < x < r \tag{1}$$

$$\alpha_1 \phi(l) - \alpha_2 \phi'(l) = 0 \tag{2}$$

$$\beta_1 \phi(r) + \beta_2 \phi'(r) = 0 \tag{3}$$

are orthogonal with weight function $p(x)$:

$$\int_l^r p(x)\phi_n(x)\phi_m(x)\,dx = 0, \qquad n \neq m \tag{4}$$

and it should be clear, from the way in which the question of orthogonality arose, that we are interested in expressing functions in terms of eigenfunction series.

Suppose that a function $f(x)$ is given in the interval $l < x < r$ and that we wish to express $f(x)$ in terms of the eigenfunctions $\phi_n(x)$ of Eqs. (1)–(3). That is, we wish to have

$$f(x) = \sum_1^\infty c_n \phi_n(x), \qquad l < x < r. \tag{5}$$

The orthogonality relation Eq. (4) clearly tells us how to compute the coefficients. Multiplying both sides of the proposed Eq. (5) by $\phi_m(x)p(x)$ (where m is a fixed integer) and integrating from l to r yields

$$\int_l^r f(x)\phi_m(x)p(x)\,dx = \sum_{n=1}^\infty c_n \int_l^r \phi_n(x)\phi_m(x)p(x)\,dx.$$

The orthogonality relation says that all the terms in the series, except that one in which $n = m$, must disappear. Thus

$$\int_l^r f(x)\phi_m(x)p(x)\,dx = c_m \int_l^r \phi_m{}^2(x)p(x)\,dx$$

gives a formula for choosing c_m.

We can now cite a convergence theorem for expansion in terms of eigenfunctions. Notice the similarity to the Fourier series convergence theorem. Of course, the Fourier sine or cosine series are series of eigenfunctions of a regular Sturm–Liouville problem in which the weight function $p(x)$ is 1.

Theorem Let ϕ_1, ϕ_2, ... be eigenfunctions of a regular Sturm–Liouville problem Eqs. (1)–(3), in which the αs and βs are not negative.
If $f(x)$ is sectionally smooth on the interval $l < x < r$, then

$$\frac{f(x+) + f(x-)}{2} = \sum c_n\phi_n(x), \qquad l < x < r \tag{6}$$

where

$$c_n = \frac{\int_l^r f(x)\phi_n(x)p(x)\,dx}{\int_l^r \phi_n{}^2(x)p(x)\,dx}.$$

Furthermore, if the series

$$\sum |c_n| \left[\int_l^r \phi_n{}^2(x)p(x)\,dx\right]^{1/2}$$

converges, then the series Eq. (6) converges uniformly, $l \le x \le r$.

Exercises

1. Verify that

$$\lambda_n{}^2 = \left(\frac{n\pi}{\ln b}\right)^2, \qquad \phi_n = \sin(\lambda_n \ln x)$$

are the eigenvalues and eigenfunctions of

$$(x\phi')' + \lambda^2\left(\frac{1}{x}\right)\phi = 0, \qquad 1 < x < b$$

$$\phi(1) = 0, \qquad \phi(b) = 0.$$

Find the expansion of the function $f(x) = x$ in terms of these eigenfunctions. To what values does the series converge at $x = 1$ and $x = b$?

2. If $\phi_1, \phi_2, \ldots$ are the eigenfunctions of a regular Sturm–Liouville problem and are orthogonal with weight function $p(x)$ on $l < x < r$, and if $f(x)$ is sectionally smooth, then

$$\int_l^r f^2(x)p(x)\, dx = \sum a_n c_n{}^2$$

where

$$a_n = \int_l^r \phi_n{}^2(x)p(x)\, dx$$

and c_n is the coefficient of f as given in the theorem. Show why this should be true and conclude that $c_n \sqrt{a_n} \to 0$ as $n \to \infty$.

3. Verify that the eigenvalues and eigenfunctions of the problem

$$(e^x \phi')' + e^x \gamma^2 \phi = 0, \qquad 0 < x < a$$
$$\phi(0) = 0, \qquad \phi(a) = 0$$

are

$$\gamma_n{}^2 = \left(\frac{n\pi}{a}\right)^2 + \frac{1}{4}, \qquad \phi_n(x) = \exp\left(-\frac{x}{2}\right)\sin\left(\frac{n\pi x}{a}\right).$$

Find the coefficients for the expansion of the function $f(x) = 1, 0 < x < a$, in terms of the ϕ_n.

4. If $\phi_1, \phi_2, \ldots$ are eigenfunctions of a regular Sturm–Liouville problem, the numbers $\sqrt{a_n}$ are called normalizing constants, and the functions $\psi_n = \phi_n/\sqrt{a_n}$ are called normalized eigenfunctions. Show that

$$\int_l^r \psi_n{}^2(x)p(x)\, dx = 1, \qquad \int_l^r \psi_n(x)\psi_m(x)p(x)\, dx = 0, \qquad n \neq m.$$

5. Find the formula for the coefficients of a sectionally smooth function $f(x)$ in the series

$$f(x) = \sum b_n \psi_n(x), \qquad l < x < r$$

where the ψ_n are normalized eigenfunctions.

6. Show that, for the function in Exercise 5

$$\int_l^r f^2(x)p(x)\, dx = \sum b_n{}^2.$$

7. What are the normalized eigenfunctions of the following problem?

$$\phi'' + \lambda^2 \phi = 0, \qquad 0 < x < 1$$
$$\phi'(0) = 0, \qquad \phi'(1) = 0$$

$$\frac{\partial}{\partial x}\left[e^{-x}\frac{\partial u(x,t)}{\partial x}\right] = \frac{1}{h}e^x\frac{\partial u}{\partial t} \quad (0<x<\ln 2\,;\, t>0)$$

$$u(0,t) = 1\,;\quad \frac{\partial u}{\partial x}(\ln 2, t) = 2\,;\quad f(x) = 0$$

9 Generalities on the heat conduction problem

On the basis of the information we have about the Sturm–Liouville problem, we can make some observations on a fairly general heat conduction problem. We take as a physical model a rod whose lateral surface is insulated. In order to simplify slightly, we will assume that no heat is generated inside the rod.

Since material properties may vary with position, the partial differential equation which governs the temperature $u(x, t)$ in the rod will be

$$\frac{\partial}{\partial x}\left(\kappa(x)\frac{\partial u}{\partial x}\right) = \rho(x)c(x)\frac{\partial u}{\partial t}, \qquad l < x < r, \quad 0 < t. \tag{1}$$

Any of the three types of boundary conditions may be imposed at either boundary, so we use as boundary conditions

$$\alpha_1 u(l, t) - \alpha_2 \frac{\partial u}{\partial x}(l, t) = c_1, \qquad t > 0 \tag{2}$$

$$\beta_1 u(r, t) + \beta_2 \frac{\partial u}{\partial x}(r, t) = c_2, \qquad t > 0. \tag{3}$$

If the temperature is fixed, the coefficient of $\partial u/\partial x$ is zero. If the boundary is insulated, the coefficient of u is zero, and the right-hand side is zero also. If there is convection at a boundary, both coefficients will be positive, and the signs will be as shown.

We already know that in the case of two insulated boundaries, the steady-state solution has some peculiarities, so we set this aside as a special case. Assume, then, that either α_1 or β_1 or both are positive. Finally we need an initial condition in the form

$$u(x, 0) = f(x), \qquad l < x < r. \tag{4}$$

Equations (1)–(4) make up an initial value–boundary value problem.

Assuming that c_1 and c_2 are constants, we must first find the steady-state solution

$$v(x) = \lim_{t \to \infty} u(x, t).$$

The function $v(x)$ satisfies the boundary value problem

$$\frac{d}{dx}\left(\kappa(x)\frac{dv}{dx}\right) = 0, \qquad l < x < r \tag{5}$$

$$\alpha_1 v(l) - \alpha_2 v'(l) = c_1 \tag{6}$$

$$\beta_1 v(r) + \beta_2 v'(r) = c_2. \tag{7}$$

Since we have assumed that at least one of α_1 or β_1 is positive, this problem can be solved. In fact, it is possible to give a formula for $v(x)$ in terms of the function (see Exercise 1)

$$\int_l^x \frac{d\xi}{\kappa(\xi)} = I(x). \tag{8}$$

Before proceeding further, it is convenient to introduce some new functions. Let $\bar{\kappa}$, $\bar{\rho}$, and $\bar{c}$ indicate average values of the functions $\kappa(x)$, $\rho(x)$, and $c(x)$. We shall define dimensionless functions $s(x)$ and $p(x)$ by

$$\kappa(x) = \bar{\kappa} s(x), \qquad \rho(x)c(x) = \bar{\rho}\bar{c}p(x).$$

Also we define the transient temperature to be

$$w(x, t) = u(x, t) - v(x).$$

By direct computation, using the fact that $v(x)$ is a solution of Eqs. (5)–(7), we can show that $w(x, t)$ satisfies the initial value–boundary value problem

$$\frac{d}{\partial x}\left(s(x)\frac{\partial w}{\partial x}\right) = \frac{1}{k}p(x)\frac{\partial w}{\partial t}, \qquad l < x < r, \qquad 0 < t \tag{9}$$

$$\alpha_1 w(l, t) - \alpha_2 \frac{\partial w}{\partial x}(l, t) = 0, \qquad\qquad 0 < t \tag{10}$$

$$\beta_1 w(r, t) + \beta_2 \frac{\partial w}{\partial x}(r, t) = 0, \qquad\qquad 0 < t \tag{11}$$

$$w(x, 0) = f(x) - v(x) = g(x), \qquad\qquad l < x < r \tag{12}$$

which has homogeneous boundary conditions. The constant k is defined to be $\bar{\kappa}/\bar{\rho}\bar{c}$.

Now we use our method of separation of variables to find w. If w has the form $w(x, t) = \phi(x)T(t)$, the differential equation becomes

$$T(t)(s(x)\phi'(x))' = \frac{1}{k}p(x)\phi(x)T'(t)$$

and, on dividing through by $p\phi T$, we find the separated equation

$$\frac{(s\phi')'}{p\phi} = \frac{T'}{kT}, \qquad l < x < r, \qquad 0 < t.$$

As before, the equality between a function of x and a function of t can hold only if their mutual value is constant. Furthermore, we expect the constant to be negative, so we put

$$\frac{(s\phi')'}{p\phi} = \frac{T'}{kT} = -\lambda^2$$

and separate two ordinary equations

$$T' + \lambda^2 kT = 0, \qquad 0 < t$$

$$(s\phi')' + \lambda^2 p\phi = 0, \qquad l < x < r.$$

The boundary conditions, being linear and homogeneous, can also be changed into conditions on ϕ. For instance, Eq. (10) becomes

$$[\alpha_1\phi(l) - \alpha_2\phi'(l)]T(t) = 0, \qquad 0 < t$$

and since $T(t) = 0$ makes $w(x, t) = 0$, we take the other factor to be zero. We have, then, the eigenvalue problem

$$(s\phi')' + \lambda^2 p\phi = 0, \qquad l < x < r \tag{13}$$

$$\alpha_1\phi(l) - \alpha_2\phi'(l) = 0 \tag{14}$$

$$\beta_1\phi(r) + \beta_2\phi'(r) = 0. \tag{15}$$

Since s and p are related to the physical properties of the rod, they should be positive. We suppose also that s, s', and p are continuous. Then Eqs. (13)–(15) is a regular Sturm–Liouville problem, and we know that

1. There are an infinite number of eigenvalues

$$0 < \lambda_1{}^2 < \lambda_2{}^2 < \cdots$$

2. To each eigenvalue corresponds just one eigenfunction (give or take a constant multiplier).

3. The eigenfunctions are orthogonal with weight $p(x)$:

$$\int_l^r \phi_n(x)\phi_m(x)p(x)\,dx = 0, \qquad n \neq m.$$

The function $T_n(t)$ that accompanies $\phi_n(x)$ is given by

$$T_n(t) = \exp(-\lambda_n{}^2 kt).$$

We now begin to assemble the solution. For each $n = 1, 2, 3, \ldots$, $w_n(x, t) = \phi_n(x)T_n(t)$ satisfies Eqs. (9)–(11). As these are all linear homogeneous equations, any linear combination of solutions is again a solution. Thus the transient temperature has the form

$$w(x, t) = \sum_{n=1}^{\infty} a_n \phi_n(x) \exp(-\lambda_n^2 kt).$$

The initial condition Eq. (12) will be satisfied if we choose the a_n so that

$$w(x, 0) = \sum_{1}^{\infty} a_n \phi_n(x) = g(x), \qquad l < x < r.$$

The convergence theorem tells us that the equality will hold, except possibly at a finite number of points, if $f(x)$—and therefore $g(x)$—is sectionally smooth. Thus $w(x, t)$ is the solution of its problem, if we choose

$$a_n = \frac{\int_l^r g(x)\phi_n(x)p(x)\ dx}{\int_l^r \phi_n^2(x)p(x)\ dx}.$$

Finally, we can write the complete solution of Eqs. (1)–(4) in the form

$$u(x, t) = v(x) + \sum a_n \phi_n(x) \exp(-\lambda_n^2 kt). \tag{16}$$

Working from the representation Eq. (16) we can draw some conclusions about the solution of Eqs. (1)–(4):

1. Since all the λ^2 are positive, $u(x, t)$ does tend to $v(x)$ as $t \to \infty$.

2. For any $t_1 > 0$, the series for $u(x, t_1)$ converges uniformly in $l \le x \le r$ because of the exponential factors; therefore $u(x, t_1)$ is a continuous function of x. Any discontinuity in the initial condition is immediately eliminated.

3. For large enough values of t, we can approximate $u(x, t)$ by

$$v(x) + a_1 \phi_1(x) \exp(-\lambda_1^2 kt)$$

(to judge how large t must be, one needs to know something about the a_n and the λ_n). Since $\phi_1(x)$ is of one sign on the interval $l < x < r$ ($\phi_1(x) > 0$ or $\phi_1(x) < 0$ for all x between l and r) the graph of the approximation above will lie either above or below the graph of $v(x)$, but will not cross it (provided that $a_1 \ne 0$).

Exercises

1. Find the explicit form for $v(x)$ in terms of the function in Eq. (8) assuming

   a. $\alpha_1 = \beta_1 = 0$, $c_1 = c_2 = 0$
   b. $\alpha_1 > 0$ or $\beta_1 > 0$, and no coefficient negative.

   Why are these two cases separate?

2. Justify each of the conclusions.

3. Derive the general form of $u(x,t)$ if the boundary conditions are $\partial u/\partial x = 0$ at both ends. In this case, $\lambda^2 = 0$ is an eigenvalue.

10 Semi-infinite rod

Up to this point we have only seen problems over finite intervals. Frequently, however, it is justifiable and useful to assume that an object is infinite in length. (Sometimes this assumption is used to disguise ignorance of a boundary condition or to suppress the influence of a complicated condition.)

If the rod we have been studying is very long, we may treat it as "semi-infinite" in length. (Sometimes this assumption is used to disguise ignorance equation governing the temperature $u(x, t)$ remains

$$\frac{\partial^2 u}{\partial x^2} = \frac{1}{k}\frac{\partial u}{\partial t}, \qquad 0 < x, \quad 0 < t \tag{1}$$

if the properties are uniform.

Suppose that at $x = 0$ the temperature is held constant, say

$$u(0, t) = 0, \qquad t > 0 \tag{2}$$

in some temperature scale. In the absence of another boundary, there is no other boundary condition. However, it is desirable that $u(x, t)$ remain finite—less than some fixed bound—as $x \to \infty$. As usual, an initial condition is necessary

$$u(x, 0) = f(x), \qquad 0 < x. \tag{3}$$

Attacking Eqs. (1)–(3) by separation of variables, we assume $u(x, t) = \phi(x)T(t)$, so the partial differential equation can be separated into two ordinary equations as usual

$$T' \pm \lambda^2 kT = 0, \qquad 0 < t \tag{4}$$

$$\phi'' \pm \lambda^2\phi = 0, \qquad 0 < x. \tag{5}$$

There is just one boundary condition on u, which requires that $\phi(0) = 0$. The boundedness condition also requires that $\phi(x)$ remain finite as $x \to \infty$. The differential equation (5) has the solution

$$\phi(x) = A \cosh \lambda x + B \sinh \lambda x$$

if the negative signs are chosen in Eqs. (5) and (4). But $\phi(0) = 0$ requires that $A = 0$, leaving $\phi(x) = B \sinh \lambda x$, a function that increases without bound as $x \to \infty$. Thus we must choose the positive signs. In this case, the solution of Eq. (5) is

$$\phi(x) = A \cos \lambda x + B \sin \lambda x$$

which does remain bounded as $x \to \infty$. Application of the boundary condition at $x = 0$ shows $\phi(0) = A = 0$, leaving $\phi(x) = \sin \lambda x$. The solution of Eq. (4) (also with the $+$ sign) is

$$T(t) = \exp(- \lambda^2 kt).$$

For any value of λ^2, the function

$$u(x, t; \lambda) = \sin \lambda x \exp(- \lambda^2 kt)$$

satisfies Eqs. (1) and (2) and the boundedness condition. Equation (1) and the boundary condition Eq. (2) are homogeneous; therefore any linear combination of solutions is a solution. Since the parameter λ may take on any value, the appropriate linear combination is an integral. So u should have the form

$$u(x, t) = \int_0^\infty B(\lambda) \sin \lambda x \exp(- \lambda^2 kt) \, d\lambda. \tag{6}$$

(We need not include negative values of λ. They give no new solutions.) The initial condition will be satisfied if $B(\lambda)$ is chosen to make

$$u(x, 0) = \int_0^\infty B(\lambda) \sin \lambda x \, d\lambda = f(x), \qquad 0 < x.$$

We recognize this as a Fourier integral; $B(\lambda)$ is to be chosen as

$$B(\lambda) = \frac{2}{\pi} \int_0^\infty f(x) \sin \lambda x \, dx.$$

If $B(\lambda)$ exists, then Eq. (6) is the solution of the problem. Notice that when $t > 0$, the exponential function makes the improper integral in Eq. (6) converge very rapidly.

Some care must be taken in the interpretation of our solution. If the rod really is finite (say length L) the expression in Eq. (6) is, of course, meaningless for x greater than L. The presence of a boundary condition at $x \doteq L$ would influence temperatures nearby, so Eq. (6) can be considered a valid approximation only for $x \ll L$.

Exercises

1. Find $u(x, t)$ if

$$f(x) = \begin{cases} 1, & 0 < x < b \\ 0, & b < x. \end{cases}$$

2. Verify that $u(x, t)$ as given by Eq. (6) is a solution of Eqs. (1)–(3). What is the steady-state temperature distribution?

3. Find the solution of Eqs. (1)–(3) if $f(x) = T_0 e^{-ax}$, $x > 0$.

4. Find a formula for the solution of the problem

$$\frac{\partial^2 u}{\partial x^2} = \frac{1}{k} \frac{\partial u}{\partial t}, \qquad 0 < x, \quad 0 < t$$

$$\frac{\partial u}{\partial x}(0, t) = 0, \qquad 0 < t$$

$$u(x, 0) = f(x), \qquad 0 < x.$$

5. Determine the solution of Exercise 4 if $f(x)$ is the function given in Exercise 1.

6. Penetration of heat into the earth. Assume that the earth is flat, occupying the region $x > 0$. If the temperature at the surface is $u(0, t) = \sin \omega t$, show that

$$u(x, t) = \exp\left(-\sqrt{\frac{\omega}{2k}}\, x\right) \sin\left(\omega t - \sqrt{\frac{\omega}{2k}}\, x\right)$$

is a solution of the heat equation which satisfies the boundary condition. Sketch $u(x, t)$ for $x = 0$, 1, and 2 meters on the same graph, taking $\omega = 2 \times 10^{-7}$ rad/sec ($\simeq 2\pi$/year), $k = 0.5 \times 10^{-6}$ m^2/sec.

7. Consider the problem

$$\frac{\partial^2 u}{\partial x^2} = \frac{1}{k} \frac{\partial u}{\partial t}, \qquad 0 < x, \quad 0 < t$$

$$u(0, t) = T_0, \qquad 0 < t$$

$$u(x, 0) = f(x), \qquad 0 < x.$$

Show that, for our method of solution to work, it is necessary to have $T_0 = \lim_{x \to \infty} f(x)$. Find a formula for $u(x, t)$.

11   Infinite rod

If we wish to study heat conduction in the center of a very long rod, we may assume that it extends from $-\infty$ to ∞. Then there are no boundary conditions, and the problem to be solved is

$$\frac{\partial^2 u}{\partial x^2} = \frac{1}{k}\frac{\partial u}{\partial t}, \qquad -\infty < x < \infty, \quad 0 < t$$

$$u(x, 0) = f(x), \qquad -\infty < x < \infty$$

$$|u(x, t)| \quad \text{bounded} \quad \text{as} \quad x \to \pm \infty.$$

Employing the same techniques as before, we have

$$u(x, t) = \phi(x)T(t),$$
$$T' + \lambda^2 kT = 0, \qquad 0 < t$$
$$\phi'' + \lambda^2 \phi = 0, \qquad -\infty < x < \infty.$$

The signs have been chosen to make $\phi(x)$ satisfy the boundedness condition, for now

$$\phi(x) = A \cos \lambda x + B \sin \lambda x$$
$$T(t) = \exp(-\lambda^2 kt).$$

We combine the solutions $\phi(x)T(t)$ in the form of an integral to obtain

$$u(x, t) = \int_0^\infty (A(\lambda) \cos \lambda x + B(\lambda) \sin \lambda x) \exp(-\lambda^2 kt) \, d\lambda. \qquad (1)$$

The initial condition is satisfied by choosing

$$\left.\begin{matrix} A(\lambda) \\ B(\lambda) \end{matrix}\right\} = \frac{1}{\pi}\int_{-\infty}^\infty f(x)\begin{Bmatrix}\cos \lambda x \\ \sin \lambda x\end{Bmatrix} dx \qquad (2)$$

for, when $t = 0$, the initial condition has the form of a Fourier integral:

$$f(x) = \int_0^\infty (A(\lambda) \cos \lambda x + B(\lambda) \sin \lambda x) \, d\lambda.$$

In this case, we can derive some very interesting results. Change the variable of integration in Eq. (2) to ξ and substitute the formulas for $A(\lambda)$ and $B(\lambda)$ into Eq. (1):

$$u(x, t) = \frac{1}{\pi}\int_0^\infty \left[\int_{-\infty}^\infty f(\xi) \cos \lambda \xi \, d\xi \cos \lambda x \right.$$

$$\left. + \int_{-\infty}^\infty f(\xi) \sin \lambda \xi \, d\xi \sin \lambda x\right]\exp(-\lambda^2 kt) \, d\lambda.$$

Combining terms, we find

$$u(x, t) = \frac{1}{\pi} \int_0^\infty \int_{-\infty}^\infty f(\xi)[\cos \lambda \xi \cos \lambda x + \sin \lambda \xi \sin \lambda x] \, d\xi \exp(-\lambda^2 kt) \, d\lambda$$

$$= \frac{1}{\pi} \int_0^\infty \int_{-\infty}^\infty f(\xi) \cos \lambda(\xi - x) \, d\xi \exp(-\lambda^2 kt) \, d\lambda.$$

If the order of integration may be reversed, we may write

$$u(x, t) = \frac{1}{\pi} \int_{-\infty}^\infty f(\xi) \int_0^\infty \cos \lambda(\xi - x) \exp(-\lambda^2 kt) \, d\lambda \, d\xi.$$

The inner integral can be computed by complex methods of integration. It is known to equal (Misc. Ex. 32, Chapter 1)

$$\sqrt{\frac{\pi}{4kt}} \exp\left[\frac{-(\xi - x)^2}{4kt}\right], \qquad t > 0.$$

This gives us, finally, a new form for the temperature distribution:

$$u(x, t) = \frac{1}{\sqrt{4k\pi t}} \int_{-\infty}^\infty f(\xi) \exp\left[\frac{-(\xi - x)^2}{4kt}\right] d\xi. \tag{3}$$

Of the two formulas, Eqs. (1) and (3), for the solution $u(x, t)$, each has its advantages. For simple problems we may be able to evaluate the coefficients $A(\lambda)$ and $B(\lambda)$ in Eq. (2). But it is a rare case indeed when the integral in Eq. (1) can be evaluated analytically. The same is true for the integral in Eq. (3). Thus, if the value of u at a specific x and t is needed, either integral would be calculated numerically. For large values of kt, the exponential factor in the integrand of Eq. (1) will be nearly zero, except for small λ. Thus, Eq. (1) is approximately

$$u(x, t) \cong \int_0^\Lambda (A(\lambda) \cos \lambda x + B(\lambda) \sin \lambda x) \exp(-\lambda^2 kt) \, d\lambda$$

for Λ not large, and the right-hand side may be found to high accuracy with little effort.

On the other hand, if kt is small, the exponential in the integrand of Eq. (3) will be nearly zero, except for ξ near x. The approximation

$$u(x, t) \cong \frac{1}{\sqrt{4k\pi t}} \int_{x-h}^{x+h} f(\xi) \exp\left(\frac{-(\xi - x)^2}{4kt}\right) d\xi$$

is satisfactory for h not large, and again numerical techniques are easily applied to the right-hand side.

The expression in Eq. (3) has a number of other advantages also. It requires no intermediate integrations (compare Eq. (2)). It shows directly the influence of initial conditions on the solution. Moreover, the function $f(x)$ need not satisfy the restriction

$$\int_{-\infty}^{\infty} |f(x)| \, dx < \infty$$

in order for Eq. (3) to satisfy the original problem.

Exercises

1. Find $u(x, t)$ using

    a.   Eq. (1)
    b.   Eq. (3)

    if $f(x)$ is

    $$f(x) = \begin{cases} 1, & -a < x < a \\ 0, & |x| > a. \end{cases}$$

2. Verify that Eq. (3) satisfies the heat equation.

3. Verify that the function

    $$u(x, t) = \frac{1}{\sqrt{4k\pi t}} \exp\left[-\frac{x^2}{4kt}\right]$$

    is a solution of the heat equation

    $$\frac{\partial^2 u}{\partial x^2} = \frac{1}{k}\frac{\partial u}{\partial t}, \qquad 0 < t, \quad -\infty < x < \infty.$$

    What can be said about u at $x = 0$? at $t = 0+$? What is $\lim_{t \to 0+} u(0, t)$? Sketch $u(x, t)$ for various fixed values of t.

4. Suppose that $f(x)$ is an odd periodic function with period $2a$. Show that $u(x, t)$ defined by Eq. (3) also has these properties.

5. If $f(x) = 1$ for all x, the solution of our heat conduction problem is $u(x, t) = 1$. Use this fact to show that

    $$1 = \frac{1}{\sqrt{4\pi kt}} \int_{-\infty}^{\infty} \exp\left[\frac{-(\xi - x)^2}{4kt}\right] d\xi.$$

6. The "error function" is defined by the equation

$$\text{erf}(q) = \frac{2}{\sqrt{\pi}} \int_0^q e^{-z^2}\, dz.$$

Prove these properties of the error function:

a. $\text{erf}(-q) = -\text{erf}(q)$

b. $\dfrac{d}{dq}\text{erf}(q) = \dfrac{2}{\sqrt{\pi}} e^{-q^2}$

c. $\lim\limits_{q \to \infty} \text{erf}(q) = 1.$ (Hint: see Exercise 5.)

7. Solve the problem below using Eq. (3), and express the solution in terms of the error function.

$$\frac{\partial^2 u}{\partial x^2} = \frac{1}{k}\frac{\partial u}{\partial t}, \qquad -\infty < x < \infty, \quad 0 < t$$

$$u(x, 0) = \begin{cases} 1, & x > 0 \\ -1, & x < 0 \end{cases}$$

8. Show that $u(x, t) = \text{erf}(x/\sqrt{4kt})$ is the solution of the problem

$$\frac{\partial^2 u}{\partial x^2} = \frac{1}{k}\frac{\partial u}{\partial t}, \qquad 0 < x, \quad 0 < t$$

$$u(0, t) = 0, \qquad 0 < t$$

$$u(x, 0) = 1, \qquad 0 < x.$$

12 Comments and references

In about 1810, Fourier made an intensive study of heat conduction problems, in which he used the product method of solution and developed the idea of Fourier series. His book, *The Analytical Theory of Heat* [15], originally published in 1822, is still available. (Sturm and Liouville made their clear and simple generalization of Fourier series in the 1830's.) Among modern works, *Conduction of Heat in Solids* by Carslaw and Jaeger [5] is the standard reference. *The Mathematics of Diffusion*, by Crank [9], is a more advanced treatise. The newest reference in this field is *The Heat Equation* by D. V. Widder [33].

Although our study has been motivated by heat conduction, many other physical phenomena are described by the heat or diffusion equation: for example, voltage and current in an inductance-free cable, vorticity transport in fluid flow, and the diffusion of a solute in a solvent. In the last case, one of great importance in chemical engineering, the governing relations are the conservation of mass and "Fick's law," expressed, respectively, as

$$-\frac{\partial v}{\partial x} = \frac{\partial c}{\partial t}, \qquad v = -k\frac{\partial c}{\partial x}$$

where v is the mass flow rate, c the concentration, and k the diffusivity. Of course, diffusion is of great importance in the physiology of the cell. Explanations and references will be found in Rashevsky's *Mathematical Biophysics* [24].

The diffusion equation also turns up in some classical problems of probability theory, especially the description of Brownian motion. Suppose a particle moves exactly one step of length Δx in each time interval Δt. The step may be either to the left or right, each equally likely. Let $u_i(m)$ denote the probability that, at time $m\,\Delta t$, the particle is at point $i\,\Delta x$ ($m = 0, 1, 2, \ldots$, $i = 0, \pm 1, \pm 2, \ldots$). In order to arrive to point $i\,\Delta x$ at time $(m + 1)\,\Delta t$, the particle must have been at one of the adjacent points $(i \pm 1)\,\Delta x$ at the preceeding time $m\,\Delta t$, and must have moved toward $i\,\Delta x$. From this, we see that the probabilities are related by the equation

$$u_i(m + 1) = \tfrac{1}{2}u_{i-1}(m) + \tfrac{1}{2}u_{i+1}(m).$$

The us are completely determined, once an initial probability distribution is given. For instance, if the particle is initially at point zero ($u_0(0) = 1$, $u_i(0) = 0$, for $i \neq 0$), the $u_i(m)$ are formed by successive applications of the difference equation, as shown in Table 2.1.

Table 2.1

i \\ m	-3	-2	-1	0	1	2	3
0	0	0	0	1	0	0	0
1	0	0	0.5	0	0.5	0	0
2	0	0.25	0	0.5	0	0.25	0
3	0.125	0	0.375	0	0.375	0	0.125

We may modify the difference equation by subtracting $u_i(m)$ from both sides:

$$u_i(m + 1) - u_i(m) = \tfrac{1}{2}(u_{i+1}(m) - 2u_i(m) + u_{i-1}(m)).$$

A further modification brings it into nearly recognizable form.

$$\frac{u_i(m+1) - u_i(m)}{\Delta t} \frac{2\,\Delta t}{(\Delta x)^2} = \frac{u_{i+1}(m) - 2u_i(m) + u_{i-1}(m)}{(\Delta x)^2}.$$

If both the time interval Δt and the step length Δx are small, we may think of $u_i(m)$ as being the value of a continuous function $u(x, t)$ at $x = i\,\Delta x$, $t = m\,\Delta t$.

In the limit, the difference quotient on the left approaches $\partial u/\partial t$. The right-hand side, being a difference of differences, approaches $\partial^2 u/\partial x^2$. The heat equation thus results if, in the simultaneous limit as Δx and Δt tend to zero, the quantity $2\,\Delta t/(\Delta x)^2$ approaches a finite, nonzero limit. In this context, the heat equation is called the Fokker–Planck equation. More details and references may be found in Feller [16], *An Introduction to Probability Theory and Its Applications.*

We have used the phrase "linear partial differential equation" several times. The most general such equation, of second order in two independent variables, can be put in the form

$$A\frac{\partial^2 u}{\partial x^2} + B\frac{\partial^2 u}{\partial x\,\partial t} + C\frac{\partial^2 u}{\partial t^2} + D\frac{\partial u}{\partial x} + E\frac{\partial u}{\partial t} + Fu + G = 0$$

where $A, B, \ldots, G$ are known—perhaps functions of x and t but not of u or its derivatives. If G is identically zero, the equation is homogeneous. Of course, the ordinary heat equation has this form if we take $A = 1, E = -1/k$, and all other coefficients equal to zero.

Some astute students will have wondered why we should seek solutions in product form. The rudest answer is that it works, in many cases. A more subtle rationale is that of seeking solutions which are geometrically similar functions of x at different times. The idea of similarity—related to dimensional analysis—has been most fruitful in the mechanics of fluids.

Miscellaneous exercises

Find the steady-state solution, the associated eigenvalue problem, and the complete solution for each of the heat conduction problems below.

1. $\dfrac{\partial^2 u}{\partial x^2} = \dfrac{1}{k}\dfrac{\partial u}{\partial t}, \quad 0 < x < a, \quad 0 < t$

$u(0, t) = T_0, \quad u(a, t) = T_0, \quad 0 < t$

$u(x, 0) = T_1, \quad 0 < x < a$

2. $\dfrac{\partial^2 u}{\partial x^2} - \gamma^2 u = \dfrac{1}{k}\dfrac{\partial u}{\partial t}, \quad 0 < x < a, \quad 0 < t$

$u(0, t) = T_0, \quad u(a, t) = T_0, \quad 0 < t$

$u(x, 0) = T_1, \quad 0 < x < a$

3. $\dfrac{\partial^2 u}{\partial x^2} - r = \dfrac{1}{k}\dfrac{\partial u}{\partial t}, \quad 0 < x < a, \quad 0 < t$

$u(0, t) = T_0, \quad u(a, t) = T_0, \quad 0 < t$

$u(x, 0) = T_1, \quad 0 < x < a \quad (r \text{ a constant})$

4. $\dfrac{\partial^2 u}{\partial x^2} = \dfrac{1}{k}\dfrac{\partial u}{\partial t}, \quad 0 < x < a, \quad 0 < t$

$u(0, t) = T_0, \quad \dfrac{\partial u}{\partial x}(a, t) = 0, \quad 0 < t$

$u(x, 0) = T_1 x/a, \quad 0 < x < a$

5. $\dfrac{\partial^2 u}{\partial x^2} - \gamma^2 u = \dfrac{1}{k}\dfrac{\partial u}{\partial t}, \quad 0 < x < a, \quad 0 < t$

$\dfrac{\partial u}{\partial x}(0, t) = 0, \quad \dfrac{\partial u}{\partial x}(a, t) = 0, \quad 0 < t$

$u(x, 0) = T_1 x/a, \quad 0 < x < a$

6. $\dfrac{\partial^2 u}{\partial x^2} = \dfrac{1}{k}\dfrac{\partial u}{\partial t}, \quad 0 < x < a, \quad 0 < t$

$u(0, t) = 0, \quad\quad u(a, t) = T_0, \quad 0 < t$

$u(x, 0) = 0, \quad 0 < x < a$

7. $\dfrac{\partial^2 u}{\partial x^2} = \dfrac{1}{k}\dfrac{\partial u}{\partial t}, \quad 0 < x < a, \quad 0 < t$

$u(0, t) = T_0, \quad u(a, t) = T_0$

$u(x, 0) = T_0, \quad 0 < x < a$

8. $\dfrac{\partial^2 u}{\partial x^2} = \dfrac{1}{k}\dfrac{\partial u}{\partial t}, \quad 0 < x < a, \quad 0 < t$

$\dfrac{\partial u}{\partial x}(0, t) = \dfrac{\Delta T}{a}, \dfrac{\partial u}{\partial x}(a, t) = \dfrac{\Delta T}{a}, \quad 0 < t$

$u(x, 0) = T_0, \quad 0 < x < a$

9. $\dfrac{\partial^2 u}{\partial x^2} = \dfrac{1}{k}\dfrac{\partial u}{\partial t}$, $\quad 0 < x < a, \quad 0 < t$

$u(0, t) = T_0$, $\quad \dfrac{\partial u}{\partial x}(a, t) = 0, \quad 0 < t$

$u(x, 0) = T_1$, $\quad 0 < x < a$

10. $\dfrac{\partial^2 u}{\partial x^2} = \dfrac{1}{k}\dfrac{\partial u}{\partial t}$, $\quad 0 < x < a, \quad 0 < t$

$\dfrac{\partial u}{\partial x}(0, t) = 0$, $\quad \dfrac{\partial u}{\partial x}(a, t) = 0, \quad 0 < t$

$u(x, 0) = \begin{cases} T_0, & 0 < x < a/2 \\ T_1, & a/2 < x < a \end{cases}$

11. $\dfrac{\partial^2 u}{\partial x^2} = \dfrac{1}{k}\dfrac{\partial u}{\partial t}$, $\quad 0 < x < \infty, \quad 0 < t$

$u(0, t) = T_0$, $\quad 0 < t$

$u(x, 0) = T_0(1 - e^{-\alpha x})$, $\quad 0 < x$

12. $\dfrac{\partial^2 u}{\partial x^2} = \dfrac{1}{k}\dfrac{\partial u}{\partial t}$, $\quad 0 < x < \infty, \quad 0 < t$

$u(0, t) = 0$, $\quad 0 < t$

$u(x, 0) = \begin{cases} T_0, & 0 < x < a \\ 0, & a < x \end{cases}$

13. $\dfrac{\partial^2 u}{\partial x^2} = \dfrac{1}{k}\dfrac{\partial u}{\partial t}$, $\quad 0 < x < \infty, \quad 0 < t$

$\dfrac{\partial u}{\partial x}(0, t) = 0$, $\quad 0 < t$

$u(x, 0) = \begin{cases} T_0, & 0 < x < a \\ 0, & a < x \end{cases}$

14. $\dfrac{\partial^2 u}{\partial x^2} = \dfrac{1}{k}\dfrac{\partial u}{\partial t}$, $\quad -\infty < x < \infty, \quad 0 < t$

$u(x, 0) = \exp\{-\alpha |x|\}$, $\quad -\infty < x < \infty$

15. $\dfrac{\partial^2 u}{\partial x^2} = \dfrac{1}{k}\dfrac{\partial u}{\partial t}$, $-\infty < x < \infty$, $0 < t$

$$u(x, 0) = \begin{cases} 0, & -\infty < x < 0 \\ T_0, & 0 < x < a \\ 0, & a < x < \infty \end{cases}$$

16. Solve the eigenvalue problem by setting $\phi(\rho) = \psi(\rho)/\rho$:

$$\dfrac{1}{\rho^2}\,(\rho^2\phi')' + \lambda^2\phi = 0, \qquad 0 < \rho < a$$

$$\phi(0)\ \text{bounded}, \qquad \phi(a) = 0.$$

Is this a regular Sturm–Liouville problem? Are the eigenfunctions orthogonal?

17. Solve this problem for heat conduction in a sphere. (Hint: let $u(\rho, t) = v(\rho, t)/\rho$.)

$$\dfrac{1}{\rho^2}\dfrac{\partial}{\partial \rho}\left(\rho^2\dfrac{\partial u}{\partial \rho}\right) = \dfrac{1}{k}\dfrac{\partial u}{\partial t}, \qquad 0 < \rho < a,\ \ 0 < t$$

$$u(0, t)\ \text{bounded}, \qquad u(a, t) = 0, \qquad 0 < t$$

$$u(\rho, 0) = T_0, \qquad 0 < \rho < a$$

18. State and solve the eigenvalue problem associated with

$$e^{-x}\dfrac{\partial}{\partial x}\left(e^x\dfrac{\partial u}{\partial x}\right) = \dfrac{1}{k}\dfrac{\partial u}{\partial t}, \qquad 0 < x < a,\ \ 0 < t$$

$$u(0, t) = 0, \qquad \dfrac{\partial u}{\partial x}(a, t) = 0.$$

19. Find the steady-state solution of

$$\dfrac{\partial^2 u}{\partial x^2} + \gamma^2(u - T) = \dfrac{1}{k}\dfrac{\partial u}{\partial t}, \qquad 0 < x < a,\ \ 0 < t$$

$$u(0, t) = T_0, \qquad \dfrac{\partial u}{\partial x}(a, t) = 0, \qquad 0 < t$$

where $T(x) = T_0 + \Delta T(x/a)$, $0 < x < a$.

20. Suppose that $u(x, t)$ is a positive function that satisfies

$$\dfrac{\partial^2 u}{\partial x^2} = \dfrac{\partial u}{\partial t}.$$

Show that the function

$$w(x, t) = -\frac{2}{u}\frac{\partial u}{\partial x}$$

satisfies the nonlinear partial differential equation called Burgers' equation:

$$\frac{\partial w}{\partial t} + w\frac{\partial w}{\partial x} = \frac{\partial^2 w}{\partial x^2}.$$

21. Find a solution of the Burgers' equation that satisfies the conditions

$$w(0, t) = 0, \qquad w(1, t) = 0, \qquad 0 < t$$
$$w(x, 0) = 1, \qquad 0 < x < 1.$$

22. Taking the function $u(x, t)$ given below as a solution of the heat equation (with $k = 1$), find a solution w of Burgers' equation. Verify that w satisfies Burgers' equation.

$$u(x, t) = \frac{1}{\sqrt{4\pi t}}\exp\left(\frac{-x^2}{4t}\right)$$

23. A metal bar and water in a water jacket exchange heat by convection. Heat balances give these two differential equations:

$$c_1\frac{du_1}{dt} = h(u_2 - u_1)$$

$$c_2\frac{du_2}{dt} = h(u_1 - u_2).$$

Find temperatures $u_1(t)$ and $u_2(t)$ assuming initial conditions $u_1(0) = 1$, $u_2(0) = 0$.

chapter 3
The
Wave
Equation

1 The vibrating string

For a simple example of the wave equation, we consider the motion of a
string which is fixed at its ends. We set up coordinates as shown in Fig. 3.1;

Figure 3.1 String fixed at ends.

the displacement $u(x, t)$ of the string is the unknown. In order to find the
equation of motion of the string, we consider a short piece whose ends are at
x and $x + \Delta x$ and apply Newton's second law of motion to it. The portions
of the string to the right and left of our element exert forces on it which cause
acceleration.

143

Assumption 1: *The string is perfectly flexible,* offers no resistance to bending. This means that the forces exerted on the element are tangent to its midline at the points where they act.

On the basis of this assumption, the forces on the element are as seen in Fig. 3.2. Applying Newton's second law in the x-direction (and assuming that gravity acts perpendicular to the x-direction) we have for the sum of forces

$$-T(x) \cos \alpha + T(x + \Delta x) \cos \beta.$$

Figure 3.2 Section of string showing forces exerted on it.

Assumption 2: *A point on the string moves only in the vertical direction.* This means that the sum of forces above equals zero. Equivalently, we may assume that the horizontal component of the tension is constant. In either case we find

$$-T(x) \cos \alpha + T(x + \Delta x) \cos \beta = 0$$
$$T(x) \cos \alpha = T(x + \Delta x) \cos \beta = T.$$

In the absence of external forces other than gravity, Newton's law in the vertical direction yields

$$-T(x) \sin \alpha + T(x + \Delta x) \sin \beta - mg = m \frac{\partial^2 u}{\partial t^2}(x, t).$$

(Since u measures the displacement in the vertical direction, $\partial^2 u / \partial t^2$ is the acceleration.)

In working with a string, it is convenient to introduce the linear density ρ, $[\rho] = m/L$. Then the mass of the element we are considering is $\rho \, \Delta x$. Now

$$T(x) = \frac{T}{\cos \alpha}, \qquad T(x + \Delta x) = \frac{T}{\cos \beta}$$

where T is constant. Hence, the equation above can be written

$$-T \tan \alpha + T \tan \beta - \rho \, \Delta x g = \rho \, \Delta x \frac{\partial^2 u}{\partial t^2}.$$

The quantity $\tan \alpha$ is the slope of the string at x, and $\tan \beta$ the slope of the string at $x + \Delta x$; that is,

$$\tan \alpha = \frac{\partial u}{\partial x}(x, t), \qquad \tan \beta = \frac{\partial u}{\partial x}(x + \Delta x, t).$$

Substituting these into the equation above, we have

$$T\left(\frac{\partial u}{\partial x}(x + \Delta x, t) - \frac{\partial u}{\partial x}(x, t)\right) = \rho \Delta x\left(\frac{\partial^2 u}{\partial t^2} + g\right).$$

On dividing through by Δx, we see a difference quotient on the left:

$$\frac{T}{\Delta x}\left(\frac{\partial u}{\partial x}(x + \Delta x, t) - \frac{\partial u}{\partial x}(x, t)\right) = \rho\left(\frac{\partial^2 u}{\partial t^2} + g\right).$$

In the limit as $\Delta x \to 0$, the difference quotient becomes a partial derivative with respect to x, leaving Newton's second law in the form

$$T \frac{\partial^2 u}{\partial x^2} = \rho \frac{\partial^2 u}{\partial t^2} + \rho g \tag{1}$$

$$\frac{\partial^2 u}{\partial x^2} = \frac{1}{c^2}\frac{\partial^2 u}{\partial t^2} + \frac{1}{c^2}g \tag{2}$$

where $c^2 = T/\rho$. If c^2 is very large (usually on the order of $10^5 \text{ m}^2/\text{sec}^2$), we neglect the last term, giving the equation of the vibrating string

$$\frac{\partial^2 u}{\partial x^2} = \frac{1}{c^2}\frac{\partial^2 u}{\partial t^2}, \qquad 0 < x < a, \quad 0 < t. \tag{3}$$

In describing the motion of an object, one must specify not only the equation of motion, but also both the initial position and velocity of the object. The initial conditions for the string, then, must state the initial displacement of every particle—that is $u(x, 0)$—and the initial velocity of every particle, $\frac{\partial u}{\partial t}(x, 0)$.

For the vibrating string as we have described it, the boundary conditions are zero displacement at the ends, so the boundary value–initial value problem for the string is

$$\frac{\partial^2 u}{\partial x^2} = \frac{1}{c^2}\frac{\partial^2 u}{\partial t^2}, \qquad\qquad 0 < x < a, \quad 0 < t \tag{4}$$

$$u(0, t) = 0, \qquad u(a, t) = 0, \quad 0 < t \tag{5}$$

$$u(x, 0) = f(x), \qquad\qquad 0 < x < a \tag{6}$$

$$\frac{\partial u}{\partial t}(x, 0) = g(x), \qquad\qquad 0 < x < a \tag{7}$$

under the assumptions noted plus the assumption that gravity is negligible.

Exercises

1. Find the dimensions of each of the following quantities, using the facts that force is equivalent to mL/t^2, and that the dimension of tension is F: u, $\partial^2 u/\partial x^2$, $\partial^2 u/\partial t^2$, c, g/c^2. Check the dimension of each term in Eq. (2).

2. Suppose a distributed vertical force $F(x, t)$ (positive upwards) acts on the string. Derive the equation of motion:

$$\frac{\partial^2 u}{\partial x^2} = \frac{1}{c^2}\frac{\partial^2 u}{\partial t^2} - \frac{1}{T}F(x, t).$$

The dimension of a distributed force are F/L. (If the weight of the string is considered as a distributed force, and is the only one, then we would have $F(x, t) = -\rho g$. Check dimensions and signs.)

3. Find a solution $v(x)$ of Eq. (2) with boundary conditions Eqs. (5) which is independent of time. (This corresponds to a "steady-state solution," but the words "steady-state" are no longer appropriate. "Equilibrium solution" is more accurate.)

2   Solution of the vibrating string problem

The initial value–boundary value problem which describes the displacement of the vibrating string,

$$\frac{\partial^2 u}{\partial x^2} = \frac{1}{c^2}\frac{\partial^2 u}{\partial t^2}, \qquad\qquad 0 < x < a, \quad 0 < t \qquad (1)$$

$$u(0, t) = 0, \qquad u(a, t) = 0, \qquad 0 < t \qquad\qquad (2)$$

$$u(x, 0) = f(x), \qquad\qquad 0 < x < a \qquad\qquad (3)$$

$$\frac{\partial u}{\partial t}(x, 0) = g(x), \qquad\qquad 0 < x < a \qquad\qquad (4)$$

contains a linear, homogeneous partial differential equation and linear, homogeneous boundary conditions. Thus we may apply the method of separation of variables with hope of success. If we assume that* $u(x, t) = \phi(x)T(t)$, Eq. (1) becomes

$$\phi''T = \frac{1}{c^2}\phi T''.$$

---

* T no longer symbolizes tension.

Dividing through by ϕT, we get

$$\frac{\phi''(x)}{\phi(x)} = \frac{T''(t)}{c^2 T(t)}, \qquad 0 < x < a, \quad 0 < t.$$

For the equality to hold, both members of this equation must be constant. We write the constant as $-\lambda^2$ and separate the equation above into two ordinary differential equations linked by the common parameter λ^2.

$$T'' + \lambda^2 c^2 T = 0, \qquad 0 < t \tag{5}$$

$$\phi'' + \lambda^2 \phi = 0, \qquad 0 < x < a. \tag{6}$$

The boundary conditions become

$$\phi(0)T(t) = 0, \qquad \phi(a)T(t) = 0, \qquad 0 < t$$

and, since $T(t) = 0$ gives a trivial solution for $u(x, t)$, we must have

$$\phi(0) = 0, \qquad \phi(a) = 0. \tag{7}$$

The eigenvalue problem Eqs. (6) and (7) is exactly the same one which we have seen and solved before. (See Chapter 2, Section 3.) We know that the eigenvalues and eigenfunctions are

$$\lambda_n^2 = \left(\frac{n\pi}{a}\right)^2, \qquad \phi_n(x) = \sin\left(\frac{n\pi x}{a}\right), \qquad n = 1, 2, 3, \ldots .$$

Equation (5) is also of a familiar type, and its solution is known to be

$$T_n(t) = a_n \cos \lambda_n ct + b_n \sin \lambda_n ct$$

where a_n and b_n are arbitrary. (In other words, there are two independent solutions.) Note, however, that there is a substantial difference between the T which arises here and the T which we found in the heat conduction problem. The most important difference is the behavior as t tends to infinity. In the heat conduction problem T tends to 0; whereas here T has no limit but oscillates periodically.

For each $n = 1, 2, 3, \ldots$ we now have a function

$$u_n(x, t) = \sin \lambda_n x[a_n \cos \lambda_n ct + b_n \sin \lambda_n ct]$$

which, for any choice of a_n and b_n, is a solution of the partial differential Eq. (1) and also satisfies the boundary conditions Eq. (2). Therefore linear combinations of the $u_n(x, t)$ also satisfy both Eqs. (1) and (2). In making our linear combinations, we need no new constants, since the a_n and b_n are arbitrary. We have, then,

$$u(x, t) = \sum_{n=1}^{\infty} \sin \lambda_n x[a_n \cos \lambda_n ct + b_n \sin \lambda_n ct]. \tag{8}$$

The initial conditions, which remain to be satisfied, have the form

$$u(x, 0) = \sum a_n \sin\left(\frac{n\pi x}{a}\right) = f(x), \qquad 0 < x < a$$

$$\frac{\partial u}{\partial t}(x, 0) = \sum b_n \frac{n\pi}{a} c \sin\left(\frac{n\pi x}{a}\right) = g(x), \qquad 0 < x < a.$$

(Here we have assumed that

$$\frac{\partial u}{\partial t}(x, t) = \sum \sin \lambda_n x[-a_n \lambda_n c \sin \lambda_n ct + b_n \lambda_n c \cos \lambda_n ct].$$

In other words, we assume that the series for u may be differentiated term-by-term.) Both of the initial conditions have the form of Fourier sine series. By applying the principle of orthogonality, we determine that

$$a_n = \frac{2}{a} \int_0^a f(x) \sin\left(\frac{n\pi x}{a}\right) dx \qquad (9)$$

$$b_n \frac{n\pi}{a} c = \frac{2}{a} \int_0^a g(x) \sin\left(\frac{n\pi x}{a}\right) dx$$

or

$$b_n = \frac{2}{n\pi c} \int_0^a g(x) \sin\left(\frac{n\pi x}{a}\right) dx. \qquad (10)$$

If the functions $f(x)$ and $g(x)$ are sectionally smooth on the interval $0 < x < a$, then we know that the initial conditions really are satisfied, except possibly at points of discontinuity of f or g. By the nature of the problem, however, one would expect that f, at least, would be continuous and would satisfy $f(0) = f(a) = 0$. Thus we expect the series for f to converge uniformly.

Let us take specific initial conditions to see an example solution. If the string is lifted in the middle and then released, appropriate initial conditions are:

$$u(x, 0) = f(x) = \begin{cases} h \cdot \dfrac{2x}{a}, & 0 < x < \dfrac{a}{2} \\ h\left(2 - \dfrac{2x}{a}\right), & \dfrac{a}{2} < x < a \end{cases}$$

$$\frac{\partial u}{\partial t}(x, 0) = 0, \qquad 0 < x < a.$$

Then $b_n = 0$ for $n = 1, 2, 3, \ldots$, and

$$a_n = \frac{2}{a}\left[\int_0^{a/2} h \cdot \frac{2x}{a} \sin\left(\frac{n\pi x}{a}\right) dx + \int_{a/2}^a h\left(2 - \frac{2x}{a}\right)\sin\left(\frac{n\pi x}{a}\right) dx\right]$$

$$= \frac{8h}{\pi^2}\frac{\sin(n\pi/2)}{n^2}.$$

Therefore the complete solution is

$$u(x, t) = \frac{8h}{\pi^2} \sum_{n=1}^{\infty} \frac{\sin(n\pi/2)}{n^2} \sin\left(\frac{n\pi x}{a}\right)\cos\left(\frac{n\pi ct}{a}\right). \tag{11}$$

Although the solution above may be considered valid, it is difficult to see, in the present form, what shape the string will take at various times. However, because of the simplicity of the sines and cosines, it is possible to rewrite the solution in such a way that $u(x, t)$ may be determined without summing a series.

By applying the trigonometric identity

$$\sin A \cos B = \tfrac{1}{2}[\sin(A - B) + \sin(A + B)]$$

we can express $u(x, t)$ as

$$u(x, t) = \frac{1}{2}\left[\frac{8h}{\pi^2} \sum_{n=1}^{\infty} \frac{\sin(n\pi/2)}{n^2} \sin\frac{n\pi(x - ct)}{a} + \frac{8h}{\pi^2} \sum_{n=1}^{\infty} \frac{\sin(n\pi/2)}{n^2} \sin\frac{n\pi(x + ct)}{a}\right].$$

We know that the series

$$\frac{8h}{\pi^2} \sum_{n=1}^{\infty} \frac{\sin(n\pi/2)}{n^2} \sin\left(\frac{n\pi x}{a}\right)$$

actually converges to the odd periodic extension, with period $2a$, of the function $f(x)$. Let us designate this extension by $\bar{f}(x)$ and note that it is defined for all values of its argument. Using this observation, we can express $u(x, t)$ more simply as

$$u(x, t) = \tfrac{1}{2}[\bar{f}(x - ct) + \bar{f}(x + ct)]. \tag{12}$$

In this form, the solution $u(x, t)$ can easily be sketched for various values of t. In Fig. 3.3 are graphs of $\bar{f}(x + ct)$, $\bar{f}(x - ct)$, and

$$u(x, t) = \tfrac{1}{2}[\bar{f}(x + ct) + \bar{f}(x - ct)]$$

for the particular example discussed here and for various values of t. The displacement $u(x, t)$ is periodic in time with period $2a/c$. During the second half period (not shown) the string returns to its initial position through the positions shown. The horizontal portions of the string have a nonzero velocity.

Equation (12) can also be used to find $u(x, t)$ for any given x and t. For instance, if we take $x = 0.2a$ and $t = 0.9a/c$, we find that

$$u(0.2a, 0.9a/c) = \tfrac{1}{2}[\bar{f}(-0.7a) + \bar{f}(1.1a)]$$
$$= \tfrac{1}{2}[(-0.6h) + (-0.2h)]$$
$$= -0.4h.$$

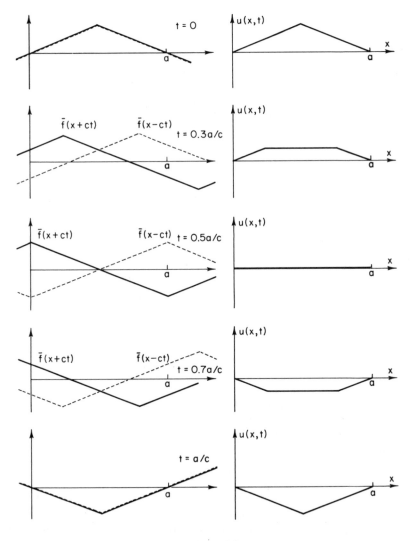

Figure 3.3

Exercises

1. Solve Eqs. (1)–(4) with the conditions

$$u(x, 0) = 0, \qquad 0 < x < a$$

$$\frac{\partial u}{\partial t}(x, 0) = 1, \qquad 0 < x < a.$$

Sketch the solution $u(x, t)$ for various values of t.

2. If

$$\sum a_n \sin\left(\frac{n\pi x}{a}\right) = \bar{f}(x), \qquad \sum b_n \cos\left(\frac{n\pi x}{a}\right) = \bar{G}(x)$$

show that $u(x, t)$ as given in Eq. (8) may be written

$$u(x, t) = \tfrac{1}{2}(\bar{f}(x - ct) + \bar{f}(x + ct)) + \tfrac{1}{2}(\bar{G}(x - ct) - \bar{G}(x + ct)).$$

Here, $\bar{f}(x)$ and $\bar{G}(x)$ are periodic with period $2a$.

3. The pressure of the air in an organ pipe satisfies the equation

$$\frac{\partial^2 p}{\partial x^2} = \frac{1}{c^2}\frac{\partial^2 p}{\partial t^2}, \qquad 0 < x < a, \quad 0 < t$$

with the boundary conditions

a. $p(0, t) = p_0$, $p(a, t) = p_0$ if the pipe is open, or

b. $p(0, t) = p_0$, $\dfrac{\partial p}{\partial x}(a, t) = 0$ if the pipe is closed at $x = a$.

Find the eigenvalues and eigenfunctions associated with the wave equation for each of these sets of boundary conditions.

4. Find the lowest frequency of vibration of the air in the organ pipes referred to in Exercise 3a and b.

5. If a string vibrates in a medium which resists the motion, the problem for the displacement of the string is

$$\frac{\partial^2 u}{\partial x^2} = \frac{1}{c^2}\frac{\partial^2 u}{\partial t^2} + k\frac{\partial u}{\partial t}, \qquad 0 < x < a, \quad 0 < t$$

$$u(0, t) = 0, \qquad u(a, t) = 0, \qquad 0 < t$$

plus initial conditions. Find eigenfunctions, eigenvalues, and product solutions for this problem. (k is small and positive.)

6. Sketch $u_1(x, t)$ and $u_2(x, t)$ as functions of x for several values of t. Assume a_1 and $a_2 = 1$, b_1 and $b_2 = 0$. (The solutions $u_n(x, t)$ are called "standing waves.")

7. The displacements $u(x, t)$ of a uniform thin beam satisfy

$$\frac{\partial^4 u}{\partial x^4} = \frac{-1}{c^2} \frac{\partial^2 u}{\partial t^2}, \qquad 0 < x < a, \quad 0 < t.$$

If the beam is simply supported at the ends, the boundary conditions are

$$u(0, t) = 0, \qquad \frac{\partial^2 u}{\partial x^2}(0, t) = 0, \qquad u(a, t) = 0, \qquad \frac{\partial^2 u}{\partial x^2}(a, t) = 0.$$

Find product solutions to this problem. What are the frequencies of vibration?

8. Verify that $u_n(x, t)$ satisfies Eqs. (1) and (2).

9. Does the series Eq. (11) converge uniformly?

3   D'Alembert's solution

We have seen a simple way of expressing the solution to the vibrating string problem in a particular case. (In fact, the one-dimensional wave equation is one of the few important partial differential equations for which a simple solution is known.) It appears that $w = x + ct$ and $z = x - ct$ are significant variables, so we shall express the wave equation in terms of these new variables. Let $u(x, t) = v(w, z)$. By the chain rule, we calculate

$$\frac{\partial u}{\partial x} = \frac{\partial v}{\partial w} \frac{\partial w}{\partial x} + \frac{\partial v}{\partial z} \frac{\partial z}{\partial x} = \frac{\partial v}{\partial w} + \frac{\partial v}{\partial z}$$

$$\frac{\partial^2 u}{\partial x^2} = \frac{\partial}{\partial w}\left(\frac{\partial v}{\partial w} + \frac{\partial v}{\partial z}\right)\frac{\partial w}{\partial x} + \frac{\partial}{\partial z}\left(\frac{\partial v}{\partial w} + \frac{\partial v}{\partial z}\right)\frac{\partial z}{\partial x}$$

$$= \frac{\partial^2 v}{\partial w^2} + 2\frac{\partial^2 v}{\partial w\, \partial z} + \frac{\partial^2 v}{\partial z^2}$$

and similarly

$$\frac{\partial^2 u}{\partial t^2} = c^2\left(\frac{\partial^2 v}{\partial w^2} - 2\frac{\partial^2 v}{\partial w\, \partial z} + \frac{\partial^2 v}{\partial z^2}\right).$$

(We have assumed that the two mixed partials $\partial^2 v/\partial z\, \partial w$ and $\partial^2 v/\partial w\, \partial z$ are equal.) If $u(x, t)$ satisfies the wave equation, then

$$\frac{\partial^2 u}{\partial x^2} = \frac{1}{c^2} \frac{\partial^2 u}{\partial t^2}.$$

In terms of the function v and the new independent variables this equation becomes

$$\frac{\partial^2 v}{\partial w^2} + 2\frac{\partial^2 v}{\partial w\, \partial z} + \frac{\partial^2 v}{\partial z^2} = \frac{\partial^2 v}{\partial w^2} - 2\frac{\partial^2 v}{\partial w\, \partial z} + \frac{\partial^2 v}{\partial z^2}$$

or, simply,

$$\frac{\partial^2 v}{\partial w\, \partial z} = 0.$$

It is actually possible to find the general solution of this last equation. Put in another form it says

$$\frac{\partial}{\partial z}\left(\frac{\partial v}{\partial w}\right) = 0$$

which means that $\partial v/\partial w$ is independent of z, or

$$\frac{\partial v}{\partial w} = \theta(w).$$

Integrating this equation, we find that

$$v = \int \theta(w)\, dw + \phi(z).$$

Here, $\phi(z)$ plays the role of an integration "constant." Since the integral of $\theta(w)$ is also a function of w, we may write the *general* solution of the partial differential equation above as

$$v(w, z) = \psi(w) + \phi(z)$$

where ψ and ϕ are arbitrary functions with continuous derivatives. Transforming back to our original variables, we get

$$u(x, t) = \psi(x + ct) + \phi(x - ct) \tag{1}$$

as a form for the general solution of the one-dimensional wave equation. This is known as d'Alembert's solution.

Now let us look at the vibrating string problem:

$$\frac{\partial^2 u}{\partial x^2} = \frac{1}{c^2}\frac{\partial^2 u}{\partial t^2}, \qquad\qquad 0 < x < a, \quad 0 < t \tag{2}$$

$$u(0, t) = 0, \qquad u(a, t) = 0, \qquad 0 < t \tag{3}$$

$$u(x, 0) = f(x), \qquad\qquad 0 < x < a \tag{4}$$

$$\frac{\partial u}{\partial t}(x, 0) = g(x), \qquad\qquad 0 < x < a. \tag{5}$$

We already know a form for u. The problem is to choose ψ and ϕ in such a way that the initial and boundary conditions are satisfied. We assume then that

$$u(x, t) = \psi(x + ct) + \phi(x - ct).$$

The initial conditions are

$$\psi(x) + \phi(x) = f(x), \qquad 0 < x < a \qquad (6)$$
$$c\psi'(x) - c\phi'(x) = g(x), \qquad 0 < x < a.$$

If we divide through the second equation by c and integrate, it becomes

$$\psi(x) - \phi(x) = G(x) + A, \qquad 0 < x < a \qquad (7)$$

where $G(x)$ stands for

$$G(x) = \frac{1}{c} \int_0^x g(y)\, dy \qquad (8)$$

and A is an arbitrary constant. Equations (6) and (7) can now be solved simultaneously to determine

$$\psi(x) = \tfrac{1}{2}(f(x) + G(x) + A), \qquad 0 < x < a$$
$$\phi(x) = \tfrac{1}{2}(f(x) - G(x) - A), \qquad 0 < x < a.$$

These equations give ψ and ϕ only for values of the argument between 0 and a. But $x \pm ct$ may take on any value whatsoever, so we must extend these functions to define them for arbitrary values of their arguments, and in such a way that the boundary conditions are satisfied. The boundary conditions are

$$u(0, t) = \psi(ct) + \phi(-ct) = 0, \qquad t > 0 \qquad (9)$$
$$u(a, t) = \psi(a + ct) + \phi(a - ct) = 0, \qquad t > 0. \qquad (10)$$

The first equation says that

$$\bar{f}(ct) + \bar{G}(ct) + A + \bar{f}(-ct) - \bar{G}(-ct) - A = 0$$
$$\bar{f}(ct) + \bar{f}(-ct) + \bar{G}(ct) - \bar{G}(-ct) = 0$$

($\bar{f}$ and $\bar{G}$ are the extensions of f and G that we are looking for). Since these equations must be true for arbitrary functions f and G (because the two functions are not interdependent), we must have individually

$$\bar{f}(ct) = -\bar{f}(-ct), \qquad \bar{G}(ct) = \bar{G}(-ct). \qquad (11)$$

That is, $\bar{f}$ is odd and $\bar{G}$ is even.

At the second endpoint, a similar calculation shows that

$$\bar{f}(a + ct) + \bar{f}(a - ct) + \bar{G}(a + ct) - \bar{G}(a - ct) = 0.$$

Once again, the independence of f and G implies that

$$\bar{f}(a + ct) = -\bar{f}(a - ct), \qquad \bar{G}(a + ct) = \bar{G}(a - ct). \tag{12}$$

The oddness of $\bar{f}$ and evenness of $\bar{G}$ can be used to transform the right-hand sides. Then

$$\bar{f}(a + ct) = \bar{f}(-a + ct), \qquad \bar{G}(a + ct) = \bar{G}(-a + ct).$$

These equations say that $\bar{f}$ and $\bar{G}$ are both periodic with period $2a$. The explicit expressions for ϕ and ψ are

$$\psi(x + ct) = \tfrac{1}{2}(\bar{f}(x + ct) + \bar{G}(x + ct) + A)$$
$$\phi(x - ct) = \tfrac{1}{2}(\bar{f}(x - ct) - \bar{G}(x - ct) - A)$$

where $\bar{f}$ is the *odd* periodic extension of f (period $2a$) and $\bar{G}$ is the *even* periodic extension of G (period $2a$).

Finally, we arrive at an expression for the solution $u(x, t)$:

$$u(x, t) = \tfrac{1}{2}[\bar{f}(x + ct) + \bar{f}(x - ct)] + \tfrac{1}{2}[\bar{G}(x + ct) - \bar{G}(x - ct)]. \tag{13}$$

This form of the solution of Eqs. (2)–(5) allows us to see directly how the initial data influence the solution at later times. From a practical point of view, it also permits us to calculate $u(x, t)$ at any x and t and even to sketch u as a function of one independent variable for a fixed value of the other. The following procedure is helpful in sketching $u(x, t)$ as a function of x for a fixed $t = t^*$, when the initial condition (5) has $g(x) \equiv 0$. It is easily adapted to other cases.

1. Sketch the odd periodic extension of f; call this $\bar{f}(q)$. (We use q as the independent variable to avoid the association with x or t.)

2. Sketch $\bar{f}(x + ct^*)$ against x; this is just the graph of $\bar{f}$ shifted ct^* units to the *left*.

3. Sketch $\bar{f}(x - ct^*)$ against x on the same axes. This graph is the same as that of $\bar{f}$ but shifted ct^* units to the *right*.

4. Average graphically the graphs made in the two preceding steps. Check that the boundary conditions are satisfied.

Exercises

1. Let $u(x, t)$ be a solution of Eqs. (2)–(5) with $g(x) \equiv 0$ and $f(x)$ a function whose graph is an isosceles triangle of width a and height h. Find $u(x, t)$ for $x = 0.25a$ and $0.5a$, and for $t = 0, 0.2a/c, 0.4a/c, 0.8a/c, 1.4a/c$.

2. Sketch $u(x, t)$ of Exercise 1 as a function of x for the times given. Compare your results with Fig. 3.3.

3. Let $u(x, t)$ be a solution of Eqs. (2)–(5) with $f(x) \equiv 0$ and $g(x) = \alpha c$, $0 < x < a$. Find $u(x, t)$ at: $x = 0$, $t = 0.5a/c$; $x = 0.2a$, $t = 0.6a/c$; $x = 0.5a$, $t = 1.2a/c$.

4. Sketch $u(x, t)$ of Exercise 3 as a function of x for times $t = 0, 0.25a/c$, $0.5a/c, a/c$.

5. Find the function $G(x)$ corresponding to (see Eq. (8))

$$g(x) = \begin{cases} 0, & 0 < x < 0.4a \\ 5c, & 0.4a < x < 0.6a \\ 0, & 0.6a < x < a. \end{cases}$$

6. Let $u(x, t)$ be a solution of Eqs. (2)–(5) with $f(x) \equiv 0$ and $g(x)$ as given in Exercise 5. Sketch $u(x, t)$ as a function of x for various times and as a function of t for various places.

7. Let $u(x, t)$ be the function referred to in Exercise 1. Sketch x and t axes; locate the region $0 < x < a$, $0 < t$; sketch some level curves of $u(x, t)$. (A level curve is the locus of points (x, t) where $u(x, t)$ has a fixed value.)

8. The equation for the forced vibrations of a string is

$$\frac{\partial^2 u}{\partial x^2} - \frac{1}{c^2} \frac{\partial^2 u}{\partial t^2} = \frac{-1}{T} F(x, t) \qquad (*)$$

(see Section 1, Exercise 2). Changing variables to

$$w = x + ct, \qquad z = x - ct, \qquad u(x, t) = v(w, z), \qquad f(w, z) = F(x, t)$$

this equation becomes

$$\frac{\partial^2 v}{\partial w \, \partial z} = \frac{-1}{4T} f(w, z).$$

Show that the general solution of this equation is

$$v(w, z) = \frac{-1}{4T} \iint f(w, z) \, dw \, dz + \psi(w) + \phi(z).$$

9. Find the general solution of $(*)$ in terms of x and t if $F(x, t) = T \cos t$.

10. Verify directly that $u(x, t)$ as given by Eq. (1) is a solution of the wave equation (2) if ϕ and ψ have at least two derivatives.

4   Generalities on the one-dimensional wave equation

As for the one-dimensional heat equation, we can make some comments for a generalized one-dimensional wave equation. For the sake of generality, we assume that some nonuniform properties are present. For the sake of simplicity, we assume that the equation is homogeneous, and free of u. Our initial value–boundary value problem will be

$$\frac{\partial}{\partial x}\left(s(x)\frac{\partial u}{\partial x}\right) = \frac{p(x)}{c^2}\frac{\partial^2 u}{\partial t^2}, \qquad l < x < r, \quad 0 < t \qquad (1)$$

$$\alpha_1 u(l, t) - \alpha_2 \frac{\partial u}{\partial x}(l, t) = c_1, \qquad\qquad 0 < t \qquad (2)$$

$$\beta_1 u(r, t) + \beta_2 \frac{\partial u}{\partial x}(r, t) = c_2, \qquad\qquad 0 < t \qquad (3)$$

$$u(x, 0) = f(x), \qquad\qquad l < x < r \qquad (4)$$

$$\frac{\partial u}{\partial t}(x, 0) = g(x), \qquad\qquad l < x < r. \qquad (5)$$

We assume that the functions $s(x)$ and $p(x)$ are positive for $l \le x \le r$, since they represent physical properties; that s, s', and p are all continuous, and that s and p have no dimensions. Also suppose that none of the coefficients α_1, α_2, β_1, β_2 is negative.

In order to get homogeneous boundary conditions we can write

$$u(x, t) = v(x) + w(x, t)$$

just as before. In the wave equation, however, neither of the names "steady-state solution" nor "transient solution" is appropriate; for as we shall see, there is no steady state, or limiting case, nor is there a part of the solution which tends to zero as t tends to infinity. Nevertheless, v represents an equilibrium solution and, more important, it is a useful mathematical device to consider u in the form above.

The function $v(x)$ is required to satisfy the conditions

$$(sv')' = 0, \qquad l < x < r$$
$$\alpha_1 v(l) - \alpha_2 v'(l) = c_1$$
$$\beta_1 v(r) + \beta_2 v'(r) = c_2.$$

Thus $v(x)$ is exactly equivalent to the "steady-state solution" discussed for the heat equation.

The function $w(x, t)$, being the difference between $u(x, t)$ and $v(x)$, satisfies the initial value–boundary value problem

$$\frac{\partial}{\partial x}\left(s(x)\frac{\partial w}{\partial x}\right) = \frac{p(x)}{c^2}\frac{\partial^2 w}{\partial t^2}, \qquad l < x < r, \quad 0 < t \tag{6}$$

$$\alpha_1 w(l, t) - \alpha_2 \frac{\partial w}{\partial x}(l, t) = 0, \qquad 0 < t \tag{7}$$

$$\beta_1 w(r, t) + \beta_2 \frac{\partial w}{\partial x}(r, t) = 0, \qquad 0 < t \tag{8}$$

$$w(x, 0) = f(x) - v(x), \qquad l < x < r \tag{9}$$

$$\frac{\partial w}{\partial t}(x, 0) = g(x), \qquad l < x < r. \tag{10}$$

Since the equation and the boundary conditions are homogeneous and linear, we attempt a solution by separation of variables. If $w(x, t) = \phi(x)T(t)$, we find in the usual way that the factor functions ϕ and T must satisfy

$$T'' + c^2\lambda^2 T = 0, \qquad 0 < t \tag{11}$$
$$(s\phi')' + p\lambda^2\phi = 0, \qquad l < x < r \tag{12}$$
$$\alpha_1\phi(l) - \alpha_2\,\phi'(l) = 0 \tag{13}$$
$$\beta_1\phi(r) + \beta_2\,\phi'(r) = 0. \tag{14}$$

The eigenvalue problem represented in the last three lines is a regular Sturm–Liouville problem, because of the assumptions we have made about s, p, and the coefficients. We know that there are an infinite number of non-negative eigenvalues $\lambda_1^2, \lambda_2^2, \ldots$, and corresponding eigenfunctions $\phi_1, \phi_2, \ldots$ which have the orthogonality property

$$\int_l^r \phi_n(x)\phi_m(x)p(x)\,dx = 0, \qquad n \neq m.$$

The solution of the equation for T is

$$T_n(t) = a_n \cos \lambda_n ct + b_n \sin \lambda_n ct.$$

Having solved the subsidiary problems which arose after separation of variables, we can begin to resassemble the solution. The function w will have the form

$$w(x, t) = \sum_{n=1}^{\infty} \phi_n(x)(a_n \cos \lambda_n ct + b_n \sin \lambda_n ct)$$

and its two initial conditions, yet to be satisfied, are

$$w(x, 0) = \sum_{n=1}^{\infty} a_n \phi_n(x) = f(x) - v(x), \qquad l < x < r$$

$$\frac{\partial w}{\partial t}(x, t) = \sum_{n=1}^{\infty} b_n \lambda_n c \phi_n(x) = g(x), \qquad l < x < r.$$

By employing the orthogonality of the ϕ_n, we determine that the coefficients a_n and b_n are given by

$$a_n = \int_l^r [f(x) - v(x)] \phi_n(x) p(x) \, dx / I_n$$

$$b_n = \int_l^r g(x) \phi_n(x) p(x) \, dx / I_n \lambda_n c$$

where

$$I_n = \int_l^r \phi_n^2(x) p(x) \, dx.$$

Finally, $u(x, t) = v(x) + w(x, t)$ is the solution of the original problem, and each of its parts is completely specified. From the form of $w(x, t)$, we can make certain observations about u:

1. $u(x, t)$ does not have a limit as $t \to \infty$. Each term of the series form of w is periodic in time and thus does not die away.

2. Except in very special cases, the eigenvalues λ_n^2 are not closely related to each other. So in general, if u causes acoustic vibrations, the result will not be musical to the ear. (A sound would be musical if, for instance, $\lambda_n = n\lambda_1$, as in the case of the uniform string.)

3. In general, $u(x, t)$ is not even periodic in time. Although each term in the series for w is periodic, the terms do not have a *common* period (except in special cases), and so the sum is not periodic.

Exercises

1. Verify the formulas for the a_n and b_n. Under what conditions on f and g can we say that the initial conditions are satisfied?

2. Check the statement that $v(x)$ is the same for the heat conduction problem and for the problem considered here.

3. Identify the period of $T_n(t)$ and the associated frequency.

4. Although $u(x, t)$ has no limit as $t \to \infty$, show that the following generalized limit is valid:

$$v(x) = \lim_{T \to \infty} \frac{1}{T} \int_0^T u(x, t)\, dt.$$

(Hint: do the integration and limiting term by term.)

5. Formally solve the problem

$$\frac{\partial}{\partial x}\left(s\frac{\partial u}{\partial x}\right) = \frac{p}{c^2}\left[\frac{\partial^2 u}{\partial t^2} + \gamma\frac{\partial u}{\partial t}\right] + qu, \qquad l < x < r, \quad 0 < t$$

with the boundary conditions Eqs. (2) and (3) and initial conditions Eqs. (4) and (5), taking γ to be constant.

6. Verify that $w(x, t)$ satisfies the differential equation and the boundary conditions Eqs. (6)–(8).

5 Estimation of eigenvalues

In many instances, one is interested not in the full solution to the wave equation, but only in the possible frequencies of vibration which may occur. For example, it is of great importance that bridges, airplane wings, and other structures should not vibrate; so it is important to know the frequencies at which a structure can vibrate, in order to avoid them. By inspecting the solution of the generalized wave equation, which we investigated in the preceding section, we can see that the frequencies of vibration are $\lambda_n c/2\pi$, $n = 1, 2, 3, \ldots$. Thus we must find the eigenvalues λ_n^2 in order to identify the frequencies of vibration.

Consider the following Sturm–Liouville problem:

$$(s\phi')' - q\phi + \lambda^2 p\phi = 0, \qquad l < x < r \tag{1}$$

$$\phi(l) = 0, \qquad \phi(r) = 0 \tag{2}$$

where s, s', q, and p are continuous, and s and p are positive for $l \le x \le r$. (Note that we have a rather general differential equation, but very special boundary conditions.)

If ϕ_1 is the eigenfunction corresponding to the smallest eigenvalue λ_1^2, then ϕ_1 satisfies Eq. (1) for $\lambda = \lambda_1$. Alternatively, we can write

$$-(s\phi_1')' + q\phi_1 = \lambda_1^2 p\phi_1, \qquad l < x < r.$$

Multiplying through this equation by ϕ_1 and integrating from l to r, we get

$$\int_l^r -(s\phi_1')'\phi_1 \, dx + \int_l^r q\phi_1{}^2 \, dx = \lambda_1{}^2 \int_l^r p\phi_1{}^2 \, dx.$$

If the first integral is integrated by parts, it becomes

$$-s\phi_1'\phi_1 \Big|_l^r + \int_l^r s\phi_1'\phi_1' \, dx.$$

But $\phi_1(l) = \phi_1(r) = 0$, so the first term vanishes and we are left with the equality

$$\int_l^r s[\phi_1']^2 \, dx + \int_l^r q\phi_1{}^2 \, dx = \lambda_1{}^2 \int_l^r p\phi_1{}^2 \, dx.$$

Since $p(x)$ is positive for $l \le x \le r$, the integral on the right is positive and we may define $\lambda_1{}^2$ as

$$\lambda_1{}^2 = \frac{\int_l^r s[\phi_1']^2 \, dx + \int_l^r q\phi_1{}^2 \, dx}{\int_l^r p\phi_1{}^2 \, dx} = \frac{N(\phi_1)}{D(\phi_1)}. \tag{3}$$

It can be shown in the calculus of variations that, if $y(x)$ is any function with two continuous derivatives ($l \le x \le r$) which satisfies $y(l) = y(r) = 0$, then

$$\lambda_1{}^2 \le \frac{N(y)}{D(y)}. \tag{4}$$

Thus, by choosing any convenient function y, which satisfies these conditions, we can find an estimate for $\lambda_1{}^2$. Ordinarily the estimate is surprisingly good. One should keep in mind that $\phi_1(x)$ does not cross the x-axis between l and r— so y should not cross the axis either.

Example

$$\phi'' + \lambda^2\phi = 0, \qquad 0 < x < 1$$
$$\phi(0) = \phi(1) = 0.$$

Let us try $y(x) = x(1 - x)$. Then $y'(x) = 1 - 2x$ and

$$N(y) = \int_0^1 [y'(x)]^2 \, dx = \int_0^1 (1 - 2x)^2 \, dx = \frac{1}{3},$$

$$D(y) = \int_0^1 y^2(x) \, dx = \int_0^1 x^2(1 - x)^2 \, dx = \frac{1}{30}.$$

Therefore $N(y)/D(y) = 10$. We know, of course, that $\phi_1(x) = \sin \pi x$, and

$$N(\phi_1) = \int_0^1 \pi^2 \cos^2 \pi x \, dx = \frac{\pi^2}{2}$$

$$D(\phi_1) = \int_0^1 \sin^2 \pi x \, dx = \frac{1}{2}$$

so $N(\phi_1)/D(\phi_1) = \lambda_1^2 = \pi^2 < 10$.

Example

$$(x\phi')' + \lambda^2 \frac{1}{x} \phi = 0, \qquad 1 < x < 2$$

$$\phi(1) = \phi(2) = 0.$$

The integrals to be calculated are

$$N(y) = \int_1^2 x(y')^2 \, dx, \qquad D(y) = \int_1^2 \frac{1}{x} y^2 \, dx.$$

The tabulation below gives results for several trial functions. It is known that the first eigenvalue and eigenfunction are

$$\lambda_1^2 = \left(\frac{\pi}{\ln 2} \right)^2 = 20.5423$$

$$\phi_1(x) = \sin\left(\frac{\pi \ln x}{\ln 2} \right).$$

The error for the best of the trial functions is about 1.44%:

$y(x)$	$\sqrt{x(2-x)(x-1)}$	$(2-x)(x-1)$	$\dfrac{(2-x)(x-1)}{x}$
$\dfrac{N(y)}{D(y)}$	23.7500	22.1349	20.8379

This method of estimating the first eigenvalue is called Rayleigh's method, and the ratio $N(y)/D(y)$ is called the Rayleigh quotient. In some mechanical systems the Rayleigh quotient may be interpreted as the ratio between potential and kinetic energy. There are many other methods for estimating eigenvalues and for systematically improving the estimates.

Exercises

1.   Using Eq. (3), show that if $q \geq 0$, then $\lambda_1{}^2 \geq 0$ also.

2.   Verify the results for at least one of the trial functions used in the second example.

3.   Estimate the first eigenvalue of the problem

$$\phi'' + \lambda^2(1 + x)\phi = 0, \qquad 0 < x < 1$$
$$\phi(0) = \phi(1) = 0.$$

6   Wave equation in unbounded regions

When the wave equation is to be solved for $x > 0$ or for $-\infty < x < \infty$, we can proceed as we did for the solution of the heat equation in these unbounded regions. That is to say, we separate variables and use a Fourier integral to combine the product solutions.

Consider the problem

$$\frac{\partial^2 u}{\partial x^2} = \frac{1}{c^2}\frac{\partial^2 u}{\partial t^2}, \qquad 0 < t, \quad 0 < x \tag{1}$$

$$u(x, 0) = f(x), \qquad 0 < x \tag{2}$$

$$\frac{\partial u}{\partial t}(x, 0) = g(x), \qquad 0 < x \tag{3}$$

$$u(0, t) = 0, \qquad 0 < t. \tag{4}$$

We require in addition that the solution $u(x, t)$ be bounded as $x \to \infty$.

On separating variables, we make $u(x, t) = \phi(x)T(t)$ and find that the factors satisfy

$$T'' + \lambda^2 c^2 T = 0, \qquad 0 < t$$
$$\phi'' + \lambda^2 \phi = 0, \qquad 0 < x$$
$$\phi(0) = 0, \qquad |\phi(x)| \text{ bounded.}$$

The solutions are easily found to be

$$\phi(x; \lambda) = \sin \lambda x, \qquad T(t) = A \cos \lambda ct + B \sin \lambda ct$$

and we combine the products $\phi(x; \lambda)T(t)$ in a Fourier integral

$$u(x, t) = \int_0^\infty (A(\lambda) \cos \lambda ct + B(\lambda) \sin \lambda ct) \sin \lambda x \, d\lambda. \tag{5}$$

The initial conditions become

$$u(x, 0) = f(x) = \int_0^\infty A(\lambda) \sin \lambda x \, d\lambda, \qquad 0 < x$$

$$\frac{\partial u}{\partial t}(x, 0) = g(x) = \int_0^\infty \lambda c B(\lambda) \sin \lambda x \, d\lambda, \qquad 0 < x.$$

Both of these equations are Fourier integrals. Thus the coefficient functions are given by

$$A(\lambda) = \frac{2}{\pi} \int_0^\infty f(x) \sin \lambda x \, dx, \qquad B(\lambda) = \frac{2}{\pi \lambda c} \int_0^\infty g(x) \sin \lambda x \, dx.$$

It is sufficient to demand that $\int_0^\infty |f(x)| \, dx$ and $\int_0^\infty |g(x)| \, dx$ both be finite in order to guarantee the existence of A and B.

The deficiency of the Fourier integral form of the solution given in Eq. (5) is that one has no idea of what $u(x, t)$ looks like. The d'Alembert solution of the wave equation can come to our aid again here. We know that the solution of Eq. (1) has the form

$$u(x, t) = \psi(x + ct) + \phi(x - ct).$$

The two initial conditions boil down to

$$\psi(x) + \phi(x) = f(x), \qquad 0 < x$$
$$\psi(x) - \phi(x) = G(x) + A, \qquad 0 < x.$$

As in the finite case, we have defined

$$G(x) = \frac{1}{c} \int_0^x g(y) \, dy$$

and A is any constant.

From the two initial conditions we get

$$\psi(x) = \tfrac{1}{2}(f(x) + G(x) + A), \qquad x > 0$$
$$\phi(x) = \tfrac{1}{2}(f(x) - G(x) - A), \qquad x > 0.$$

Both f and G are known for $x > 0$. Thus

$$\psi(x + ct) = \tfrac{1}{2}(f(x + ct) + G(x + ct) + A)$$

is defined for all $x > 0$ and $t \geq 0$. But $\phi(x - ct)$ is not yet defined for $x - ct < 0$. That means that we must extend the functions f and G in such a way as to define ϕ for negative arguments and also satisfy the boundary condition. The sole boundary condition is Eq. (4), which becomes

$$u(0, t) = 0 = \psi(ct) + \phi(-ct).$$

In terms of $\bar{f}$ and $\bar{G}$, extensions of f and G, this is

$$0 = f(ct) + G(ct) + A + \bar{f}(-ct) - \bar{G}(-ct) - A.$$

Since f and G are not dependent on each other, we must have individually

$$f(ct) + \bar{f}(-ct) = 0, \qquad G(ct) - \bar{G}(-ct) = 0.$$

That is, $\bar{f}$ is the odd extension of f, and $\bar{G}$ is the even extension of G.
Finally we arrive at a formula for the solution

$$u(x, t) = \tfrac{1}{2}[f(x + ct) + G(x + ct)] + \tfrac{1}{2}[\bar{f}(x - ct) - \bar{G}(x - ct)]. \quad (6)$$

Now given the functions $f(x)$ and $g(x)$, it is a simple matter to construct $\bar{f}$ and $\bar{G}$ and thus to graph $u(x, t)$ as a function of either variable or to evaluate it for specific values of x and t. By way of illustration, Fig. 3.4 shows the solution of Eqs. (1)–(4) as a function of x at various times, for $f(x)$ as shown and $g(x) \equiv 0$.

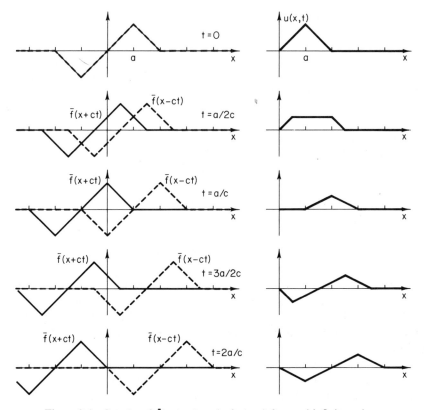

Figure 3.4   Graphs of $\bar{f}(x + ct)$ and of $u(x, t)$ for semi-infinite string.

Another interesting problem which can be treated by the d'Alembert method is one in which the boundary condition is a function of time. For simplicity, we take zero initial conditions. Our problem becomes

$$\frac{\partial^2 u}{\partial x^2} = \frac{1}{c^2}\frac{\partial^2 u}{\partial t^2} \qquad 0 < x, \qquad 0 < t \tag{7}$$

$$u(x, 0) = 0, \qquad 0 < x \tag{8}$$

$$\frac{\partial u}{\partial t}(x, 0) = 0, \qquad 0 < x \tag{9}$$

$$u(0, t) = h(t), \qquad 0 < t. \tag{10}$$

As u is to be a solution of the wave equation, it must have the form

$$u(x, t) = \psi(x + ct) + \phi(x - ct). \tag{11}$$

The two initial conditions, Eqs. (8) and (9), can be treated exactly as in the first problem. Of course, $G(x) \equiv 0$, and the constant A, being arbitrary, may be taken as 0. The conclusion is that

$$\psi(x) = 0, \qquad \phi(x) = 0, \quad 0 < x.$$

Since both x and t are positive in this problem, we see that $\psi(x + ct) = 0$ always, so that Eq. (11) may be simplified to

$$u(x, t) = \phi(x - ct). \tag{12}$$

The boundary condition Eq. (10) will tell us how to evaluate ϕ for negative arguments. The equation is

$$u(0, t) = \phi(-ct) = h(t), \qquad 0 < t. \tag{13}$$

We now put together what we know of the function ϕ:

$$\phi(q) = \begin{cases} 0, & q > 0 \\ h\left(-\dfrac{q}{c}\right), & q < 0. \end{cases} \tag{14}$$

The argument q is a dummy, used to avoid association with either x or t. Equations (12) and (14) now specify the solution $u(x, t)$ completely.

For a specific example, we take $h(t)$ as shown in Fig. 3.5. There also is a graph of $\phi(q)$. Note that the graph of ϕ for negative argument is that of h, reflected.

The graphs in Fig. 3.6 show $u(x, t)$ as a function of x for various values of t. It is clear from both the graphs and the formula that the disturbance caused by the variable boundary condition arrives at a fixed point x at time x/c. Thus the disturbance travels with the velocity c, the wave speed.

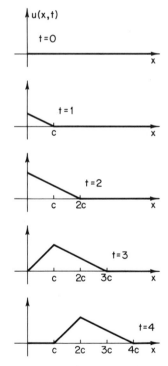

Figure 3.5 Graphs of $h(t)$ and $\phi(q)$ for semi-infinite string with time-varying boundary condition.

A wave equation accompanied by nonzero initial conditions and time-varying boundary conditions can be solved by breaking it into two problems, one like Eqs. (1)–(4) with zero boundary condition, and the other like Eqs. (7)–(10) with zero initial conditions.

Figure 3.6 Graphs of $u(x, t)$ vs. x for the semi-infinite string with time-varying boundary condition.

Exercises

1. Derive a formula similar to Eq. (6) for the case in which the boundary condition Eq. (4) is replaced by

$$\frac{\partial u}{\partial x}(0, t) = 0, \qquad 0 < t.$$

2. Derive Eq. (6) from Eq. (5) by using trigonometric identities for $\sin \lambda x \cdot \cos \lambda ct$, and so forth, and recognizing certain Fourier integrals.

3. Sketch the solution of Eqs. (1)–(4) as a function of x at times $t = 0$, $1/2c$, $1/c$, $2/c$, $3/c$, if $g(x) = 0$ everywhere and $f(x)$ is the rectangular pulse

$$f(x) = \begin{cases} 0, & 0 < x < 1 \\ 1, & 1 < x < 2 \\ 0, & 2 < x. \end{cases}$$

4. Same as Exercise 3, but $f(x) = 0$ and $g(x)$ is the rectangular pulse

$$g(x) = \begin{cases} 0, & 0 < x < 1 \\ c, & 1 < x < 2 \\ 0, & 2 < x. \end{cases}$$

5. Sketch the solution of Eqs. (7)–(10) at times $t = 0$, $\pi/2$, π, $3\pi/2$, 2π, $5\pi/2$, if $h(t) = \sin t$.

6. Sketch the solution of Eqs. (7)–(10) at times $t = 0$, $1/2$, $3/2$, and $5/2$, if

$$h(t) = \begin{cases} 0, & 0 < t < 1 \\ 1, & 1 < t < 2 \\ 0, & 2 < t. \end{cases}$$

7. Use the d'Alembert solution of the wave equation to solve the problem

$$\frac{\partial^2 u}{\partial x^2} = \frac{1}{c^2}\frac{\partial^2 u}{\partial t^2}, \qquad -\infty < x < \infty, \quad 0 < t$$

$$u(x, 0) = f(x), \qquad -\infty < x < \infty$$

$$\frac{\partial u}{\partial t}(x, 0) = g(x), \qquad -\infty < x < \infty.$$

7 References

The wave equation is one of the oldest equations of mathematical physics. Euler, Bernoulli, and d'Alembert all solved the problem of the vibrating string about 1750, using either separation of variables or what we called d'Alembert's method. This latter is, in fact, a very special case of the method of characteristics, in essence a way of identifying new independent variables having special significance. Dennemeyer's *Introduction to Partial Differential Equations and Boundary Value Problems* [11] has a good introduction to the method of characteristics.

Because the wave equation has been around so long, there has been established a large vocabulary of special terms, many based on physical phenomena. Brief explanations of these concepts may be found in *Elementary Partial Differential Equations* by Berg and McGregor [3] or *Partial Differential Equations* by Weinberger [32]. A number of physical examples, especially from electrical engineering, are presented in *Partial Differential Equations* by Moon and Spencer [19]. The Scientific American article, "'Second Sound' in Solid Helium" by Bertman and Sandiford [4] discusses an interesting phenomenon relating heat flow and wave effects.

Many nonmechanical phenomena can be described by the wave equation. Among the most important are electrical and magnetic fields represented by special cases of the Maxwell field equations. Discussions will be found in Duff and Naylor [12] and in Moon and Spencer [19].

The potential difference V between the interior and exterior of a nerve axon can be modeled approximately by the Fitzhugh–Nagumo equations,

$$\frac{\partial V}{\partial t} = \frac{\partial^2 V}{\partial x^2} + V - \frac{1}{3} V^3 - R$$

$$\frac{\partial R}{\partial t} = k(V + a - bR).$$

Here R represents a restoring effect and a, b, and k are constants. At first glance, one would expect V to behave like the solution of a heat equation. But a traveling-wave solution

$$V(x, t) = F(x - ct), \qquad R(x, t) = G(x - ct)$$

of these equations can be found that shows many important features of nerve impulses. Further details will be found in *Biological Engineering* [28]. Other biological applications will be found in *Partial Differential Equations in Biology*, by Peskin [22].

More information about the Rayleigh quotient and estimation of eigenvalues is in *Boundary and Eigenvalue Problems in Mathematical Physics*, by Sagan [27].

Miscellaneous exercises

Exercises 1–5 refer to the problem

$$\frac{\partial^2 u}{\partial x^2} = \frac{1}{c^2}\frac{\partial^2 u}{\partial t^2}, \qquad\qquad 0 < x < a, \quad 0 < t$$

$$u(0, t) = 0, \; u(a, t) = 0, \qquad\qquad 0 < t$$

$$u(x, 0) = f(x), \frac{\partial u}{\partial t}(x, 0) = g(x), \qquad 0 < x < a.$$

1. Take $f(x) = 1$, $0 < x < a$ and $g(x) \equiv 0$. (This is rather unrealistic if $u(x, t)$ is the displacement of a vibrating string.) Find a series (separation of variables) solution.

2. Sketch $u(x, t)$ of Exercise 1 as a function of x at various times throughout one period.

3. Make a sketch of the region $0 < x < a$, $0 < t$ and find the locus of points where $u(x, t) = 1, 0$ or -1; u is the function of Exercise 1.

4. Take $g(x) = 0$ and $f(x)$ to be this function:

$$f(x) = \begin{cases} 3hx/2a, & 0 < x < 2a/3 \\ 3h(a - x)/2a, & 2a/3 < x < a. \end{cases}$$

    The graph of f is triangular, with peak at $x = 2a/3$. Find a series solution for $u(x, t)$.

5. Sketch $u(x, t)$ of Exercise 4 as a function of x at times 0 to a/c in steps of $a/(6c)$.

6. Find an analytic (integral) solution of this wave problem

$$\frac{\partial^2 u}{\partial x^2} = \frac{1}{c^2}\frac{\partial^2 u}{\partial t^2}, \qquad\qquad -\infty < x < \infty, \quad 0 < t$$

$$u(x, 0) = f(x), \quad \frac{\partial u}{\partial t}(x, 0) = g(x), \qquad -\infty < x < \infty,$$

    with $g(x) = 0$ and

$$f(x) = \begin{cases} h, & |x| < \varepsilon \\ 0, & |x| > \varepsilon. \end{cases}$$

7. Sketch the solution of the problem in Exercise 6 at times $t = 0$, ε/c, $2\varepsilon/c$, $3\varepsilon/c$.

8. Same as Exercise 7, but $f(x) = 0$ and

$$g(x) = \begin{cases} c, & |x| < \varepsilon \\ 0, & |x| > \varepsilon. \end{cases}$$

9. Let $u(x, t)$ be the solution of the problem

$$\frac{\partial^2 u}{\partial x^2} = \frac{1}{c^2}\frac{\partial^2 u}{\partial t^2}, \qquad\qquad 0 < x, \quad 0 < t$$

$$u(0, t) = 0, \qquad\qquad 0 < t$$

$$u(x, 0) = f(x), \qquad \frac{\partial u}{\partial t}(x, 0) = g(x), \qquad 0 < x.$$

Sketch the solution $u(x, t)$ as a function of x at times $t = 0$, $a/6c$, $a/2c$, $5a/6c$, $7a/6c$. Use $g(x) = 0$ and

$$f(x) = \begin{cases} 3hx/2a, & 0 < x < 2a/3 \\ 3h(a - x)/a, & 2a/3 < x < a \\ 0, & a < x. \end{cases}$$

10. Same task as in Exercise 9 but

$$f(x) = \begin{cases} \sin x, & 0 < x < \pi \\ 0, & \pi < x \end{cases}$$

and $g(x) = 0$. Sketch at times $t = 0$, $\pi/4c$, $\pi/2c$, $3\pi/4c$, π/c, $2\pi/c$.

11. Let $u(x, t)$ be the solution of the wave equation on the semi-infinite interval $0 < x < \infty$, with both initial conditions equal to zero, but with the time-varying boundary condition

$$u(0, t) = \begin{cases} \sin ct/a, & 0 < t < \pi a/c \\ 0, & \pi a/c < t. \end{cases}$$

Sketch $u(x, t)$ as a function of x at various times.

12. Same as Exercise 11, but the boundary condition is $u(0, t) = h$, for all $t > 0$.

13. Same as Exercise 11, but the boundary condition is

$$u(0, t) = \begin{cases} hct/a, & 0 < t < a/c \\ h(2a - ct)/a, & a/c < t < 2a/c \\ 0, & 2a/c < t. \end{cases}$$

14. Estimate the lowest eigenvalue of the problem

$$(e^{ax}\phi')' + \lambda^2 e^{ax}\phi = 0, \qquad 0 < x < 1$$
$$\phi(0) = 0, \qquad \phi(1) = 0.$$

(This problem can be solved exactly.)

15. Estimate the lowest eigenvalue of the problem

$$\phi'' - \frac{3}{4x^2}\phi' + \lambda^2\phi = 0, \qquad \frac{1}{4} < x < \frac{5}{4},$$

$$\phi\left(\frac{1}{4}\right) = 0, \qquad \phi\left(\frac{5}{4}\right) = 0.$$

16. Show that the nonlinear wave equation

$$\frac{\partial u}{\partial t} + u\frac{\partial u}{\partial x} + \frac{\partial^3 u}{\partial x^3} = 0$$

(the Korteweg–DeVries equation) has, as one solution,

$$u(x, t) = 12a^2\,\text{sech}^2(ax - 4a^3 t).$$

17. The solution in Exercise 16 is of the form $u(x, t) = f(x - ct)$. What is the function f, and what is the wave speed c?

18. For $t < 0$, water flows steadily through a long pipe connected at $x = 0$ to a large reservoir and open at $x = a$ to the air. At time $t = 0$ a valve at $x = a$ is suddenly closed. Reasonable expressions for the conservation of momentum and of mass are

$$\frac{\partial u}{\partial t} = -\frac{\partial p}{\partial x}, \qquad 0 < x < a, \quad 0 < t \qquad\qquad \text{(A)}$$

$$\frac{\partial p}{\partial t} = -\beta\frac{\partial u}{\partial x}, \qquad 0 < x < a, \quad \text{all } t \qquad\qquad \text{(B)}$$

where p is gauge pressure, u is the mass flow rate, and β (constant) is the compressibility of water. Show that both u and p satisfy the wave equation with $c^2 = \beta$.

19. Introduce a function v with the definition $u = \partial v/\partial x$, $p = -\partial v/\partial t$. Show that (A) becomes an identity and that (B) becomes the wave equation for v.

20. Reasonable boundary and initial conditions for u and p are

$$u(x, 0) = U_0 \text{ (constant)}, \qquad 0 < x < a$$
$$p(x, 0) = -kx, \qquad 0 < x < a$$
$$p(0, t) = 0, \qquad \text{all } t$$
$$u(a, t) = 0, \qquad t > 0.$$

Restate these as conditions on v. Show that the first and third equations may be replaced by

$$v(x, 0) = U_0 x, \qquad 0 < x < a$$
$$v(0, t) = 0, \qquad \text{all } t.$$

21. Solve the problem in Exercise 20; find a series form for $v(x, t)$.

22. In many problems involving fluid flow, the combination

$$\frac{\partial u}{\partial t} + V \frac{\partial u}{\partial x}$$

(called the Stokes derivative) appears. Here V is the speed of the fluid in the x direction. If $u = V$, or if V is a function of u, this operator is very nonlinear. But if V is not a function of u, x, or t, it is a linear operator. Suppose that this is the case. Define new variables

$$\xi = x + Vt, \qquad \tau = x - Vt, \qquad u(x, t) = v(\xi, \tau).$$

Show that

$$\frac{\partial u}{\partial t} + V \frac{\partial u}{\partial x} = 2V \frac{\partial v}{\partial \xi}.$$

23. Assume that $u(x, y, t)$ has the product form shown below. Separate the variables in the given partial differential equation.

$$u(x, y, t) = \psi(x + Vt)\phi(x - Vt)\,Y(y), \qquad \frac{\partial^2 u}{\partial y^2} = \frac{1}{k}\left(\frac{\partial u}{\partial t} + V \frac{\partial u}{\partial x}\right).$$

24. A fluid flows between two parallel plates held at temperature 0. At the inlet, fluid temperature is T_0 and initially the fluid is at temperature T_1. If V is the speed of the fluid in the x direction, a problem describing the temperature $u(x, y, t)$ is

$$\frac{\partial^2 u}{\partial y^2} = \frac{1}{k}\left(\frac{\partial u}{\partial t} + V \frac{\partial u}{\partial x}\right), \qquad 0 < y < b, \quad 0 < x, \quad 0 < t$$

$$u(x, 0, t) = 0, \ u(x, b, t) = 0, \qquad 0 < x, \quad 0 < t$$
$$u(0, y, t) = T_0, \qquad 0 < y < b, \quad 0 < t$$
$$u(x, y, 0) = T_1, \qquad 0 < x, \quad 0 < y < b.$$

Make a separation of variables as in Exercise 23. State and solve the eigenvalue problem for Y. Show that

$$u_n(x, y, t) = \phi_n(x - Vt) \exp(-\lambda_n{}^2 k(x + Vt)/2V) \sin \lambda_n y$$

satisfies partial differential equation and boundary conditions at $y = 0$ and $y = b$.

25. Satisfy the initial and inlet conditions in the problem of Exercise 24, forming a sum of product solutions and correctly choosing the ϕ_n.

26. In a uniform electric transmission line, extending along the x axis from $x = 0$ to $x = a$, the current and voltage to ground are represented by $i(x, t)$ and $e(x, t)$. The line is characterized by these parameters, all per unit length: resistance R, inductance L, leakage conductance G, and capacitance C. A section of the line lying between x and $x + \Delta x$ has the equivalent circuit shown in Fig. 3.7. Kirchhoff's laws applied to this

Figure 3.7 Equivalent circuit for a section of transmission line.

circuit yield the equations

$$i(x, t) = i(x + \Delta x, t) + G \,\Delta x e(x, t) + C \,\Delta x \frac{\partial e}{\partial t}(x, t)$$

$$e(x, t) = e(x + \Delta x, t) + R \,\Delta x i(x, t) + L \,\Delta x \frac{\partial i}{\partial t}(x, t).$$

By rearranging terms and taking a limit as $\Delta x \to 0$, derive the simultaneous partial differential equations

$$\frac{\partial i}{\partial x} + Ge + C\frac{\partial e}{\partial t} = 0 \tag{C}$$

$$\frac{\partial e}{\partial x} + Ri + L\frac{\partial i}{\partial t} = 0. \tag{D}$$

27. In telegraph lines, the leakage conductance and the inductance are negligible. Taking $G = 0$ and $L = 0$, combine equations (C) and (D) so as to get a single second-order partial differential equation for $i(x, t)$. Do the same for $e(x, t)$.

28. In other cases the resistance and leakage conductance are negligible. Taking $G = 0$ and $R = 0$, show that both e and i satisfy the wave equation ("high frequency equations") with

$$c = \sqrt{1/LC}.$$

29. Find solutions (in series form) of the equations (C) and (D) with $G = 0$ and $R = 0$ and with the auxiliary conditions

$$e(x, 0) = V, \qquad i(x, 0) = 0, \qquad 0 < x < a$$
$$e(0, t) = 0, \qquad e(a, t) = 0, \qquad 0 < t.$$

(The boundary conditions are referred to as short-circuit conditions.)

30. Obtain a solution by d'Alembert's method for the problem in Exercise 29.

31. Find series solutions of equations (C) and (D) for the case where $G = 0$ and $L = 0$, with the initial and boundary conditions

$$e(0, t) = V_0, \qquad e(a, t) = 0, \qquad 0 < t$$
$$e(x, 0) = V_0, \qquad i(x, 0) = 0, \qquad 0 < x < a.$$

32. Solve the equations (C) and (D) with $R = 0$ and $G = 0$, if initially both voltage and current are zero, and if at time $t = 0$ a constant voltage V is applied at $x = 0$, while the end at $x = a$ is short-circuited.

33. Solve equations (C) and (D) with $R = 0$ and $G = 0$, on the assumption that the line extends from $x = 0$ to infinity. Take the initial and boundary conditions

$$e(x, 0) = Ve^{-ax}, \qquad i(x, 0) = 0, \quad 0 < x$$
$$e(0, t) = V, \qquad\qquad 0 < t.$$

chapter 4
The
Potential
Equation

1   Potential equation

The equation for the steady-state temperature distribution in two dimensions (see Chapter 5) is

$$\frac{\partial^2 u}{\partial y^2} + \frac{\partial^2 u}{\partial x^2} = 0.$$

The same equation describes the equilibrium (time independent) displacements of a two-dimensional membrane, and so is an important common part of both the heat and wave equations in two dimensions. Many other physical phenomena—gravitational and electrostatic potentials, certain fluid flows—and an important class of functions are described by this equation, thus making it one of the most important of mathematics, physics, and engineering. The analogous equation in three dimensions is

$$\frac{\partial^2 u}{\partial x^2} + \frac{\partial^2 u}{\partial y^2} + \frac{\partial^2 u}{\partial z^2} = 0.$$

Either equation may be written $\nabla^2 u = 0$ and is commonly called Laplace's or the potential equation.

The solutions of Laplace's equation (called harmonic functions) have many interesting properties. One important one, which can be understood intuitively, is the maximum principle: if $\nabla^2 u = 0$ in a region, then u cannot have a relative maximum or minimum inside the region unless u is constant. (Thus, if $\partial u/\partial x$ and $\partial u/\partial y$ are both zero at some point, it is a saddle point.) If u is thought of as the steady-state temperature distribution in a metal plate, it is clear that the temperature cannot be greater at one point than at all other nearby points. For, if such were the case, heat would flow away from the hot point to cooler points nearby, thus reducing the temperature at the hot point. But then the temperature would not be unchanging with time.

A complete boundary value problem consists of the potential equation in a region plus boundary conditions. These may be of any of the three types

$$u \text{ given,} \qquad \frac{\partial u}{\partial n} \text{ given,} \qquad \text{or} \quad u + \gamma \frac{\partial u}{\partial n} \text{ given}$$

along any section of the boundary. (By $\partial u/\partial n$, we mean the directional derivative in the direction normal or perpendicular to the boundary.) When u is specified along the whole boundary, the problem is called Dirichlet's problem; if $\partial u/\partial n$ is specified along the whole boundary, it is Neumann's problem. The solutions of Neumann's problem are not unique, for if u is a solution, so is u plus a constant.

It is often useful to consider Laplace's equation in other coordinate systems. One of the most important is the polar coordinate system, in which the variables are

$$r = \sqrt{x^2 + y^2}, \qquad x = r \cos \theta$$

$$\theta = \tan^{-1}\left(\frac{y}{x}\right), \qquad y = r \sin \theta.$$

By convention we require $r \geq 0$. We shall define

$$u(x, y) = u(r \cos \theta, r \sin \theta) = v(r, \theta)$$

and find an expression for the "Laplacian operator"

$$\frac{\partial^2 u}{\partial x^2} + \frac{\partial^2 u}{\partial y^2}$$

in terms of v and its derivatives by using the chain rule. The calculations are elementary, but tedious. The results are

$$\frac{\partial^2 u}{\partial x^2} = \cos^2 \theta \, \frac{\partial^2 v}{\partial r^2} - \frac{2 \sin \theta \cos \theta}{r} \frac{\partial^2 v}{\partial \theta \, \partial r} + \frac{\sin^2 \theta}{r^2} \frac{\partial^2 v}{\partial \theta^2}$$

$$+ \frac{\sin^2 \theta}{r} \frac{\partial v}{\partial r} + \frac{2 \sin \theta \cos \theta}{r^2} \frac{\partial v}{\partial \theta}$$

$$\frac{\partial^2 u}{\partial y^2} = \sin^2 \theta \, \frac{\partial^2 v}{\partial r^2} + \frac{2 \sin \theta \cos \theta}{r} \frac{\partial^2 v}{\partial \theta \, \partial r} + \frac{\cos^2 \theta}{r^2} \frac{\partial^2 v}{\partial \theta^2}$$

$$+ \frac{\cos^2 \theta}{r} \frac{\partial v}{\partial r} - \frac{2 \sin \theta \cos \theta}{r^2} \frac{\partial v}{\partial \theta}.$$

From these equations we easily find that the potential equation in polar coordinates is

$$\frac{\partial^2 v}{\partial r^2} + \frac{1}{r} \frac{\partial v}{\partial r} + \frac{1}{r^2} \frac{\partial^2 v}{\partial \theta^2} = \frac{1}{r} \frac{\partial}{\partial r} \left(r \frac{\partial v}{\partial r} \right) + \frac{1}{r^2} \frac{\partial^2 v}{\partial \theta^2} = 0.$$

In cylindrical (r, θ, z) coordinates, the potential equation is

$$\frac{1}{r} \frac{\partial}{\partial r} \left(r \frac{\partial v}{\partial r} \right) + \frac{1}{r^2} \frac{\partial^2 v}{\partial \theta^2} + \frac{\partial^2 v}{\partial z^2} = 0.$$

Exercises

1. Find a relation among the coefficients of the polynomial

$$p(x, y) = a + bx + cy + dx^2 + exy + fy^2$$

which makes it satisfy the potential equation. Choose a specific polynomial and show that, if $\partial p/\partial x$ and $\partial p/\partial y$ are both zero at some point, the surface there is saddle shaped.

2. Show that $u(x, y) = x^2 - y^2$ and $u(x, y) = xy$ are solutions of Laplace's equation. Sketch the surfaces $z = u(x, y)$. What boundary conditions do these functions fulfill on the lines $x = 0$, $x = a$, $y = 0$, $y = b$?

3. If a solution of Laplace's equation in the square $0 < x < 1$, $0 < y < 1$ has the form $u(x, y) = Y(y) \sin \pi x$, of what form is the function Y? Find a function Y which makes $u(x, y)$ satisfy the boundary conditions $u(x, 0) = 0$, $u(x, 1) = \sin \pi x$.

4. Find a function $u(x)$ independent of y, which satisfies the potential equation.

5. What functions $v(r)$, independent of θ and z, satisfy the potential equation in polar coordinates?

6. Show that $r^n \sin n\theta$ and $r^n \cos n\theta$ both satisfy the potential equation in polar coordinates ($n = 0, 1, 2, \ldots$).

7. Find expressions for the partial derivatives of u with respect to x and y in terms of derivatives of v with respect to r and θ.

8. If u and v are the x- and y-components of the velocity in a fluid, it can be shown (under certain assumptions) that these functions satisfy the equations

$$\frac{\partial u}{\partial x} + \frac{\partial v}{\partial y} = 0 \tag{A}$$

$$\frac{\partial u}{\partial y} - \frac{\partial v}{\partial x} = 0. \tag{B}$$

Show that the definition of a velocity potential function by the equations

$$u = \frac{\partial \phi}{\partial x}, \qquad v = \frac{\partial \phi}{\partial y}$$

causes (B) to be identically satisfied and turns (A) into the potential equation.

2 Potential in a rectangle

One of the simplest and most important problems in mathematical physics is Dirichlet's problem in a rectangle. To take an easy case, we consider a problem with just two nonzero boundary conditions:

$$\frac{\partial^2 u}{\partial x^2} + \frac{\partial^2 u}{\partial y^2} = 0, \qquad 0 < x < a, \quad 0 < y < b \tag{1}$$

$$u(x, 0) = f_1(x), \qquad 0 < x < a \tag{2}$$

$$u(x, b) = f_2(x), \qquad 0 < x < a \tag{3}$$

$$u(0, y) = 0, \qquad 0 < y < b \tag{4}$$

$$u(a, y) = 0, \qquad 0 < y < b. \tag{5}$$

If we assume that $u(x, y)$ has a product form $u = X(x) Y(y)$, then Eq. (1) becomes

$$X'' Y + X Y'' = 0.$$

This equation can be separated by dividing through by XY yielding:

$$\frac{X''}{X} = -\frac{Y''}{Y}.\tag{6}$$

The nonhomogeneous conditions Eqs. (2) and (3) will not, in general, become conditions on X or Y, but the homogeneous conditions Eqs. (4) and (5), as usual, require that

$$X(0) = 0, \qquad X(a) = 0.\tag{7}$$

Now, both sides of Eq. (6) must be constant, but the sign of the constant is not obvious. If we try a positive constant (say μ^2), Eq. (6) represents two ordinary equations

$$X'' - \mu^2 X = 0, \qquad Y'' + \mu^2 Y = 0.$$

The solutions of these equations are

$$X = A \cosh \mu x + B \sinh \mu x, \qquad Y = C \cos \mu y + D \sin \mu y.$$

In order to make X satisfy the boundary conditions Eq. (7), both A and B must be zero, leading to a solution $u(x, y) = 0$. Thus we try the other possibility for sign, taking both members in Eq. (6) to equal $-\lambda^2$.

Under the new assumption, Eq. (6) separates into

$$X'' + \lambda^2 X = 0, \qquad Y'' - \lambda^2 Y = 0.\tag{8}$$

The first of these equations, along with the boundary conditions, is recognizable as an eigenvalue problem, whose solutions are

$$X_n(x) = \sin \lambda_n x, \qquad \lambda_n^2 = \left(\frac{n\pi}{a}\right)^2.$$

The functions Y which accompany the Xs are

$$Y_n = a_n \cosh \lambda_n y + b_n \sinh \lambda_n y.$$

The as and bs are for the moment unknown.

We see that $X_n Y_n$ is a solution of the potential Eq. (1) which satisfies the homogeneous conditions Eqs. (4) and (5). A sum of these functions should satisfy the same conditions and equation, so u may have the form

$$u(x, y) = \sum_{n=1}^{\infty} (a_n \cosh \lambda_n y + b_n \sinh \lambda_n y) \sin \lambda_n x.\tag{9}$$

The nonhomogeneous boundary conditions Eqs. (2) and (3) are yet to be satisfied. If u is to be of the form above, the boundary condition Eq. (2) becomes

$$u(x, 0) = \sum_{n=1}^{\infty} a_n \sin\left(\frac{n\pi x}{a}\right) = f_1(x), \qquad 0 < x < a.\tag{10}$$

We recognize a problem in Fourier series immediately. The a_n must be the Fourier coefficients of $f_1(x)$

$$a_n = \frac{2}{a} \int_0^a f_1(x) \sin\left(\frac{n\pi x}{a}\right) dx.$$

The second boundary condition reads

$$u(x, b) = \sum_1^\infty (a_n \cosh \lambda_n b + b_n \sinh \lambda_n b) \sin\left(\frac{n\pi x}{a}\right)$$

$$= f_2(x), \qquad 0 < x < a.$$

This also is a problem in Fourier series, but not as neat. The constant

$$a_n \cosh \lambda_n b + b_n \sinh \lambda_n b$$

must be the nth Fourier sine coefficient of f_2, and a_n is known; thus, b_n can be determined from the following computations:

$$a_n \cosh \lambda_n b + b_n \sinh \lambda_n b = \frac{2}{a} \int_0^a f_2(x) \sin \lambda_n x \, dx = c_n$$

$$b_n = \frac{c_n - a_n \cosh \lambda_n b}{\sinh \lambda_n b}.$$

If we use this last expression for b_n and substitute into Eq. (9), we find the solution

$$u(x, y) =$$

$$\sum_{n=1}^\infty \left\{ c_n \frac{\sinh(n\pi y/a)}{\sinh(n\pi b/a)} + a_n \left[\cosh\left(\frac{n\pi y}{a}\right) - \frac{\cosh(n\pi b/a)}{\sinh(n\pi b/a)} \sinh\left(\frac{n\pi y}{a}\right) \right] \right\} \sin\left(\frac{n\pi x}{a}\right).$$

$$(11)$$

Notice that the function multiplying c_n is 0 at $y = 0$ and is 1 at $y = b$. Similarly, the function multiplying a_n is 1 at $y = 0$ and 0 at $y = b$. An easier way to write this latter function is

$$\frac{\sinh \lambda_n(b - y)}{\sinh \lambda_n b}$$

as can readily be found from hyperbolic identities.

Let us take a specific example in order to see more clearly how the solution looks. Suppose f_1 and f_2 are the same,

$$f_1(x) = f_2(x) = \begin{cases} \dfrac{2x}{a}, & 0 < x < \dfrac{a}{2} \\[2mm] 2\left(\dfrac{a-x}{a}\right), & \dfrac{a}{2} < x < a. \end{cases}$$

Then

$$c_n = a_n = \frac{8}{\pi^2} \frac{\sin(n\pi/2)}{n^2}.$$

The solution of the potential equation for these boundary conditions is

$$u(x, y) = \frac{8}{\pi^2} \sum_{n=1}^{\infty} \frac{\sin(n\pi/2)}{n^2} \left\{ \frac{\sinh[(n\pi/a)y] + \sinh[(n\pi/a)(b - y)]}{\sinh[(n\pi/a)b]} \right\} \cdot \sin\left(\frac{n\pi x}{a}\right)$$

In Fig. 4.1 is a graph of some level curves, $u(x, y) = $ constant, for the case where $a = b$.

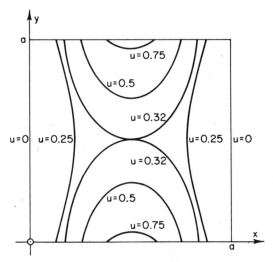

Figure 4.1

Now we have seen a solution of Dirichlet's problem in a rectangle with homogeneous conditions on two parallel sides. In general, of course, the boundary conditions will be nonhomogeneous on all four sides of the rectangle. But this more general problem can be broken down into two problems like the one we have solved.

Consider the problem

$$\nabla^2 u = 0, \qquad 0 < x < a, \quad 0 < y < b \qquad (12)$$

$$u(x, 0) = f_1(x), \qquad 0 < x < a \qquad (13)$$

$$u(x, b) = f_2(x), \qquad 0 < x < a \qquad (14)$$

$$u(0, y) = g_1(y), \qquad 0 < y < b \qquad (15)$$

$$u(a, y) = g_2(y), \qquad 0 < y < b. \qquad (16)$$

Let $u(x, y) = u_1(x, y) + u_2(x, y)$. We will put conditions on u_1 and u_2 so that they can readily be found, and from them u can be put together. The most obvious conditions are the following:

$$\nabla^2 u_1 = 0, \qquad\qquad \nabla^2 u_2 = 0$$
$$u_1(x, 0) = f_1(x), \qquad u_2(x, 0) = 0$$
$$u_1(x, b) = f_2(x), \qquad u_2(x, b) = 0$$
$$u_1(0, y) = 0, \qquad\qquad u_2(0, y) = g_1(y)$$
$$u_1(a, y) = 0, \qquad\qquad u_2(a, y) = g_2(y).$$

It is evident that $u_1 + u_2$ is the solution of the original problem Eqs. (12)–(16). Also, each of the functions u_1 and u_2 has homogeneous conditions on parallel boundaries. We already have determined the form of u_1. The other function would be of the form

$$u_2(x, y) = \sum_{n=1}^{\infty} \sin \mu_n y \left[\frac{A_n \sinh \mu_n x + B_n \sinh \mu_n(a - x)}{\sinh \mu_n a} \right] \qquad (17)$$

where $\mu_n = n\pi/b$, and

$$A_n = \frac{2}{b} \int_0^b g_2(y) \sin \mu_n y \, dy$$

$$B_n = \frac{2}{b} \int_0^b g_1(y) \sin \mu_n y \, dy.$$

In the individual problems for u_1 and u_2, the technique of separation of variables works, because the homogeneous conditions on parallel sides of the rectangle can be translated into conditions on one of the factor functions.

Exercises

1. Show that $\sinh \lambda y$ and $\sinh \lambda(b - y)$ are independent solutions of $Y'' - \lambda^2 Y = 0$. Thus a combination of these two functions may replace a combination of sinh and cosh as the general solution of this differential equation.

2. Show that the solution of the example problem may be written

$$u(x, y) = \frac{8}{\pi^2} \sum_{n=1}^{\infty} \frac{\sin(n\pi/2)}{n^2} \frac{\cosh[(n\pi/a)(y - \tfrac{1}{2}b)]}{\cosh(n\pi b/2a)} \cdot \sin\left(\frac{n\pi x}{a}\right).$$

3. Use the form above to compute u in the center of the rectangle in the three cases: $b = a$, $b = 2a$, $b = a/2$. (Hint: Check the magnitude of the terms.)

4. Verify that each term of Eq. (9) is a solution of Eqs. (1), (4), and (5).

5. Solve the problem:

$$\nabla^2 u = 0, \qquad\qquad\qquad 0 < x < a, \qquad 0 < y < b$$

$$u(0, y) = 0, \qquad u(a, y) = 0, \qquad 0 < y < b$$

$$u(x, 0) = 0, \qquad u(x, b) = f(x), \qquad 0 < x < a$$

where f is the same as in the example. Sketch some level curves of $u(x, y)$.

6. Verify that Eq. (17) is a solution of its problem (that is, it satisfies the partial differential equation and the boundary conditions).

7. Solve the problem of Laplace's equation in the rectangle $0 < x < a$, $0 < y < b$, for each of the following sets of boundary conditions:

a. $\dfrac{\partial u}{\partial x}(0, y) = 0$; $u = 1$ on the remainder of the boundary;

b. $\dfrac{\partial u}{\partial x} = 0$ at $x = 0$ and $x = a$; $u(x, 0) = 0$, $u(x, b) = 1$;

c. $\dfrac{\partial u}{\partial y}(x, 0) = 0$, $u(x, b) = 0$; $u(0, y) = 1$, $u(a, y) = 0$.

8. Solve the problem for u_2. (That is, derive Eq. (17).)

3 Potential in a slot

The potential equation, as well as the heat and wave equations, can be solved in unbounded regions. Consider the following problem in which the region involved is half a vertical strip, or a slot:

$$\frac{\partial^2 u}{\partial x^2} + \frac{\partial^2 u}{\partial y^2} = 0, \qquad 0 < x < a, \quad 0 < y$$

$$u(x, 0) = f(x), \qquad 0 < x < a$$

$$u(0, y) = g_1(y), \qquad 0 < y$$

$$u(a, y) = g_2(y), \qquad 0 < y.$$

As usual, we require that $u(x, y)$ remain bounded as $y \to \infty$.

In order to make the separation of variables work, we must break this up into two problems. Following the model of the preceding section, we set

$u(x, y) = u_1(x, y) + u_2(x, y)$ and require that the parts satisfy the two solvable problems:

$$\nabla^2 u_1 = 0, \qquad \nabla^2 u_2 = 0, \qquad 0 < x < a, \quad 0 < y$$
$$u_1(x, 0) = f(x), \qquad u_2(x, 0) = 0, \qquad 0 < x < a$$
$$u_1(0, y) = 0, \qquad u_2(0, y) = g_1(y), \qquad 0 < y$$
$$u_1(a, y) = 0, \qquad u_2(a, y) = g_2(y), \qquad 0 < y.$$

We attack the problem for u_1 by assuming the product form and separating variables:

$$u_1(x, y) = X(x)Y(y), \qquad \frac{X''}{X} = -\frac{Y''}{Y} = -\lambda^2.$$

The sign of the constant $-\lambda^2$ is determined by the boundary conditions at $x = 0$ and $x = a$, which become homogeneous conditions on the factor $X(x)$:

$$X(0) = 0, \qquad X(a) = 0. \tag{1}$$

(We also can see that the condition to be satisfied along $y = 0$ demands functions of x which permit a representation of an arbitrary function.)

The boundary conditions, Eq. (1), together with the differential equation

$$X'' + \lambda^2 X = 0 \tag{2}$$

that comes from the separation of variables, constitute a familiar eigenvalue problem whose solution is

$$X_n(x) = \sin\left(\frac{n\pi x}{a}\right), \qquad \lambda_n^2 = \left(\frac{n\pi}{a}\right)^2.$$

The equation for Y is

$$Y'' - \lambda^2 Y = 0, \qquad 0 < y.$$

In addition to satisfying this differential equation, Y must remain bounded as $y \to \infty$. The solutions of the equation are $e^{\lambda y}$ and $e^{-\lambda y}$. Of these, the first is unbounded, so

$$Y_n(y) = \exp(-\lambda_n y).$$

Finally, we can write the solution of the first problem as

$$u_1(x, y) = \sum_{n=1}^{\infty} a_n \sin\left(\frac{n\pi x}{a}\right) \exp\left(\frac{-n\pi y}{a}\right). \tag{3}$$

The constants a_n are to be determined from the condition at $y = 0$.

The solution of the second problem is somewhat different. Again for solutions in the product form $u_2(x, y) = X(x)Y(y)$, Laplace's equation becomes

$$\frac{X''}{X} = -\frac{Y''}{Y} = \mu^2. \tag{4}$$

Here we have chosen the sign of the separation constant to be opposite that in the first case. The requirements on Y are that $Y(0) = 0$ and $Y(y)$ remain bounded as $y \to \infty$. We easily find that the solution of $Y'' + \mu^2 Y = 0$ which satisfies these conditions is

$$Y(y) = \sin \mu y$$

valid for all $\mu > 0$. The solution of the differential equation in X is:

$$X(x) = A \frac{\sinh \mu x}{\sinh \mu a} + B \frac{\sinh \mu(a - x)}{\sinh \mu a}.$$

We have chosen this special form on the basis of our experience in solving Laplace's equation in the rectangle.

Since μ is a continuous parameter, we combine our product solutions by means of an integral, finding

$$u_2(x, y) = \int_0^\infty \left[A(\mu) \frac{\sinh \mu x}{\sinh \mu a} + B(\mu) \frac{\sinh \mu(a - x)}{\sinh \mu a} \right] \sin \mu y \, d\mu. \tag{5}$$

The nonhomogeneous boundary conditions at $x = 0$ and $x = a$ are satisfied if

$$u_2(0, y) = \int_0^\infty B(\mu) \sin \mu y \, d\mu = g_1(y), \qquad 0 < y$$

$$u_2(a, y) = \int_0^\infty A(\mu) \sin \mu y \, d\mu = g_2(y), \qquad 0 < y.$$

Obviously these two equations are Fourier integral problems, so we know how to determine the coefficients $A(\mu)$ and $B(\mu)$.

Laplace's equation can also be solved in a strip $(0 < x < a, -\infty < y < \infty)$, a quarter plane $(0 < x, 0 < y)$, or a half plane $(0 < x, -\infty < y < \infty)$. Along each boundary line, a boundary condition is imposed, and the solution is required to remain bounded in remote portions of the region considered. In general, a Fourier integral is employed in the solution, because the separation constant is a continuous parameter, as in the second problem here.

Exercises

1. Find a formula for the constants a_n in Eq. (3).

2. Verify that $u_1(x, y)$ in the form given in Eq. (3) satisfies the potential equation and the homogeneous boundary conditions.

3. Find formulas for $A(\mu)$ and $B(\mu)$ of Eq. (5).

4. Verify that $u_2(x, y)$ as given in Eq. (5) satisfies the potential equation and the boundary condition at $y = 0$.

5. Solve the potential equation in the strip $0 < x < a$, $-\infty < y < \infty$, subject to the boundary conditions

$$u(0, y) = g_1(y), \qquad u(a, y) = g_2(y).$$

(Note that u must remain bounded as $y \to \pm \infty$.)

6. State the complete boundary value problem for Laplace's equation in a quarter plane. Indicate how to solve the problem. (Hint: break it up into two simpler problems.)

7. Solve Laplace's equation in the quarter plane $x > 0$, $y > 0$, with the boundary conditions $u(0, y) = e^{-y}$, $u(x, 0) = e^{-x}$.

8. Show that, if the separation constant had been chosen as $-\mu^2$ instead of μ^2 in solving for u_2 (leading to $Y'' - \mu^2 Y = 0$), then the only function that satisfies the differential equation, the condition $Y(0) = 0$, and remains bounded as $y \to \infty$, is $Y(y) \equiv 0$.

9. Find the solution of the problem stated in this section, taking $f(x) = 0$, $g_1(y) = 0$, $g_2(y) = e^{-y}$.

10. Show that $u(x, y) = x$ is the solution of Laplace's equation in a slot under the boundary conditions: $f(x) = x$, $g_1(y) = 0$, $g_2(y) = a$. Can this solution be found by the method of this section?

11. Solve the problem of potential in a slot under the boundary conditions

$$u(x, 0) = 1, \qquad u(0, y) = u(a, y) = e^{-y}.$$

12. Show that the function $v(x, y)$ given below satisfies the potential equation and the boundary conditions on the "long" sides in Exercise 11, provided that $\cos(a/2) \neq 0$.

$$v(x, y) = \frac{\cos(x - \tfrac{1}{2}a)}{\cos(\tfrac{1}{2}a)} e^{-y}$$

What partial differential equation and boundary conditions are satisfied by $w(x, y) = u(x, y) - v(x, y)$, if u is the function of Exercise 11?

13. Solve Laplace's equation in the slot $0 < y < b$, $0 < x$ for each of the following sets of boundary conditions:

a. $u(0, y) = 0$, $u(x, 0) = 0$, $u(x, b) = \begin{cases} 1, & 0 < x < a \\ 0, & a < x \end{cases}$

b. $u(0, y) = 0$, $u(x, 0) = e^{-x}$, $u(x, b) = 0$.

4 Potential in a disk

If we need to solve the potential equation in a circular disk $x^2 + y^2 < c^2$, it is natural to use polar coordinates r, θ, in terms of which the disk is described by $0 < r < c$. The Dirichlet problem on the disk is

$$\frac{1}{r} \frac{\partial}{\partial r}\left(r \frac{\partial v}{\partial r}\right) + \frac{1}{r^2} \frac{\partial^2 v}{\partial \theta^2} = 0, \qquad 0 < r < c, \quad -\pi < \theta \le \pi \qquad (1)$$

$$v(c, \theta) = f(\theta), \qquad -\pi < \theta \le \pi. \qquad (2)$$

Since θ and $\theta + 2\pi$ refer to the same angle, really, we restrict θ to an interval from $-\pi$ to π. However, the ray $\theta = \pi$ is not a "boundary" in the sense we have been using, since the outer world does not affect v at that point. In order to insure that the introduction of the false boundary does not cause a discontinuity in v and its derivatives, we require that

$$v(r, \pi) = v(r, -\pi), \qquad 0 < r < c \qquad (3)$$

$$\frac{\partial v}{\partial \theta}(r, \pi) = \frac{\partial v}{\partial \theta}(r, -\pi), \qquad 0 < r < c. \qquad (4)$$

Alternately, we may allow θ to assume any value but require $v(r, \theta)$ and $f(\theta)$ to be periodic in θ with period 2π.

By assuming $v(r, \theta) = R(r)\Theta(\theta)$ we can separate variables. The potential equation becomes

$$\frac{1}{r}(rR')'\Theta + \frac{1}{r^2} R\Theta'' = 0.$$

Separation is effected by dividing through by $R\Theta/r^2$:

$$\frac{r(rR')'}{R} + \frac{\Theta''}{\Theta} = 0.$$

Evidently the boundary condition Eq. (2) will have to be satisfied with a linear combination of solutions; hence we choose $\Theta''/\Theta = -\lambda^2$. Conditions (3) and (4) become conditions on Θ. The individual problems for the functions R and Θ are

$$\Theta'' + \lambda^2\Theta = 0, \qquad -\pi < \theta \le \pi \tag{5}$$

$$\Theta(-\pi) = \Theta(\pi) \tag{6}$$

$$\Theta'(-\pi) = \Theta'(\pi) \tag{7}$$

$$r(rR')' - \lambda^2 R = 0, \qquad 0 < r < c. \tag{8}$$

The solutions of Eq. (5) all have the form

$$\Theta(\theta) = A \cos \lambda\theta + B \sin \lambda\theta. \tag{9}$$

If we write Eqs. (6) and (7) in terms of the function Θ of Eq. (9) and use the properties of sine and cosine, we have

$$A \cos \lambda\pi - B \sin \lambda\pi = A \cos \lambda\pi + B \sin \lambda\pi$$

$$\lambda A \sin \lambda\pi + \lambda B \cos \lambda\pi = -\lambda A \sin \lambda\pi + \lambda B \cos \lambda\pi.$$

After simple manipulations, these equations reduce to

$$B \sin \lambda\pi = 0, \qquad \lambda A \sin \lambda\pi = 0.$$

Not both A and B should be zero, for then Θ would be identically zero. Thus either $\lambda = 0$ and $B = 0$, or $\sin \lambda\pi = 0$. The possibilities for nonzero solutions are thus

$$\lambda = 0, \qquad\qquad \Theta = 1$$

$$\lambda = 1, 2, 3, \ldots, \qquad \Theta = \cos \lambda\theta \qquad \text{or} \qquad \sin \lambda\theta.$$

Now we have found that the eigenvalue $\lambda^2 = 0$ corresponds to the eigenfunction $\Theta = 1$ (constant), and the eigenvalues $\lambda_n^2 = n^2$ ($n = 1, 2, 3, \ldots$) each correspond to *two* independent eigenfunctions: $\cos n\theta$ and $\sin n\theta$.

Knowing that $\lambda_n^2 = n^2$, we can easily find $R(r)$. The equation for R becomes

$$r^2 R'' + rR' - n^2 R = 0, \qquad 0 < r < c$$

when the indicated differentiations are carried out. This is a Cauchy–Euler equation, whose solutions are known to have the form $R(r) = r^\alpha$, where α is constant. Substituting $R = r^\alpha$, $R' = \alpha r^{\alpha-1}$, and $R'' = \alpha(\alpha - 1)r^{\alpha-2}$ into it leaves

$$(\alpha(\alpha - 1) + \alpha - n^2)r^\alpha = 0, \qquad 0 < r < c.$$

As r^α is not zero, the constant factor in parentheses must be zero—that is, $\alpha = \pm n$. The general solution of the differential equation is any combination of r^n and r^{-n}. The latter, however, is unbounded as r approaches zero, so we discard that solution, retaining $R_n(r) = r^n$. In the special case $n = 0$, the two solutions are the constant function 1 and $\ln r$. The logarithm is discarded because of its behavior at $r = 0$.

Now we reassemble our solution. The functions

$$r^0 \cdot 1 = 1, \qquad r^n \cos n\theta, \qquad r^n \sin n\theta$$

are all solutions of Laplace's equation, so a general linear combination of these solutions will also be a solution. Thus $v(r, \theta)$ may have the form

$$v(r, \theta) = a_0 + \sum_{n=1}^{\infty} a_n r^n \cos n\theta + \sum_{n=1}^{\infty} b_n r^n \sin n\theta. \qquad (10)$$

At the true boundary $r = c$, the boundary condition reads

$$v(c, \theta) = a_0 + \sum c^n(a_n \cos n\theta + b_n \sin n\theta) = f(\theta), \qquad -\pi < \theta \le \pi.$$

This is a Fourier series problem, solved by choosing

$$a_0 = \frac{1}{2\pi} \int_{-\pi}^{\pi} f(\theta)\, d\theta,$$

$$a_n = \frac{1}{\pi c^n} \int_{-\pi}^{\pi} f(\theta) \cos n\theta\, d\theta, \qquad b_n = \frac{1}{\pi c^n} \int_{-\pi}^{\pi} f(\theta) \sin n\theta\, d\theta. \qquad (11)$$

Now that we have the form Eq. (10) of the solution of Laplace's equation, we can see some important properties of the function $v(r, \theta)$. In particular, by setting $r = 0$ we obtain

$$v(0, \theta) = a_0 = \frac{1}{2\pi} \int_{-\pi}^{\pi} f(\theta)\, d\theta = \frac{1}{2\pi} \int_{-\pi}^{\pi} v(c, \theta)\, d\theta.$$

This says that the solution of Laplace's equation at the center of a disk is equal to the average of its values around the edge of the disk. It is easy to show also that

$$v(0, \theta) = \frac{1}{2\pi} \int_{-\pi}^{\pi} v(r, \theta)\, d\theta \qquad (12)$$

for any r between 0 and c! This characteristic of solutions of the potential equation is called the *mean value property*. From the mean value property, it is just a step to prove the maximum principle mentioned in the first section. For the mean value of a function lies between the minimum and the maximum, and cannot equal either unless the function is constant.

$$v(r, \theta) = \text{Re} \sum_{n=0}^{\infty} C_n (r e^{i\theta})^n = \text{Re} \sum_0^{\infty} C_n z^n$$

$$\underset{z}{\big\downarrow}$$

converges if $\qquad |z| < \dfrac{1}{\lim\limits_{n \to \infty} \left| \dfrac{C_{n+1}}{C_n} \right|} \quad = \text{radius of convergence}$

Exercises

1. Solve Laplace's equation in the disk $0 < r < c$ if the boundary condition is $v(c, \theta) = |\theta|$, $-\pi < \theta \leq \pi$.

2. Same as Exercise 1 if $v(c, \theta) = \theta$, $-\pi < \theta < \pi$. Is the boundary condition satisfied at $\theta = \pm \pi$?

3. Compute the value of the solution of Laplace's equation at $r = 0$ for the cases given in Exercises 1 and 2.

4. Verify Eq. (12) by integrating the series Eq. (10) term-by-term.

5. If the function $f(\theta)$ in Eq. (2) is continuous, sectionally smooth, and satisfies $f(-\pi+) = f(\pi-)$, what can be said about convergence of the series for $v(c, \theta)$?

6. Show that

$$v(r, \theta) = a_0 + \sum r^{-n}(a_n \cos n\theta + b_n \sin n\theta)$$

is a solution of Laplace's equation in the region $r > c$ (exterior of a disk) and has the property that $|v(r, \theta)|$ is bounded as $r \to \infty$.

7. If the condition $v(c, \theta) = f(\theta)$ is given, what are the formulas for the as and bs in Exercise 6?

8. The solution of Eqs. (1)–(4) can be written in a single formula by the following sequence of operations:
   a. Replace θ by ϕ in Eq. (11) for the as and bs;
   b. replace the as and bs in Eq. (10) by the integrals in part (a);
   c. use the trigonometric identity

   $$\cos n\theta \cos n\phi + \sin n\theta \sin n\phi = \cos n(\theta - \phi)$$

   d. take the integral outside the series;
   e. add up the series (see Chapter 1, Section 9, Exercise 1). Then $v(r, \theta)$ is given by the single integral (Poisson integral formula)

   $$v(r, \theta) = \frac{1}{2\pi} \int_{-\pi}^{\pi} f(\phi) \frac{c^2 - r^2}{c^2 + r^2 - 2rc \cos(\theta - \phi)} \, d\phi.$$

9. Solve Laplace's equation in the sector $0 < \theta < \pi/2$, $0 < r < c$, subject to the boundary conditions $v(r, 0) = 0$, $v(r, \pi/2) = 0$, $v(c, \theta) = 1$.

$$r^n C_n = \frac{1}{2\pi} \int_{-\pi}^{\pi} f(\phi) e^{in\phi} \, d\phi$$

5   Classification of partial differential equations and limitations of the product method

By this time, we have seen a variety of equations and solutions. We have concentrated on three different, homogeneous equations (heat, wave, and potential) and have found the qualitative features summarized in the following tabulation:

Equation	Features
Heat	Exponential behavior in time. Existence of a limiting (steady-state) solution. Smooth graph for $t > 0$.
Wave	Oscillatory (not always periodic) behavior in time. Retention of discontinuities for $t > 0$.
Potential	Smooth surface. Maximum principle. Mean value property.

These three two-variable equations are the most important representatives of the three classes of second-order linear partial differential equations in two variables. The most general equation which fits this description is

$$A \frac{\partial^2 u}{\partial \xi^2} + B \frac{\partial^2 u}{\partial \xi \, \partial \eta} + C \frac{\partial^2 u}{\partial \eta^2} + D \frac{\partial u}{\partial \xi} + E \frac{\partial u}{\partial \eta} + Fu + G = 0$$

where A, B, C, and so forth, are, in general, functions of ξ and η. We use Greek letters for the independent variables to avoid implying any relations to space or time. Such an equation can be classified according to the sign of $B^2 - 4AC$:

$$B^2 - 4AC < 0 \qquad \text{elliptic}$$
$$B^2 - 4AC = 0 \qquad \text{parabolic}$$
$$B^2 - 4AC > 0 \qquad \text{hyperbolic.}$$

Since A, B, and C are functions of ξ and η (not of u), the classification of an equation may vary from point to point. It is easy to see that the heat equation is parabolic, the wave equation is hyperbolic, and the potential equation is elliptic. The classification of an equation determines important features of the solution and also dictates the method of attack when numerical techniques are used for solution.

The question naturally arises whether separation of variables works on all equations. The answer is no. For instance, the equation

$$(\xi + \eta^2)\frac{\partial^2 u}{\partial \xi^2} + \frac{\partial^2 u}{\partial \eta^2} = 0$$

does not admit separation of variables. It is difficult to say, however, just which equations can be solved by this method. However, it is necessary to have $B \equiv 0$.

The region in which the solution is to be found also limits the applicability of the method we have used. The region must be a "generalized rectangle." By this, we mean a region that can be described by inequalities whose end-points are fixed quantities. For instance, we have worked in regions described by the following sets of inequalities:

$$
\begin{aligned}
0 &< x < a, & 0 &< t \\
0 &< x, & 0 &< t \\
-\infty &< x < \infty, & 0 &< t \\
0 &< x < a, & 0 &< y < b \\
0 &< r < c, & -\pi &< \theta \le \pi.
\end{aligned}
$$

All of these are generalized rectangles, but only one is an ordinary rectangle. An L-shaped region is not a generalized rectangle, and our methods would break down if applied to, for instance, Laplace's equation there.

There are, as we know, restrictions on the kinds of boundary conditions which can be handled. From the examples in this chapter it is clear that we need homogeneous or "homogeneous-like" conditions on opposite sides of a generalized rectangle. Examples of "homogeneous-like" conditions are the requirement that a function remain bounded as some variable tends to infinity, or the "periodic conditions" at $\theta = \pm\pi$ (see Section 4). The point is that if two or more functions satisfy the conditions, so does a sum of those functions.

In spite of the limitations of the method of separation of variables, it works well on many important problems in two or more variables, and pro-vides insight into the nature of their solutions. Moreover, it is known that in those cases where separation of variables can be carried out, it will find a solution if one exists.

Exercises

1. Classify the following equations:

a. $\dfrac{\partial^2 u}{\partial x \, \partial y} = 0$

b. $\dfrac{\partial^2 u}{\partial x^2} + \dfrac{\partial^2 u}{\partial x\, \partial y} + \dfrac{\partial^2 u}{\partial y^2} = 2x$

c. $\dfrac{\partial^2 u}{\partial x^2} - \dfrac{\partial^2 u}{\partial x\, \partial y} + \dfrac{\partial^2 u}{\partial y^2} = 2u$

d. $\dfrac{\partial^2 u}{\partial x^2} - 2\dfrac{\partial^2 u}{\partial x\, \partial y} + \dfrac{\partial^2 u}{\partial y^2} = \dfrac{\partial u}{\partial y}$

e. $\dfrac{\partial^2 u}{\partial x^2} - \dfrac{\partial^2 u}{\partial y^2} - \dfrac{\partial u}{\partial y} = 0.$

2. Show that, in polar coordinates, an annulus, a sector, and a sector of an annulus are all generalized rectangles.

3. In which of the equations in Exercise 1 can the variables be separated?

4. Sketch the regions listed in the text as generalized rectangles.

5. Solve the three problems below and compare the solutions.

a. $\dfrac{\partial^2 u}{\partial x^2} + \dfrac{\partial^2 u}{\partial y^2} = 0,$ $\qquad\qquad 0 < x < 1, \quad 0 < y$

$u(x, 0) = f(x),$ $\qquad\qquad 0 < x < 1$

$u(0, y) = 0,$ $\qquad u(1, y) = 0, \quad 0 < y$

b. $\dfrac{\partial^2 u}{\partial x^2} = \dfrac{\partial^2 u}{\partial y^2},$ $\qquad\qquad 0 < x < 1, \quad 0 < y$

$u(x, 0) = f(x), \quad \dfrac{\partial u}{\partial y}(x, 0) = 0, \quad 0 < x < 1$

$u(0, y) = 0,$ $\qquad u(1, y) = 0, \quad 0 < y$

c. $\dfrac{\partial^2 u}{\partial x^2} = \dfrac{\partial u}{\partial y},$ $\qquad\qquad 0 < x < 1, \quad 0 < y$

$u(x, 0) = f(x),$ $\qquad\qquad 0 < x < 1$

$u(0, y) = 0,$ $\qquad u(1, y) = 0, \quad 0 < y$

6. Show that if $f_1, f_2, \ldots$ all satisfy the periodic boundary conditions
$$f(-\pi) = f(\pi), \qquad f'(-\pi) = f'(\pi)$$
then so does the function $c_1 f_1 + c_2 f_2 + \cdots$, where the cs are constants.

6 Comments and references

While the potential equation describes many physical phenomena, there is one which makes the solution of the Dirichlet problem very easy to visualize. Suppose a piece of wire is bent into a closed curve or frame. When the frame is held over a level surface, its projection onto the surface is a plane curve C enclosing a region R. If one forms a soap film on the frame, the height $u(x, y)$ of the film above the level surface is a function that satisfies the potential equation, if the effects of gravity are negligible (see Chapter 5). The height of the frame above the curve C gives the boundary condition on u.

It turns out that the potential equation (but not all elliptic equations) is best studied through the use of complex variables. A complex variable may be written $z = x + iy$, where x and y are real and $i^2 = -1$; similarly a function of z is denoted by $f(z) = u(x, y) + iv(x, y)$, u and v being real functions of real variables. If f has a derivative with respect to z, then both u and v satisfy the potential equation. Easy examples such as polynomials and exponentials lead to familiar solutions:

$$z^2 = (x + iy)^2 = x^2 - y^2 + i2xy$$

$$e^z = e^{x+iy} = e^x e^{iy} = e^x \cos y + ie^x \sin y.$$

An excellent, compact presentation of Laplace's equation and its relation to functions of complex variables is given in *Differential Equations of Applied Mathematics* by Duff and Naylor [12]. More detailed treatments are in Churchill, *Complex Variables and Applications* [6] and Kreyszig, *Advanced Engineering Mathematics* [18].

Two more references of interest are the book *Maximum Principles in Differential Equations*, by Protter and Weinberger (23), an elementary and elegant study of maximum principles; and a *Scientific American* article, "Brownian Motion and Potential Theory," by Hersh and Griego [17], which relates some probabilistic phenomena to Laplace's equation.

Miscellaneous exercises

1. Solve the potential equation in the rectangle $0 < x < a$, $0 < y < b$ with the boundary conditions

$$u(0, y) = 1, \quad u(a, y) = 0, \quad 0 < y < b$$

$$u(x, 0) = 0, \quad u(x, b) = 0, \quad 0 < x < a.$$

2. If $a = b$ in Exercise 1, then $u(a/2, a/2) = \frac{1}{4}$. Why?

3. Solve the potential equation on the rectangle $0 < x < a$, $0 < y < b$ with the boundary conditions

$$u(0, y) = 1, \qquad u(a, y) = 1, \qquad 0 < y < b$$

$$\frac{\partial u}{\partial y}(x, 0) = 0, \qquad \frac{\partial u}{\partial y}(x, b) = 0, \qquad 0 < x < a.$$

4. Same as Exercise 3, but the boundary conditions are

$$u(0, y) = 1, \qquad \frac{\partial u}{\partial x}(a, y) = 0, \qquad 0 < y < b$$

$$u(x, 0) = 1, \qquad \frac{\partial u}{\partial y}(x, b) = 0, \qquad 0 < x < a.$$

5. Same as Exercise 3, but the boundary conditions are

$$u(0, y) = 1, \qquad u(a, y) = 1, \qquad 0 < y < b$$

$$\frac{\partial u}{\partial y}(x, 0) = 0, \qquad u(x, b) = 0, \qquad 0 < x < a.$$

6. Same as Exercise 3, but the boundary conditions are

$$u(0, y) = 1, \qquad u(a, y) = 0, \qquad 0 < y < b$$

$$u(x, 0) = 1, \qquad u(x, b) = 0, \qquad 0 < x < a.$$

7. Same as Exercise 3, but the region is a square ($b = a$) and the boundary conditions are

$$u(0, y) = f(y), \qquad u(a, y) = 0, \qquad 0 < y < a$$

$$u(x, 0) = f(x), \qquad u(x, a) = 0, \qquad 0 < x < a,$$

where f is a function whose graph is an isosceles triangle of height h and width a.

8. Solve the potential equation in the region $0 < x < a$, $0 < y$ with the boundary conditions

$$u(x, 0) = 1, \qquad\qquad 0 < x < a$$

$$u(0, y) = 0, \qquad u(a, y) = 0, \quad 0 < y.$$

9. Find the solution of the potential equation on the strip $0 < y < b$, $-\infty < x < \infty$, subject to the conditions below. Supply boundedness

conditions as necessary.

$$u(x, 0) = \begin{cases} 1, & -a < x < a \\ 0, & a < |x| \end{cases}$$

$$u(x, b) = 0, \qquad -\infty < x < \infty$$

10. Obtain a formula for the solution of the potential equation in the upper half plane.

$$\frac{\partial^2 u}{\partial x^2} + \frac{\partial^2 u}{\partial y^2} = 0, \qquad -\infty < x < \infty, \quad 0 < y$$

$$u(x, 0) = f(x), \qquad -\infty < x < \infty$$

11. Solve the problem of Exercise 10 taking $f(x) = \exp(-\alpha|x|)$.

12. Convert your solution of Exercise 10 into the following formula. (See Exercise 8 of Section 4 and Section 11 of Chapter 2.)

$$u(x, y) = \frac{1}{\pi} \int_{-\infty}^{\infty} f(x') \frac{y}{y^2 + (x - x')^2} \, dx'$$

13. Apply the formula in Exercise 12 to the case where $f(x) \equiv 1$, $-\infty < x < \infty$.

14. Show that the function $u(x, y) = \tan^{-1}(y/x)$ is a solution of the potential equation in the first quadrant. What conditions does u satisfy along the rays $x = 0$, $y > 0$ and $x > 0$, $y = 0$?

15. Solve the potential equation in a disk of radius c with boundary condition

$$u(c, \theta) = \begin{cases} 1, & 0 < \theta < \pi \\ 0, & -\pi < \theta < 0. \end{cases}$$

16. What is the value of u at the center of the disk, in Exercise 15?

17. Same as Exercise 15, but the boundary condition is

$$u(c, \theta) = |\sin \theta|.$$

18. For the potential problem on an annular ring

$$\frac{1}{r} \frac{\partial}{\partial r} \left(r \frac{\partial u}{\partial r} \right) + \frac{1}{r^2} \frac{\partial^2 u}{\partial \theta^2} = 0, \qquad a < r < b,$$

show that product solutions have the form

$$r^n(A_n \cos n\theta + B_n \sin n\theta) + r^{-n}(C_n \cos n\theta + D_n \sin n\theta)$$

or $A + B \ln r$.

19. Solve the potential problem on the annular ring as stated in Exercise 18 with boundary conditions

$$u(a, \theta) = 1, \qquad u(b, \theta) = 0.$$

20. Find product solutions of the potential equation on a sector of a disk with zero boundary conditions on the straight edges.

$$\nabla^2 u = 0, \qquad 0 \le r < c, \quad 0 < \theta < \alpha$$

$$u(r, 0) = 0, \qquad u(r, \alpha) = 0$$

21. Solve the potential problem in a slit disk:

$$\nabla^2 u = 0, \qquad 0 \le r < c, \quad 0 < \theta < 2\pi$$

$$u(r, 0) = 0, \qquad u(r, 2\pi) = 0$$

$$u(r, \theta) = f(\theta), \qquad 0 < \theta < 2\pi.$$

22. Show that the function $u(x, y) = \sin(\pi x/a) \sinh(\pi y/a)$ satisfies the potential problem

$$\nabla^2 u = 0, \qquad\qquad 0 < x < a, \quad 0 < y$$

$$u(0, y) = 0, u(a, y) = 0, \qquad 0 < y$$

$$u(x, 0) = 0, \qquad\qquad 0 < x < a.$$

This solution is eliminated if it is also required that $u(x, y)$ be bounded as $y \to \infty$.

23. Find a relationship among the coefficients of the polynomial

$$p(x, y) = a + bx + cy + dx^2 + exy + fy^2$$

so that it satisfies the nonhomogeneous equation

$$\frac{\partial^2 p}{\partial x^2} + \frac{\partial^2 p}{\partial y^2} = -1.$$

24. Find a polynomial $p(x, y)$ that satisfies the partial differential equation of Exercise 23 and also the boundary conditions

$$p(x, 0) = 0, \qquad p(x, b) = 0.$$

25. Solve the following boundary value problem

$$\frac{\partial^2 u}{\partial x^2} + \frac{\partial^2 u}{\partial y^2} = -1, \qquad\qquad 0 < x < a, \quad 0 < y < b$$

$$u(x, 0) = 0, \qquad u(x, b) = 0, \qquad 0 < x < a$$

$$u(0, y) = 0, \qquad u(a, y) = 0, \qquad 0 < y < b.$$

26. An equation of the form $\nabla^2 u = -f$, where f is a function of the space variables, is called Poisson's equation. (See Exercises 23–25.) When f depends on just one variable, it is often easy to find a solution. Solve these problems in polar coordinates:

   a. $\nabla^2 u = -1$

   b. $\nabla^2 u = -\dfrac{1}{x^2 + y^2}.$

27. (Flow past a plate)   A fluid occupies the half plane $y > 0$ and flows past (left to right, approximately) a plate located near the x-axis. If the x and y components of velocity are $U_0 + u(x, y)$, and $v(x, y)$, respectively ($U_0 = $ constant free-stream velocity), under certain assumptions, the equations of motion, continuity, and state can be reduced to

$$\frac{\partial u}{\partial y} = \frac{\partial v}{\partial x}, \qquad (1 - M^2)\frac{\partial u}{\partial x} + \frac{\partial v}{\partial y} = 0,$$

valid for all x and $y > 0$. M is the free-stream Mach number. Define the velocity potential ϕ by the equations $u = \partial\phi/\partial x$ and $v = \partial\phi/\partial y$. Show that the first equation is automatically satisfied and the second is a partial differential equation which is elliptic if $M < 1$ or hyperbolic if $M > 1$.

28. If the plate is wavy—say its equation is $y = \varepsilon \cos \alpha x$—then the boundary conditions, that the vector velocity be parallel to the wall, is

$$v(x, \varepsilon \cos \alpha x) = -\varepsilon\alpha \sin \alpha x(U_0 + u(x, \varepsilon \cos \alpha x)).$$

This equation is impossible to use, so it is replaced by

$$v(x, 0) = -\varepsilon\alpha U_0 \sin \alpha x$$

on the assumption that ε is small and u is much smaller than U_0. Using this boundary condition, and the condition that $u(x, y) \to 0$ as $y \to \infty$, set up and solve a complete boundary value problem for ϕ, assuming $M < 1$.

29. By superposition of solutions (α ranging from 0 to ∞) find the flow past a wall whose equation is $y = f(x)$. (Hint: use the boundary condition

$$v(x, 0) = U_0 f'(x) = \int_0^\infty [A(\alpha) \cos \alpha x + B(\alpha) \sin \alpha x]\, d\alpha.)$$

chapter 5
Problems in Several Dimensions

1   Derivation of the two-dimensional wave equation

For an example of a two-dimensional wave equation, we consider a membrane which is stretched taut over a flat frame in the xy-plane (Fig. 5.1). The displacement of the membrane above the point (x, y) at time t is $u(x, y, t)$. We assume that the surface tension σ (dimensions F/L) is constant and independent of position. We also suppose that the membrane is perfectly flexible: that is, it does not resist bending. (A soap film satisfies these assumptions quite accurately.) Let us imagine that a small rectangle (of dimensions Δx by Δy aligned with the coordinate axes) is cut out of the membrane and then

Figure 5.1

apply Newton's law of motion to it. On each edge of the rectangle, the rest of the membrane exerts a distributed force of magnitude σ (symbolized by the arrows in Fig. 5.2); these distributed forces can be resolved into concentrated forces of magnitude $\sigma \, \Delta x$ or $\sigma \, \Delta y$, according to the length of the segment involved (see Fig. 5.3).

Looking at projections on the xu- and yu-planes (Figs. 5.5–5.6), we see

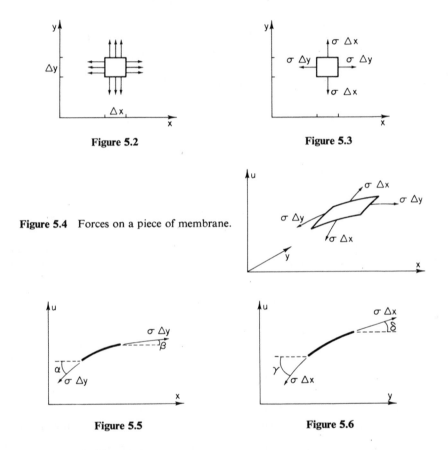

Figure 5.2

Figure 5.3

Figure 5.4   Forces on a piece of membrane.

Figure 5.5

Figure 5.6

that the sum of forces in the x-direction is $\sigma \, \Delta y \, (\cos \beta - \cos \alpha)$, and the sum of forces in the y-direction is $\sigma \, \Delta x \, (\cos \delta - \cos \gamma)$. It is desirable that both these sums be zero or at least negligible. Therefore we shall assume that α, β, γ, and δ are all small angles. Since we know that

$$\tan \alpha = \frac{\partial u}{\partial x}, \qquad \tan \gamma = \frac{\partial u}{\partial y}$$

and so forth, when the derivatives are evaluated at some appropriate point near (x, y), we are assuming that the slopes $\partial u/\partial x$ and $\partial u/\partial y$ of the membrane are very small.

Adding up forces in the vertical direction, and equating the sum to the mass times acceleration (in the vertical direction) we get

$$\sigma \, \Delta y \, (\sin \beta - \sin \alpha) + \sigma \, \Delta x \, (\sin \delta - \sin \gamma) = \rho \, \Delta x \, \Delta y \, \frac{\partial^2 u}{\partial t^2}$$

where ρ is the surface density $[m/L^2]$. Since the angles α, β, γ, and δ are small, the sine of each is approximately equal to its tangent:

$$\sin \alpha \simeq \tan \alpha = \frac{\partial u}{\partial x} (x, y, t),$$

and so forth. With these approximations used throughout, the equation above becomes

$$\sigma \, \Delta y \left(\frac{\partial u}{\partial x} (x + \Delta x \ y, t) - \frac{\partial u}{\partial x} (x, y, t) \right)$$

$$+ \sigma \, \Delta x \left(\frac{\partial u}{\partial y} (x, y + \Delta y, t) - \frac{\partial u}{\partial y} (x, y, t) \right) = \rho \, \Delta x \, \Delta y \, \frac{\partial^2 u}{\partial t^2}.$$

On dividing through by $\Delta x \, \Delta y$, we recognize two difference quotients in the left-hand member. In the limit they become partial derivatives, yielding the equation

$$\sigma \left(\frac{\partial^2 u}{\partial x^2} + \frac{\partial^2 u}{\partial y^2} \right) = \rho \, \frac{\partial^2 u}{\partial t^2}$$

or

$$\frac{\partial^2 u}{\partial x^2} + \frac{\partial^2 u}{\partial y^2} = \frac{1}{c^2} \frac{\partial^2 u}{\partial t^2}$$

if $c^2 = \sigma/\rho$. This is the two-dimensional wave equation.

If the membrane is fixed to the flat frame, the boundary condition would be

$$u(x, y, t) = 0 \qquad \text{for} \quad (x, y) \text{ on the boundary.}$$

Naturally, it is necessary to give initial conditions describing the displacement and velocity of each point on the membrane at $t = 0$:

$$u(x, y, 0) = f(x, y)$$

$$\frac{\partial u}{\partial t} (x, y, 0) = g(x, y).$$

Exercises

1. Suppose that the frame is rectangular, bounded by segments of the lines $x = 0$, $x = a$, $y = 0$, $y = b$. Write an initial value–boundary value problem, complete with inequalities, for a membrane stretched over this frame.

2. Suppose that the frame is circular, and its equation is $x^2 + y^2 = a^2$. Write an initial value–boundary value problem for a membrane on a circular frame. (Use polar coordinates.)

3. What should the three-dimensional wave equation be?

2 Derivation of the two-dimensional heat equation

Imagine a thin flat plate of heat-conducting material sandwiched between sheets of insulation. Let us suppose that a coordinate system is chosen as shown in Fig. 5.7 and that the temperature in the plate depends only on the

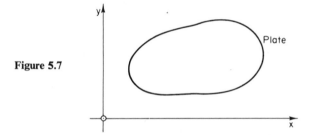

Figure 5.7

position in the xy-plane and time. We apply the law of conservation of energy (in rate form) to a small rectangle with sides of lengths Δx and Δy (Fig. 5.8).

Let $q_x(x, y)$ and $q_y(x, y)$ denote the heat flow rates in the x- and y-directions, respectively. The quantities needed for expressing the conservation of

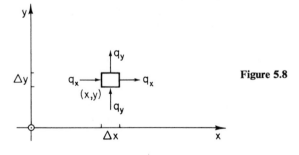

Figure 5.8

energy are:

Rate in: $q_x(x, y + \frac{1}{2}\Delta y)\theta\, \Delta y + q_y(x + \frac{1}{2}\Delta x, y)\theta\, \Delta x$

Rate out: $q_x(x + \Delta x, y + \frac{1}{2}\Delta y)\theta\, \Delta y + q_y(x + \frac{1}{2}\Delta x, y + \Delta y)\theta\, \Delta x$

Rate of generation: $r\theta\, \Delta x\, \Delta y$

Rate of storage: $\rho c \theta\, \Delta x\, \Delta y\, \dfrac{\partial u}{\partial t}.$

In these expressions, θ is the thickness of the plate, r is the rate of heat generation per unit volume, and ρ and c are density and heat capacity.

The conservation of energy requires that

rate in + rate of generation = rate out + rate of storage.

It is easier to formulate the equation as:

rate in − rate out = rate of storage − rate of generation.

Thus (after dividing through by θ) we find

$$[q_x(x, y + \tfrac{1}{2}\Delta y) - q_x(x + \Delta x, y + \tfrac{1}{2}\Delta y)]\, \Delta y +$$

$$[q_y(x + \tfrac{1}{2}\Delta x, y) - q_y(x + \tfrac{1}{2}\Delta x, y + \Delta y)]\, \Delta x = \left(-r + \rho c\, \frac{\partial u}{\partial t}\right) \Delta x\, \Delta y.$$

On dividing through by $\Delta x\, \Delta y$, we see two difference quotients on the left-hand side of the equation. In the limit as Δx and Δy tend to zero, they become partial derivatives, leaving

$$-\frac{\partial q_x}{\partial x} - \frac{\partial q_y}{\partial y} = -r + \rho c\, \frac{\partial u}{\partial t}.$$

All functions now are evaluated at the point (x, y).

We resort again to Fourier's law to eliminate the rates q_x and q_y. After doing so, the equation above becomes

$$\frac{\partial}{\partial x}\left(\kappa_x\, \frac{\partial u}{\partial x}\right) + \frac{\partial}{\partial y}\left(\kappa_y\, \frac{\partial u}{\partial y}\right) = \rho c\, \frac{\partial u}{\partial t} - r.$$

In general, κ_x and κ_y may be different functions of position. If, however, $\kappa_x = \kappa_y$ (isotropic material) and both are independent of position (uniform material), our equation reduces to

$$\kappa\left(\frac{\partial^2 u}{\partial x^2} + \frac{\partial^2 u}{\partial y^2}\right) = \rho c\, \frac{\partial u}{\partial t} - r$$

which is usually called the two-dimensional heat equation.

A two-dimensional heat conduction problem requires a description of what happens at the boundary of the plate and a specification of the initial temperature distribution.

Exercises

1. Suppose the plate lies in the rectangle $0 < x < a, 0 < y < b$. State a complete initial value–boundary value problem for temperature in the plate if:

    a. There is no heat generation;
    b. the temperature is held at T_0 along $x = a$ and $y = 0$;
    c. the edges at $x = 0$ and $y = b$ are insulated.

2. Find the dimensions of ρ, c, κ, q, and r, and verify that the dimensions of the right and left members of the heat equation are the same.

3. Derive the three-dimensional heat equation for a uniform, isotropic body.

4. Suppose that the plate is rectangular (as in Exercise 1) and that it lies between $z = 0$ and $z = \theta$. If we consider the plate as a three-dimensional body (uniform and isotropic, without generation) the equation for the · temperature $u(x, y, z, t)$ is

$$\frac{\partial^2 u}{\partial x^2} + \frac{\partial^2 u}{\partial y^2} + \frac{\partial^2 u}{\partial z^2} = \frac{1}{k}\frac{\partial u}{\partial t}$$

where $k = \kappa/\rho c$. Let

$$v(x, y, t) = \int_0^\theta u(x, y, z, t)\,dz$$

and assume that $\partial u/\partial z = 0$ at $z = 0$ and at $z = \theta$. Show that v satisfies the two-dimensional heat equation.

3   Solution of the two-dimensional heat equation

In order to see the technique of solution for a two-dimensional problem, we shall consider the diffusion of heat in a rectangular plate of uniform, isotropic material. The steady-state temperature distribution is a solution of Laplace's equation (see Exercise 6). Suppose that the initial value–boundary value problem for the transient temperature $u(x, y, t)$ is

$$\frac{\partial^2 u}{\partial x^2} + \frac{\partial^2 u}{\partial y^2} = \frac{1}{k}\frac{\partial u}{\partial t}, \qquad\qquad 0 < x < a, \quad 0 < y < b, \quad 0 < t \qquad (1)$$

$$u(x, 0, t) = 0, \qquad u(x, b, t) = 0, \qquad 0 < x < a, \quad 0 < t \qquad\qquad (2)$$

$$u(0, y, t) = 0, \qquad u(a, y, t) = 0, \qquad 0 < y < b, \quad 0 < t \qquad\qquad (3)$$

$$u(x, y, 0) = f(x, y), \qquad\qquad\qquad 0 < x < a, \quad 0 < y < b. \qquad (4)$$

This problem contains a homogeneous partial differential equation and homogeneous boundary conditions. We may thus proceed with separation of variables by seeking solutions in the form

$$u(x, y, t) = \phi(x, y)T(t).$$

On substituting u in product form into Eq. (1), we find that it becomes

$$\left(\frac{\partial^2 \phi}{\partial x^2} + \frac{\partial^2 \phi}{\partial y^2}\right)T = \frac{1}{k}\phi T'.$$

Separation can be achieved by dividing through by ϕT, which leaves

$$\left(\frac{\partial^2 \phi}{\partial x^2} + \frac{\partial^2 \phi}{\partial y^2}\right)\frac{1}{\phi} = \frac{T'}{kT}.$$

We may argue, as usual, that the mutual value of the members of this equation must be a constant, which we expect to be negative $(-\lambda^2)$. The equations that result are

$$T' + \lambda^2 kT = 0, \qquad 0 < t \tag{5}$$

$$\frac{\partial^2 \phi}{\partial x^2} + \frac{\partial^2 \phi}{\partial y^2} = -\lambda^2 \phi, \qquad 0 < x < a, \quad 0 < y < b. \tag{6}$$

In terms of the product solutions, the boundary conditions become

$$\phi(x, 0)T(t) = 0, \qquad \phi(x, b)T(t) = 0$$
$$\phi(0, y)T(t) = 0, \qquad \phi(a, y)T(t) = 0.$$

In order to satisfy all four equations, either $T(t) = 0$ for all t, or $\phi = 0$ on the boundary. We have seen many times that the choice of $T(t) = 0$ wipes out our solution completely. Therefore, we require that ϕ satisfy the conditions

$$\phi(x, 0) = 0, \qquad \phi(x, b) = 0, \qquad 0 < x < a \tag{7}$$
$$\phi(0, y) = 0, \qquad \phi(a, y) = 0, \qquad 0 < y < b \tag{8}$$

We are not yet out of difficulty, because Eqs. (6)–(8) constitute a new problem, a two-dimensional eigenvalue problem. It is evident, however, that the partial differential equation and the boundary conditions are linear and homogeneous; thus separation of variables may work again. Supposing that ϕ has the form

$$\phi(x, y) = X(x)Y(y)$$

we find that the partial differential equation (6) becomes

$$\frac{X''}{X} + \frac{Y''}{Y} = -\lambda^2, \qquad 0 < x < a, \quad 0 < y < b.$$

The sum of a function of x and a function of y can be constant only if those two functions are individually constant:

$$\frac{X''}{X} = \text{constant}, \qquad \frac{Y''}{Y} = \text{constant}.$$

Before naming the constants, let us look at the boundary conditions on $\phi = XY$:

$$X(x)Y(0) = 0, \qquad X(x)Y(b) = 0, \qquad 0 < x < a$$
$$X(0)Y(y) = 0, \qquad X(a)Y(y) = 0, \qquad 0 < y < b.$$

If either of the functions X or Y is zero throughout the whole interval of its variable, the conditions are certainly satisfied, but ϕ is identically zero. We therefore require each of the functions X and Y to be zero at the endpoints of its interval:

$$Y(0) = 0, \qquad Y(b) = 0 \tag{9}$$
$$X(0) = 0, \qquad X(a) = 0. \tag{10}$$

Now it is clear that each of the ratios X''/X and Y''/Y should be a negative constant, designated by $-\mu^2$ and $-\nu^2$, respectively. The separate equations for X and Y are

$$X'' + \mu^2 X = 0, \qquad 0 < x < a \tag{11}$$
$$Y'' + \nu^2 Y = 0, \qquad 0 < y < b. \tag{12}$$

Finally, the original separation constant $-\lambda^2$ is determined by

$$\lambda^2 = \mu^2 + \nu^2. \tag{13}$$

Now we see two independent eigenvalue problems: Eqs. (9) and (12) form one problem and Eqs. (10) and (11) the other. Each is of a very familiar form; the solutions are

$$X_m(x) = \sin\left(\frac{m\pi x}{a}\right), \qquad \mu_m^2 = \left(\frac{m\pi}{a}\right)^2$$
$$Y_n(y) = \sin\left(\frac{n\pi y}{b}\right), \qquad \nu_n^2 = \left(\frac{n\pi}{b}\right)^2.$$

Notice that the indices n and m are independent. This means that ϕ will have a double index. Specifically, the solutions of the two-dimensional eigenvalue problem (6)–(8) are

$$\phi_{mn}(x, y) = X_m(x)Y_n(y)$$
$$\lambda_{mn}^2 = \mu_m^2 + v_n^2$$

and the corresponding function T is

$$T_{mn} = \exp(-\lambda_{mn}^2 kt).$$

We now begin to assemble the solution. For each pair of indices m, n ($m = 1, 2, 3, \ldots, n = 1, 2, 3, \ldots$) there is a function

$$u_{mn}(x, y, t) = \phi_{mn}(x, y)T_{mn}(t)$$

$$= \sin\left(\frac{m\pi x}{a}\right)\sin\left(\frac{n\pi y}{b}\right)\exp(-\lambda_{mn}^2 kt)$$

which satisfies the partial differential Eq. (1) and the boundary conditions Eqs. (2) and (3). We may form linear combinations of these solutions to get other solutions. The most general linear combination would be the double series

$$u(x, y, t) = \sum_{m=1}^{\infty} \sum_{n=1}^{\infty} a_{mn} \phi_{mn}(x, y)T_{mn}(t) \tag{14}$$

and any such combination should satisfy Eqs. (1)–(3). There remains the initial condition Eq. (4) to be satisfied. If u has the form above, then the initial condition becomes

$$\sum_{m=1}^{\infty} \sum_{n=1}^{\infty} a_{mn} \phi_{mn}(x, y) = f(x, y), \qquad 0 < x < a, \quad 0 < y < b. \tag{15}$$

The idea of orthogonality is once again applicable to the problem of selecting the coefficients a_{mn}. One can show by direct computation that

$$\int_0^b \int_0^a \phi_{mn}(x, y)\phi_{pq}(x, y)\, dx\, dy = \begin{cases} \dfrac{ab}{4}, & \text{if } m = p \text{ and } n = q \\ 0, & \text{otherwise.} \end{cases} \tag{16}$$

Thus, the appropriate formula for the coefficients a_{mn} is

$$a_{mn} = \frac{4}{ab} \int_0^b \int_0^a f(x, y) \sin\left(\frac{m\pi x}{a}\right)\sin\left(\frac{n\pi y}{b}\right) dx\, dy.$$

If f is a sufficiently regular function, the series Eq. (15) will converge and equal $f(x, y)$ in the rectangular region $0 < x < a$, $0 < y < b$. We may then say that the problem is solved.

It is reassuring to notice that each term in the series Eq. (14) contains a decaying exponential, and thus, as t increases, $u(x, y, t)$ tends to zero, as expected.

Let us take the specific initial condition

$$f(x, y) = xy, \qquad 0 < x < a, \quad 0 < y < b.$$

The coefficients are easily found to be

$$a_{mn} = \frac{4ab}{\pi^2} \frac{\cos m\pi \cos n\pi}{mn}$$

whence the solution to this problem is

$$u(x, y, t) = \frac{4ab}{\pi^2} \sum_{m=1}^{\infty} \sum_{n=1}^{\infty} \frac{\cos m\pi \cos n\pi}{mn}$$

$$\times \sin\left(\frac{n\pi x}{a}\right) \sin\left(\frac{m\pi y}{b}\right) \exp(-\lambda_{mn}^2 kt). \tag{17}$$

The double series which appear here are best handled by converting them into single series. To do this, arrange the terms in order of increasing values of λ_{mn}^2. Then the first terms in the single series are the most significant, those that decay least rapidly. For example, if $a = 2b$, so that

$$\lambda_{mn}^2 = \frac{(m^2 + 4n^2)\pi^2}{a^2},$$

then the following list gives the double index (m, n) in order of increasing values of λ_{mn}^2:

$$(1, 1), (2, 1), (3, 1), (1, 2), (2, 2), (4, 1), (3, 2), \ldots.$$

Exercises

1. Write out the "first few" terms of the series Eq. (17). By "first few," we mean those for which λ_{mn}^2 is smallest. (Assume $a = b$ in determining relative magnitudes of the λ^2.)

2. Provide the details of the separation of variables by which Eqs. (9)–(13) are derived.

3. Find the frequencies of vibration of a rectangular membrane. See Section 1, Exercise 1.

4. Verify that $u_{mn}(x, y, t)$ satisfies Eqs. (1)–(3).

5. Show that $X_m = \cos(m\pi x/a)$ $(m = 0, 1, 2, \ldots)$ if the boundary conditions Eq. (3) are replaced by

$$\frac{\partial u}{\partial x}(0, y, t) = 0, \qquad \frac{\partial u}{\partial x}(a, y, t) = 0, \qquad 0 < y < b, \quad 0 < t.$$

What values will the λ_{mn}^2 have, and of what form will the solution $u(x, y, t)$ be?

6. Suppose that, instead of boundary conditions Eqs. (2) and (3), we have

$$u(x, 0, t) = f_1(x), \qquad u(x, b, t) = f_2(x), \qquad 0 < x < a, \quad 0 < t \quad (2')$$

$$u(0, y, t) = g_1(y), \qquad u(a, y, t) = g_2(y), \qquad 0 < y < b, \quad 0 < t. \quad (3')$$

Show that the steady-state solution involves Laplace's equation, and indicate how to solve it.

7. Solve the two-dimensional heat conduction problem in a rectangle if there is insulation on all boundaries and the initial condition is

a. $u(x, y, 0) = 1$
b. $u(x, y, 0) = x + y$
c. $u(x, y, 0) = xy$.

8. Verify the orthogonality relation Eq. (16) and the formula for a_{mn}.

9. Show that the separation constant $-\lambda^2$ must be negative by showing that $-\mu^2$ and $-\nu^2$ must both be negative.

4 Problems in polar coordinates

We found that the one-dimensional wave and heat problems have a great deal in common. Namely, the steady-state or time-independent solutions and the eigenvalue problems which arise are identical in both cases. Also, in problems in a rectangle, the same features are shared.

If we consider now the vibrations of a circular membrane or heat conduction in a circular plate, we shall see common features again. Below, these two problems are given side by side for the region $0 < r < a$, $-\pi < \theta \leq \pi$, $0 < t$:

Wave	Heat
$\nabla^2 v = \dfrac{1}{c^2} \dfrac{\partial^2 v}{\partial t^2}$	$\nabla^2 v = \dfrac{1}{k} \dfrac{\partial v}{\partial t}$
$v(a, \theta, t) = f(\theta)$	$v(a, \theta, t) = f(\theta)$
$v(r, \theta, 0) = g(r, \theta)$	$v(r, \theta, 0) = g(r, \theta)$
$\dfrac{\partial v}{\partial t}(r, \theta, 0) = h(r, \theta).$	

In both problems we require

$$v(r, -\pi, t) = v(r, \pi, t), \qquad 0 < r < a, \quad 0 < t$$

$$\frac{\partial v}{\partial \theta}(r, -\pi, t) = \frac{\partial v}{\partial \theta}(r, \pi, t), \qquad 0 < r < a, \quad 0 < t$$

so that no discontinuity will arise at the artificial boundaries $\theta = \pm\pi$.

Although the interpretation of the function v is different in the two cases, we see that the solution of the problem

$$\nabla^2 v = 0, \qquad v(a, \theta) = f(\theta)$$

is the rest-state or steady-state solution for both problems, and will be needed in both problems to make the boundary condition at $r = a$ homogeneous. Let us suppose that the time-independent solution has been found and subtracted: that is, we will replace $f(\theta)$ by zero. Then we have

$$\nabla^2 v = \frac{1}{c^2}\frac{\partial^2 v}{\partial t^2} \qquad \nabla^2 v = \frac{1}{k}\frac{\partial v}{\partial t}$$

$$v(a, \theta, t) = 0 \qquad\qquad v(a, \theta, t) = 0$$

plus the appropriate initial conditions. If we attempt to solve by separation of variables, setting $v(r, \theta, t) = \phi(r, \theta)T(t)$, in both cases we will find that $\phi(r, \theta)$ must satisfy

$$\nabla^2\phi = -\lambda^2\phi, \qquad 0 < r < a, \quad -\pi < \theta \leq \pi \qquad (1)$$

$$\phi(a, \theta) = 0, \qquad\qquad -\pi < \theta \leq \pi \qquad\qquad (2)$$

$$\phi(r, -\pi) = \phi(r, \pi), \qquad 0 < r < a \qquad\qquad (3)$$

$$\frac{\partial\phi}{\partial\theta}(r, -\pi) = \frac{\partial\phi}{\partial\theta}(r, \pi), \qquad 0 < r < a. \qquad\qquad (4)$$

Now we shall concentrate on the solution of this two-dimensional eigenvalue problem. Written out in polar coordinates, Eq. (1) becomes

$$\frac{1}{r}\frac{\partial}{\partial r}\left(r\frac{\partial\phi}{\partial r}\right) + \frac{1}{r^2}\frac{\partial^2\phi}{\partial\theta^2} = -\lambda^2\phi.$$

We can separate variables again by assuming that $\phi(r, \theta) = R(r)\Theta(\theta)$. After some algebra, we find that

$$\frac{(rR')'}{rR} + \frac{\Theta''}{r^2\Theta} = -\lambda^2 \qquad\qquad (5)$$

$$R(a) = 0 \qquad\qquad (6)$$

$$\Theta(-\pi) = \Theta(\pi) \qquad\qquad (7)$$

$$\Theta'(-\pi) = \Theta'(\pi). \qquad\qquad (8)$$

The ratio Θ''/Θ must be constant; otherwise, λ^2 could not be constant. Choosing $\Theta''/\Theta = -\mu^2$, we get a familiar problem

$$\Theta'' + \mu^2 \Theta = 0, \qquad -\pi < \theta \leq \pi \tag{9}$$

$$\Theta(-\pi) = \Theta(\pi) \tag{10}$$

$$\Theta'(-\pi) = \Theta'(\pi). \tag{11}$$

We found (Chapter 4) that the solutions of this problem are

$$\begin{aligned} \mu_0^2 &= 0, & \Theta(\theta) &= 1 \\ \mu_m^2 &= m^2, & \Theta(\theta) &= \cos m\theta \quad \text{or} \quad \sin m\theta \end{aligned} \tag{12}$$

where $m = 1, 2, 3, \ldots$.

There remains a problem in R:

$$(rR')' - \frac{\mu^2}{r} R + \lambda^2 rR = 0, \qquad 0 < r < a \tag{13}$$

$$R(a) = 0. \tag{14}$$

Equation (13) is called *Bessel's equation*, and we shall solve it in the next section.

Exercises

1. State the full initial value–boundary value problems which result from the problems as originally given when the steady-state or time-independent solution is subtracted from v.

2. Verify the separation of variables that leads to Eqs. (1) and (2).

3. Substitution of $v(r, \theta, t)$ in the form of a product led to the problem of Eqs. (1)–(4) for the factor $\phi(r, \theta)$. What differential equation is to be satisfied by the factor $T(t)$?

4. Solve Eqs. (9)–(11) and check the solutions given.

5. Suppose the problems originally stated were to be solved in the half-disk $0 < r < a, 0 < \theta < \pi$, with additional conditions

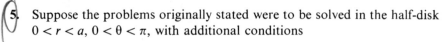

$$\begin{aligned} v(r, 0, t) &= 0, & 0 < r < a, \quad 0 < t \\ v(r, \pi, t) &= 0, & 0 < r < a, \quad 0 < t. \end{aligned}$$

What eigenvalue problem arises in place of Eqs. (9)–(11)? Solve it.

6. Suppose that the boundary condition

$$\frac{\partial v}{\partial r}(a, \theta, t) = 0, \qquad -\pi < \theta \le \pi, \quad 0 < t$$

were given instead of $v(a, \theta, t) = f(\theta)$. Carry out the steps involved in separation of variables. Show that the only change is in Eq. (14), which becomes $R'(a) = 0$.

7. One of the consequences of Green's theorem is the integral relation

$$\iint_R (f\nabla^2 g - g\nabla^2 f)\, dA = \int_C \left(f\frac{\partial g}{\partial n} - g\frac{\partial f}{\partial n} \right) ds$$

where R is a region in the plane, C is the closed curve that bounds R, and $\partial f/\partial n$ is the directional derivative in the direction normal to the curve C. Use this relation to show that eigenfunctions of the problem

$$\nabla^2 \phi = -\lambda^2 \phi, \qquad \text{in} \quad R$$
$$\phi = 0, \qquad \text{on} \quad C$$

are orthogonal if they correspond to different eigenvalues. (Hint: use $f = \phi_k$, $g = \phi_m$, $m \ne k$.)

8. Same problem as above, except that the boundary condition is

$$\phi + \gamma \frac{\partial \phi}{\partial n} = 0$$

on the boundary.

5 Bessel's equation

In order to solve the Bessel equation

$$(rR')' - \frac{\mu^2}{r} R + \lambda^2 rR = 0 \tag{1}$$

we apply the method of Frobenius. Assume that $R(r)$ has the form of a power series multiplied by an unknown power of r:

$$R(r) = r^\alpha (c_0 + c_1 r + \cdots + c_k r^k + \cdots). \tag{2}$$

When the differentiations in Eq. (1) are carried out and the equation is multiplied by r, it becomes

$$r^2 R'' + rR' - \mu^2 R + \lambda^2 r^2 R = 0.$$

We now substitute the infinite series Eq. (2) for R:

$$R(r) = c_0 r^\alpha + c_1 r^{\alpha+1} + \cdots + c_k r^{\alpha+k} + \cdots$$

and analogous forms for the derivatives of R. The four terms of the differential equation are

$$r^2 R'' = \alpha(\alpha - 1)c_0 r^\alpha + (\alpha + 1)\alpha c_1 r^{\alpha+1} + (\alpha + 2)(\alpha + 1)c_2 r^{\alpha+2} + \cdots$$
$$+ (\alpha + k)(\alpha + k - 1)c_k r^{\alpha+k} + \cdots$$

$$rR' = \qquad \alpha c_0 r^\alpha + (\alpha + 1)c_1 r^{\alpha+1} \qquad + (\alpha + 2)c_2 r^{\alpha+2} + \cdots$$
$$+ (\alpha + k)c_k r^{\alpha+k} + \cdots$$

$$-\mu^2 R = \qquad - \mu^2 c_0 r^\alpha \qquad - \mu^2 c_1 r^{\alpha+1} \qquad - \mu^2 c_2 r^{\alpha+2} - \cdots$$
$$- \mu^2 c_k r^{\alpha+k} + \cdots$$

$$\lambda^2 r^2 R = \qquad \lambda^2 c_0 r^{\alpha+2} + \cdots$$
$$+ \lambda^2 c_{k-2} r^{\alpha+k} + \cdots.$$

The expression for $\lambda^2 r^2 R$ is jogged to the right to make like powers of r line up vertically. Note that the lowest power of r present in $\lambda^2 r^2 R$ is $r^{\alpha+2}$.

Now we add the tableau vertically. The sum of the left-hand sides is, according to the differential equation, equal to zero. Therefore

$$0 = c_0(\alpha^2 - \mu^2)r^\alpha + c_1[(\alpha + 1)^2 - \mu^2]r^{\alpha+1} + [c_2((\alpha + 2)^2 - \mu^2) + \lambda^2 c_0]r^{\alpha+2}$$
$$+ \cdots + [c_k((\alpha + k)^2 - \mu^2) + \lambda^2 c_{k-2}]r^{\alpha+k} + \cdots.$$

Each term in this power series must be zero in order for the equality to hold. Therefore, each coefficient of a term must be zero:

$$c_0(\alpha^2 - \mu^2) = 0$$
$$c_1((\alpha + 1)^2 - \mu^2) = 0$$
$$\vdots$$
$$c_k((\alpha + k)^2 - \mu^2) + \lambda^2 c_{k-2} = 0, \qquad k \geq 2.$$

As a bookkeeping agreement, we take $c_0 \neq 0$. Thus $\alpha = \pm \mu$. Let us study the case $\alpha = \mu \geq 0$. The second equation becomes

$$c_1((\mu + 1)^2 - \mu^2) = 0$$

and this implies $c_1 = 0$. Now in general the relation

$$c_k = - \frac{\lambda^2 c_{k-2}}{(\mu + k)^2 - \mu^2} = -\lambda^2 \frac{c_{k-2}}{k(2\mu + k)}, \qquad k \geq 2 \qquad (3)$$

says that c_k can be found from c_{k-2}. In particular, we find

$$c_2 = -\frac{\lambda^2}{2(2\mu + 2)} c_0$$

$$c_4 = -\frac{\lambda^2}{4(2\mu + 4)} c_2 = \frac{\lambda^4}{2 \cdot 4 \cdot (2\mu + 2)(2\mu + 4)} c_0$$

and so forth. All cs with odd index are zero, since they are all multiples of c_1. The general formula for a coefficient with even index $k = 2m$ is

$$c_{2m} = \frac{(-1)^m}{m!(\mu + 1)(\mu + 2) \cdots (\mu + m)} \left(\frac{\lambda}{2}\right)^{2m} c_0. \tag{4}$$

For integral values of μ, c_0 is chosen by convention to be

$$c_0 = \left(\frac{\lambda}{2}\right)^\mu \cdot \frac{1}{\mu!}.$$

Then the solution of Eq. (1) that we have found is called the *Bessel function of the first kind of order* μ:

$$J_\mu(\lambda r) = \left[\frac{\lambda r}{2}\right]^\mu \sum_{m=0}^\infty \frac{(-1)^m}{m! \, (\mu + m)!} \left[\frac{\lambda r}{2}\right]^{2m}. \tag{5}$$

There must be a second independent solution of Bessel's equation, which can be found by using variation of parameters. This method yields a solution in the form

$$J_\mu(\lambda r) \cdot \int \frac{dr}{r J_\mu{}^2(\lambda r)}. \tag{6}$$

In its standard form, the second solution of Bessel's equation is called the *Bessel function of second kind of order* μ and is denoted by $Y_\mu(\lambda r)$.

The most important feature of the second solution is its behavior near $r = 0$. When r is very small, we can approximate $J_\mu(\lambda r)$ by the first term of its series expansion:

$$J_\mu(\lambda r) \simeq \left(\frac{\lambda}{2}\right)^\mu \frac{1}{\mu!} r^\mu, \qquad r \ll 1.$$

The solution Eq. (6) then can be approximated by

$$\text{constant} \times r^\mu \int \frac{dr}{r^{1+2\mu}} = \text{constant} \times \begin{cases} \ln r, & \text{if} \quad \mu = 0 \\ r^{-\mu}, & \text{if} \quad \mu > 0. \end{cases}$$

In either case, it is easy to see that

$$|Y_\mu(\lambda r)| \to \infty, \qquad \text{as} \quad r \to 0.$$

Both kinds of Bessel functions have an infinite number of zeros. That is, there are an infinite number of values of α (and β) for which

$$J_\mu(\alpha) = 0, \qquad Y_\mu(\beta) = 0.$$

Also, as $r \to \infty$, both $J_\mu(\lambda r)$ and $Y_\mu(\lambda r)$ tend to zero. In Fig. 5.9 are graphs of several Bessel functions, and in Table 5.1 are values of their zeros. Further information can be found in most books of tables.

Summary The differential equation

$$\frac{d}{dr}\left(r \frac{dR}{dr}\right) - \frac{\mu^2}{r} R + \lambda^2 r R = 0$$

is called Bessel's equation. Its general solution is

$$R(r) = A J_\mu(\lambda r) + B Y_\mu(\lambda r)$$

(*A* and *B* arbitrary constants). The functions J_μ and Y_μ are called Bessel functions of order μ of the first and second kinds, respectively. The Bessel function of the second kind is unbounded at the origin.

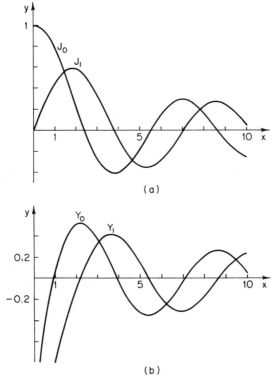

(a)

(b)

Figure 5.9 Graphs of Bessel functions.

Table 5.1   Zeros of Bessel functions[a]

m \ n	1	2	3	4
0	2.405	5.520	8.654	11.792
1	3.832	7.016	10.173	13.324
2	5.136	8.417	11.620	14.796
3	6.380	9.761	13.015	16.223

[a] The values α_{mn} satisfy the equation $J_m(\alpha_{mn}) = 0$.

Exercises

1. Find the values of the parameter λ for which the following problem has a nonzero solution:

$$\frac{1}{r}\frac{d}{dr}\left(r\frac{d\phi}{dr}\right) + \lambda^2\phi = 0, \qquad 0 < r < a$$

$$\phi(a) = 0, \qquad \phi(0) \text{ bounded.}$$

2. Sketch the first few eigenfunctions found in Exercise 1.

3. Show that

$$\frac{d}{dr}J_\mu(\lambda r) = \lambda J_\mu'(\lambda r)$$

where the prime denotes differentiation with respect to the argument.

4. Show from the series that

$$\frac{d}{dr}J_0(\lambda r) = -\lambda J_1(\lambda r).$$

5. By using Rolle's theorem, and knowing that $J_0(x) = 0$ for an infinite number of values of x, show that $J_1(x) = 0$ has an infinite number of solutions.

6. Using the infinite series representations for the Bessel functions, verify the formulas

$$\frac{d}{dx}(x^{-\mu}J_\mu(x)) = -x^{-\mu}J_{\mu+1}(x)$$

$$\frac{d}{dx}(x^\mu J_\mu(x)) = x^\mu J_{\mu-1}(x).$$

7. Use the second formula in Exercise 6 to derive the integral formula

$$\int x^\mu J_\mu(x) x \, dx = x^{\mu+1} J_{\mu+1}(x).$$

6 Temperature in a cylinder

In Section 4, we observed that both the heat and the wave equations have a great deal in common, especially the equilibrium solution and the eigenvalue problem. To reinforce that observation, we will solve a heat problem and a wave problem with analogous conditions so that the similar points may be seen. These examples illustrate another important point: that problems which would be two-dimensional in one coordinate system (rectangular) may become one-dimensional in another system (polar). In order to obtain this simplification, we will assume that the unknown function, $v(r, \theta, t)$, is actually independent of the angular coordinate θ. (We write $v(r, t)$ then.) As a consequence of this assumption, the two-dimensional Laplacian operator becomes

$$\nabla^2 v = \frac{1}{r} \frac{\partial}{\partial r} \left(r \frac{\partial v}{\partial r} \right).$$

Suppose that the temperature $v(r, t)$ in a large cylinder (radius a) satisfies the problem

$$\frac{1}{r} \frac{\partial}{\partial r} \left(r \frac{\partial v}{\partial r} \right) = \frac{1}{k} \frac{\partial v}{\partial t}, \qquad 0 < r < a, \quad 0 < t \tag{1}$$

$$v(a, t) = 0, \qquad 0 < t \tag{2}$$

$$v(r, 0) = f(r), \qquad 0 < r < a. \tag{3}$$

Since the differential equation (1) and boundary condition (2) are homogeneous, we may start the separation of variables by assuming $v(r, t) = \phi(r)T(t)$. Using this form for v, we find that the partial differential equation (1) becomes

$$\frac{1}{r} (r\phi')'T = \frac{1}{k} \phi T'.$$

After dividing through this equation by ϕT, we arrive at the equality

$$\frac{(r\phi')'}{r\phi} = \frac{T'}{kT}. \tag{4}$$

The two members of this equation must both be constant; call their mutual value $-\lambda^2$. Then we have two linked ordinary differential equations,

$$T' + \lambda^2 kT = 0, \qquad 0 < t \tag{5}$$

$$(r\phi')' + \lambda^2 r\phi = 0, \qquad 0 < r < a. \tag{6}$$

The boundary condition, Eq. (2), becomes $\phi(a)T(t) = 0$, $0 < t$. It will be satisfied by requiring that

$$\phi(a) = 0. \tag{7}$$

We can recognize Eq. (6) as Bessel's equation with $\mu = 0$. (See Summary, Section 5.) The general solution, therefore, has the form

$$\phi(r) = A J_0(\lambda r) + B Y_0(\lambda r).$$

If $B \neq 0$, $\phi(r)$ must become infinite as r approaches zero. The physical implications of this possibility are unacceptable, so we require that $B = 0$. In effect we have added the boundedness condition

$$|v(r, t)| \qquad \text{bounded at } r = 0, \tag{8}$$

which we shall employ frequently.

The function $\phi(r) = J_0(\lambda r)$ is a solution of (6), and we wish to choose λ so that (7) is satisfied. Then we must have

$$J_0(\lambda a) = 0$$

or

$$\lambda_n = \alpha_n / a \qquad n = 1, 2, \ldots$$

where α_n are the zeros of the function J_0. Thus the eigenfunctions and eigenvalues of (6), (7), and (8) are

$$\phi_n = J_0(\lambda_n r), \qquad \lambda_n{}^2 = (\alpha_n / a)^2. \tag{9}$$

Returning to Eq. (5), we determine that the time factors T_n are

$$T_n(t) = \exp(-\lambda_n{}^2 k t).$$

We may now assemble the general solution of the partial differential equation (1), under the boundary condition (2) and boundedness condition (8), as a general linear combination of our product solutions:

$$v(r, t) = \sum a_n J_0(\lambda_n r) \exp(-\lambda_n{}^2 k t). \tag{10}$$

It remains to determine the coefficients a_n so as to satisfy the initial condition (3), which now takes the form

$$v(r, 0) = \sum a_n J_0(\lambda_n r) = f(r), \qquad 0 < r < a. \tag{11}$$

While this problem is not a routine exercise in Fourier series, nor even a regular Sturm–Liouville problem (see Section 7 of Chapter 2, and especially Exercise 6 there), it is nevertheless true that the eigenfunctions of Eqs. (6) and (7) are orthogonal, as expressed by the relation

$$\int_0^a \phi_n(r) \phi_m(r) r \, dr = 0 \qquad (n \neq m)$$

or

$$\int_0^a J_0(\lambda_n r) J_0(\lambda_m r) r \, dr = 0 \qquad (n \neq m).$$

More importantly, the following theorem gives us justification for Eq. (11).

Theorem If $f(r)$ is sectionally smooth on the interval $0 < r < a$, then at every point r on that interval,

$$\sum_{n=1}^{\infty} a_n J_0(\lambda_n r) = \frac{f(r+) + f(r-)}{2}, \qquad 0 < r < a, \tag{12}$$

where the λ_n are solutions of $J_0(\lambda a) = 0$, and

$$a_n = \frac{\int_0^a f(r) J_0(\lambda_n r) r \, dr}{\int_0^a J_0^2(\lambda_n r) r \, dr}. \tag{13}$$

Now we may proceed with the problem at hand. If the function $f(r)$ in the initial condition (3) is sectionally smooth, the use of Eq. (13) to choose the coefficients a_n guarantees that Eq. (11) is satisfied (as nearly as possible) and hence the function

$$v(r, t) = \sum a_n J_0(\lambda_n r) \exp(-\lambda_n^2 k t)$$

satisfies the problem expressed by Eqs. (1), (2), (3), and (8).

By way of example, let us suppose that the function $f(r) = T_0$, $0 < r < a$. It is necessary to determine the coefficients a_n by formula (13). The numerator is the integral

$$\int_0^a T_0 J_0(\lambda_n r) r \, dr.$$

This integral is evaluated by means of the relation (see Exercise 6 of Section 5)

$$\frac{d}{dx}(x J_1(x)) = x J_0(x). \tag{14}$$

Hence, we find

$$\int_0^a J_0(\lambda_n r) r \, dr = \frac{1}{\lambda_n} r J_1(\lambda_n r) \Big|_0^a$$

$$= \frac{a}{\lambda_n} J_1(\lambda_n a) = \frac{a^2}{\alpha_n} J_1(\alpha_n). \tag{15}$$

The denominator of Eq. (13) is known to have the value (Exercise 5)

$$\int_0^a J_0^2(\lambda_n r)r\,dr = \frac{a^2}{2}J_1^2(\lambda_n a)$$

$$= \frac{a^2}{2}J_1^2(\alpha_n). \tag{16}$$

Putting together the numerator and denominator from Eqs. (15) and (16), we find that the coefficients we need are

$$a_n = \frac{2T_0}{\alpha_n J_1(\alpha_n)}. \tag{17}$$

Thus, the solution to the heat conduction problem is

$$v(r, t) = T_0 \sum_{n=1}^{\infty} \frac{2}{\alpha_n J_1(\alpha_n)} J_0(\lambda_n r) \exp(-\lambda_n^2 kt). \tag{18}$$

In Table 5.2 are listed the first few values of the ratio $2/[\alpha_n J_1(\alpha_n)]$.

Table 5.2

n	α_n	$J_1(\alpha_n)$	$2/[\alpha_n J_1(\alpha_n)]$
1	2.405	+0.5191	+1.602
2	5.520	−0.3403	−1.065
3	8.654	+0.2715	+0.8512
4	11.792	−0.2325	−0.7295

Exercises

1. Use Eq. (18) to find an expression for the function $v(0, t)/T_0$. Evaluate the function for

$$\frac{kt}{a^2} = 0.1, 0.2, 0.3.$$

(The first two terms of the series are sufficient.)

2. Write out the first three terms of the series in Eq. (18).

3. Use Exercise 7 of Section 5 to validate Eq. (16).

4. Let $\phi(r) = J_0(\lambda r)$, so that $\phi(r)$ satisfies Bessel's equation of order 0. Multiply through the differential equation by $r\phi'$ and conclude that

$$\frac{d}{dr}[(r\phi')^2] + \lambda^2 r^2 \frac{d}{dr}[\phi^2] = 0.$$

5. Assuming that λ is chosen so that $\phi(a) = 0$, integrate the equation above over the interval $0 < r < a$ to find

$$\int_0^a \phi^2(r)r \, dr = \frac{1}{2\lambda^2} \, (a\phi'(a))^2.$$

7 Vibrations of a circular membrane

We shall now attempt to solve the problem of describing the displacement of a circular membrane that is fixed at its edge. To begin with, we treat the simple case in which the initial conditions are independent of θ. Thus the displacement $v(r, t)$ satisfies the problem

$$\frac{1}{r} \frac{\partial}{\partial r} \left(r \frac{\partial v}{\partial r} \right) = \frac{1}{c^2} \frac{\partial^2 v}{\partial t^2}, \qquad 0 < r < a, \quad 0 < t \tag{1}$$

$$v(a, t) = 0, \qquad\qquad 0 < t \tag{2}$$

$$v(r, 0) = f(r), \qquad\quad 0 < r < a \tag{3}$$

$$\frac{\partial v}{\partial t}(r, 0) = g(r), \qquad 0 < r < a. \tag{4}$$

We start immediately with separation of variables, assuming $v(r, t) = \phi(r)T(t)$. The differential equation (1) becomes

$$\frac{1}{r} \, (r\phi')'T = \frac{1}{c^2} \, \phi T''$$

and the variables may be separated by dividing by ϕT. Then we find

$$\frac{(r\phi')'}{r\phi} = \frac{T''}{c^2 T}.$$

The two sides must both be equal to a constant (say $-\lambda^2$) yielding two linked, ordinary differential equations

$$T'' + \lambda^2 c^2 T = 0, \qquad 0 < t \tag{5}$$

$$(r\phi')' + \lambda^2 r\phi = 0, \qquad 0 < r < a. \tag{6}$$

The boundary condition Eq. (2) is satisfied if

$$\phi(a) = 0. \tag{7}$$

Of course, since $r = 0$ is a singular point of the differential equation (6), we add the requirement

$$|\phi(r)| \qquad \text{bounded at } r = 0, \tag{8}$$

which is equivalent to requiring that $|v(r, t)|$ be bounded at $r = 0$. We recognize that Eq. (6) is Bessel's equation, of which the function $\phi(r) = J_0(\lambda r)$ is the solution bounded at $r = 0$. In order to satisfy the boundary condition Eq. (7), we must have

$$J_0(\lambda a) = 0$$

or

$$\lambda_n = \frac{\alpha_n}{a}, \qquad n = 1, 2, \ldots \tag{9}$$

where α_n are the zeros of the function J_0. Thus the eigenfunctions and eigenvalues of Eqs. (6)–(8) are

$$\phi_n = J_0(\lambda_n r), \qquad \lambda_n^2 = \left(\frac{\alpha_n}{a}\right)^2.$$

The rest of our problem can now be dispatched easily. Returning to Eq. (5), we see that

$$T_n(t) = a_n \cos \lambda_n ct + b_n \sin \lambda_n ct,$$

and then for each $n = 1, 2, \ldots$ we have a solution of Eqs. (1), (2), and (8)

$$v_n(r, t) = \phi_n(r)T_n(t).$$

The most general linear combination of the v_n would be

$$v(r, t) = \sum J_0(\lambda_n r)[a_n \cos \lambda_n ct + b_n \sin \lambda_n ct]. \tag{10}$$

The initial conditions Eqs. (3) and (4) are satisfied if

$$v(r, 0) = \sum a_n J_0(\lambda_n r) = f(r), \qquad 0 < r < a$$

$$\frac{\partial v}{\partial t}(r, 0) = \sum b_n \lambda_n cJ_0(\lambda_n r) = g(r), \qquad 0 < r < a.$$

As in the preceding section, the coefficients of these series are to be found through the integral formulas

$$a_n = \int_0^a f(r)J_0(\lambda_n r)r \, dr/I_n$$

$$b_n = \int_0^a g(r)J_0(\lambda_n r)r \, dr/(\lambda_n cI_n)$$

$$I_n = \int_0^a [J_0(\lambda_n r)]^2 r \, dr.$$

With the coefficients determined by these formulas, the function given in Eq. (10) is the solution to the vibrating membrane problem that we started with.

Having seen the simplest case of the vibrations of a circular membrane, we return to the more general case. The full problem was

$$\frac{1}{r}\frac{\partial}{\partial r}\left(r\frac{\partial u}{\partial r}\right) + \frac{1}{r^2}\frac{\partial^2 u}{\partial\theta^2} = \frac{1}{c^2}\frac{\partial^2 u}{\partial t^2}, \qquad 0 < r < a, \quad 0 < t, \quad -\pi < \theta \le \pi. \tag{11}$$

$$u(a, \theta, t) = 0, \qquad 0 < t, \quad -\pi < \theta \le \pi \tag{12}$$

$$|u(0, \theta, t)| \quad \text{bounded}, \qquad 0 < t, \quad -\pi < \theta \le \pi \tag{13}$$

$$u(r, -\pi, t) = u(r, \pi, t), \qquad 0 < r < a, \quad 0 < t$$

$$\frac{\partial u}{\partial\theta}(r, -\pi, t) = \frac{\partial u}{\partial\theta}(r, \pi, t), \qquad 0 < r < a, \quad 0 < t \tag{14}$$

$$u(r, \theta, 0) = f(r, \theta), \qquad 0 < r < a, \quad -\pi < \theta \le \pi \tag{15}$$

$$\frac{\partial u}{\partial t}(r, \theta, 0) = g(r, \theta), \qquad 0 < r < a, \quad -\pi < \theta \le \pi. \tag{16}$$

Condition (13) has been added because $r = 0$ is a singular point of the partial differential equation.

Following the procedure suggested in Section 4, we assume that u has the product form

$$u = \phi(r, \theta)T(t)$$

and we find that the Eq. (11) separates into two linked equations

$$T'' + \lambda^2 c^2 T = 0, \qquad 0 < t \tag{17}$$

$$\frac{1}{r}\frac{\partial}{\partial r}\left(r\frac{\partial\phi}{\partial r}\right) + \frac{1}{r^2}\frac{\partial^2\phi}{\partial\theta^2} = -\lambda^2\phi, \qquad 0 < r < a, \quad -\pi < \theta \le \pi. \tag{18}$$

If we separate variables of the function ϕ by assuming $\phi(r, \theta) = R(r)\Theta(\theta)$, Eq. (18) takes the form

$$\frac{1}{r}(rR')'\Theta + \frac{1}{r^2}R\Theta'' = -\lambda^2 R\Theta.$$

The variables will separate if we multiply by r^2 and divide by $R\Theta$. Then the equation above may be put in the form

$$\frac{r(rR')'}{R} + \lambda^2 r^2 = -\frac{\Theta''}{\Theta} = \mu^2.$$

Finally we get two problems for R and θ:

$$\Theta'' + \mu^2\Theta = 0, \qquad -\pi < \theta \leq \pi$$

$$\Theta(-\pi) = \Theta(\pi) \tag{19}$$

$$\Theta'(-\pi) = \Theta'(\pi)$$

$$(rR')' - \frac{\mu^2}{r} R + \lambda^2 rR = 0, \qquad 0 < r < a$$

$$|R(0)| \quad \text{bounded} \tag{20}$$

$$R(a) = 0.$$

As we observed before, the problem (19) has the solutions

$$\mu^2 = 0, \qquad \Theta_0 = 1$$
$$\mu^2 = m^2, \qquad \Theta_m = \cos m\theta \quad \text{or} \quad \sin m\theta, \qquad m = 1, 2, 3, \dots.$$

Also the differential equation of Eq. (20) will be recognized as Bessel's equation, the general solution of which is (using $\mu = m$)

$$R(r) = CJ_m(\lambda r) + DY_m(\lambda r).$$

In order for the boundedness condition in Eq. (20) to be fulfilled, D must be zero. Then we are left with

$$R(r) = J_m(\lambda r).$$

(Since any multiple of a solution is another solution, we can drop the constant C.) The boundary condition of Eq. (20) becomes

$$R(a) = J_m(\lambda a) = 0$$

implying that λa must be a root of the equation

$$J_m(\alpha) = 0.$$

(See Table 5.1.) For each fixed integer m,

$$\alpha_{m1}, \qquad \alpha_{m2}, \qquad \alpha_{m3}, \qquad \cdots$$

are the first, second, third, $\dots$, solutions of the equation above. The values of λ for which $J_m(\lambda r)$ solves the differential equation and satisfies the boundary condition are

$$\lambda_{mn} = \frac{\alpha_{mn}}{a}, \qquad m = 0, 1, 2, \dots, \qquad n = 1, 2, 3, \dots.$$

Now that the functions R and Θ are determined, we can construct ϕ. For $m = 1, 2, 3, \ldots$ and $n = 1, 2, 3, \ldots$, both of the functions

$$J_m(\lambda_{mn} r) \cos m\theta, \qquad J_m(\lambda_{mn} r) \sin m\theta \qquad (21)$$

are solutions of the problem Eq. (18), both corresponding to the same eigenvalue λ_{mn}^2. For $m = 0$ and $n = 1, 2, 3, \ldots$, we have the functions

$$J_0(\lambda_{0n} r) \qquad (22)$$

which correspond to the eigenvalues λ_{0n}^2. (Compare with the simple case.) The function $T(t)$ which is a solution of Eq. (17) is any combination of $\cos \lambda_{mn} ct$ and $\sin \lambda_{mn} ct$.

Now the solutions of Eqs. (11)–(14) have any of the forms

$$J_m(\lambda_{mn} r) \cos m\theta \cos \lambda_{mn} ct, \qquad J_m(\lambda_{mn} r) \sin m\theta \cos \lambda_{mn} ct$$
$$J_m(\lambda_{mn} r) \cos m\theta \sin \lambda_{mn} ct, \qquad J_m(\lambda_{mn} r) \sin m\theta \sin \lambda_{mn} ct \qquad (23)$$

for $m = 1, 2, 3, \ldots$ and $n = 1, 2, 3, \ldots$. In addition, there is the special case $m = 0$, for which solutions have the form

$$J_0(\lambda_{0n} r) \cos \lambda_{0n} ct, \qquad J_0(\lambda_{0n} r) \sin \lambda_{0n} ct. \qquad (24)$$

The general solution of the problem Eqs. (11)–(14) will thus have the form of a linear combination of the solutions above. We shall use several series to form the combination:

$$u(r, \theta, t) = \sum_n a_{0n} J_0(\lambda_{0n} r) \cos \lambda_{0n} ct + \sum_{m,n} a_{mn} J_m(\lambda_{mn} r) \cos m\theta \cos \lambda_{mn} ct$$

$$+ \sum_{m,n} b_{mn} J_m(\lambda_{mn} r) \sin m\theta \cos \lambda_{mn} ct$$

$$+ \sum_n A_{0n} J_0(\lambda_{0n} r) \sin \lambda_{0n} ct + \sum_{m,n} A_{mn} J_m(\lambda_{mn} r) \cos m\theta \sin \lambda_{mn} ct$$

$$+ \sum_{m,n} B_{mn} J_m(\lambda_{mn} r) \sin m\theta \sin \lambda_{mn} ct. \qquad (25)$$

When $t = 0$, the last three sums disappear, and the cosines of t in the first three sums are all equal to 1. Thus

$$u(r, \theta, 0) = \sum_n a_{0n} J_0(\lambda_{0n} r) + \sum_{m,n} a_{mn} J_m(\lambda_{mn} r) \cos m\theta + \sum_{m,n} b_{mn} J_m(\lambda_{mn} r) \sin m\theta$$

$$= f(r, \theta), \qquad 0 < r < a, \quad -\pi < \theta \le \pi. \qquad (26)$$

We expect to fulfill this equality by choosing the as and bs according to some orthogonality principle. Since each function present in the series is an eigenfunction of the problem

$$\nabla^2 \phi = -\lambda^2 \phi, \qquad 0 < r < a, \quad -\pi < \theta \le \pi$$

$$\phi(a, \theta) = 0, \qquad -\pi < \theta \le \pi$$

$$\phi(r, -\pi) = \phi(r, \pi), \qquad \frac{\partial \phi}{\partial \theta}(r, -\pi) = \frac{\partial \phi}{\partial \theta}(r, \pi), \qquad 0 < r < a$$

we expect it to be orthogonal to each of the others (see Section 4, Exercise 7). This is indeed true: any function from one series is orthogonal to all of the functions in the other series, and also to the rest of the functions in its own series. To illustrate this orthogonality, we have

$$\iint_R J_0(\lambda_{0n} r) J_m(\lambda_{mn} r) \cos m\theta \, dA$$

$$= \int_0^a J_0(\lambda_{0n} r) J_m(\lambda_{mn} r) \int_{-\pi}^{\pi} \cos m\theta \, d\theta \, r \, dr = 0, \qquad m \ne 0. \quad (27)$$

There are two other relations like this one involving functions from two different series.

We already know that the functions within the first series are orthogonal to each other:

$$\int_{-\pi}^{\pi} \int_0^a J_0(\lambda_{0n} r) J_0(\lambda_{0q} r) r \, dr \, d\theta = 0, \qquad n \ne q.$$

Within the second series we must show that, if $m \ne p$ or $n \ne q$, then

$$0 = \int_{-\pi}^{\pi} \int_0^a J_m(\lambda_{mn} r) \cos m\theta \, J_p(\lambda_{pq} r) \cos p\theta \, r \, dr \, d\theta. \quad (28)$$

(Recall that $r \, dr \, d\theta = dA$ in polar coordinates.) Integrating with respect to θ first, we see that the integral must be zero if $m \ne p$, by the orthogonality of $\cos m\theta$ and $\cos p\theta$. If $m = p$, the integral above becomes

$$\pi \int_0^a J_m(\lambda_{mn} r) J_m(\lambda_{mq} r) r \, dr$$

after the integration with respect to θ. Finally, if $n \ne q$, this integral is zero; the demonstration follows the same lines as the usual Sturm–Liouville proof. Thus the functions within the second series are shown orthogonal to each other. For the functions of the last series, the proof of orthogonality is similar.

Equipped now with an orthogonality relation, we can determine formulas for the as and bs. For instance,

$$a_{0n} = \frac{\int_0^{2\pi} \int_0^a f(r, \theta) J_0(\lambda_{0n} r) r \, dr \, d\theta}{2\pi \int_0^a J_0{}^2(\lambda_{0n} r) \, r \, dr}. \quad (29)$$

The As and Bs are calculated from the second initial condition.

It should now be clear that, while the computation of the solution to the original problem is possible in theory, it will be very painful in practice. Worse yet, the final form of the solution Eq. (25) does not give a clear idea of what u looks like. All is not wasted, however. We can say, from an examination of the λs, that the tone produced is not musical—that is, u is not periodic in t. Also we can sketch some of the fundamental modes of vibration of the membrane corresponding to some low eigenvalues (Fig. 5.10). The curves represent points for which displacement is zero in that mode (nodal curves).

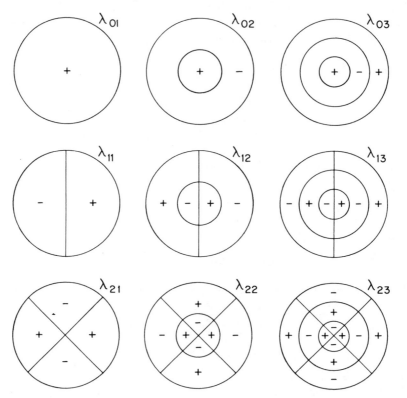

Figure 5.10 Nodal curves: The curves in these graphs represent solutions of $\phi_{mn}(r, \theta)$ $= 0$. Adjacent regions bulge up or down, according to the sign. Only those ϕs containing the factor $\cos m\theta$ have been used.

Exercises

1. Verify that each of the functions in the series Eq. (10) satisfies Eqs. (1), (2), and (8).

2. Derive the formulas for the as and bs of Eq. (10).

3. List the five lowest frequencies of vibration of a circular membrane.

4. Sketch the function $J_0(\lambda_n r)$, for $n = 1, 2, 3$.

5. What boundary conditions must the function ϕ of Eq. (18) satisfy?

6. Justify the derivation of Eqs. (19) and (20) from Eqs. (12)–(14) and (18).

7. Show that

$$\int_0^a J_m(\lambda_{mn} r) J_m(\lambda_{mq} r) r \, dr = 0, \qquad n \neq q$$

if

$$J_m(\lambda_{ms} a) = 0, \qquad s = 1, 2, \dots .$$

8. Sketch the nodal curves of the eigenfunctions Eq. (21) corresponding to $\lambda_{31}, \lambda_{32}$, and λ_{33}.

8 Some applications of Bessel functions

After the elementary functions, the Bessel functions are among the most useful in engineering and physics. One reason for their usefulness is the fact that they solve a fairly general differential equation. The general solution of

$$\phi'' + \frac{1 - 2\alpha}{x} \phi' + \left[(\lambda\gamma x^{\gamma - 1})^2 - \frac{p^2\gamma^2 - \alpha^2}{x^2} \right] \phi = 0 \tag{1}$$

is

$$\phi(x) = x^\alpha [A J_p(\lambda x^\gamma) + B Y_p(\lambda x^\gamma)].$$

Below are several problems in which the Bessel functions play an important role. The details of separation of variables, which should now be routine, are kept to a minimum.

A Laplace's equation in a cylinder

The steady-state temperature distribution in a circular cylinder with insulated surface is determined by the problem

$$\frac{1}{r} \frac{\partial}{\partial r} \left(r \frac{\partial u}{\partial r} \right) + \frac{\partial^2 u}{\partial z^2} = 0, \qquad 0 < r < a, \quad 0 < z < b \tag{2}$$

$$\frac{\partial u}{\partial r} (a, z) = 0, \qquad 0 < z < b \tag{3}$$

$$u(r, 0) = f(r), \qquad 0 < r < a \tag{4}$$

$$u(r, b) = g(r), \qquad 0 < r < a. \tag{5}$$

Here we are considering the boundary conditions to be independent of θ, so that u is independent of θ also.

Assuming that $u = R(r)Z(z)$ we find that

$$(rR')' + \lambda^2 rR = 0, \qquad 0 < r < a \tag{6}$$

$$R'(a) = 0 \tag{7}$$

$$|R(0)| \quad \text{bounded} \tag{8}$$

$$Z'' - \lambda^2 Z = 0. \tag{9}$$

Condition (8) has been added because $r = 0$ is a singular point. The solution of Eqs. (6)–(8) is

$$R_n(r) = J_0(\lambda_n r) \tag{10}$$

where the eigenvalues $\lambda_n{}^2$ are defined by the solutions of

$$R'(a) = \lambda J_0'(\lambda a) = 0. \tag{11}$$

Since $J_0' = -J_1$, the λs are related to the zeros of J_1. The first three eigenvalues are: 0, $(3.832/a)^2$, and $(7.016/a)^2$. Note that $R(0) = J_0(0) = 1$.

The solution of the problem Eqs. (2)–(5) may be put in the form

$$u(r, z) = a_0 + b_0 z + \sum_{n=1}^{\infty} J_0(\lambda_n r) \left[a_n \frac{\sinh \lambda_n z}{\sinh \lambda_n b} + b_n \frac{\sinh \lambda_n(b - z)}{\sinh \lambda_n b} \right]. \tag{12}$$

The as and bs are determined from Eqs. (4) and (5) by using the orthogonality relation

$$\int_0^a J_0(\lambda_n r) J_0(\lambda_m r) r \, dr = 0, \qquad n \neq m.$$

B Spherical waves

In spherical (ρ, θ, ϕ) coordinates, the Laplacian operator ∇^2 becomes

$$\nabla^2 u = \frac{1}{\rho^2} \frac{\partial}{\partial \rho} \left(\rho^2 \frac{\partial u}{\partial \rho} \right) + \frac{1}{\rho^2 \sin \phi} \frac{\partial}{\partial \phi} \left(\sin \phi \frac{\partial u}{\partial \phi} \right) + \frac{1}{\rho^2 \sin^2 \phi} \frac{\partial^2 u}{\partial \theta^2}.$$

Consider a wave problem in a sphere when the initial conditions depend only on the radial coordinate ρ:

$$\frac{1}{\rho^2} \frac{\partial}{\partial \rho} \left(\rho^2 \frac{\partial u}{\partial \rho} \right) = \frac{1}{c^2} \frac{\partial^2 u}{\partial t^2}, \qquad 0 < \rho < a, \quad 0 < t \tag{13}$$

$$u(a, t) = 0, \qquad 0 < t \tag{14}$$

$$u(\rho, 0) = f(\rho), \qquad 0 < \rho < a \tag{15}$$

$$\frac{\partial u}{\partial t}(\rho, 0) = g(\rho), \qquad 0 < \rho < a. \tag{16}$$

Assuming $u(\rho, t) = R(\rho)T(t)$, we separate variables and find

$$T'' + \lambda^2 c^2 T = 0 \qquad (17)$$

$$(\rho^2 R')' + \lambda^2 \rho^2 R = 0, \qquad 0 < \rho < a \qquad (18)$$

$$R(a) = 0 \qquad (19)$$

$$|R(0)| \quad \text{bounded.} \qquad (20)$$

Again, the condition (20) has been added because $\rho = 0$ is a singular point. Equation (18) may be put into the form

$$R'' + \frac{2}{\rho} R' + \lambda^2 R = 0$$

and comparison with Eq. (1) shows that $\alpha = -\frac{1}{2}$, $\gamma = 1$, and $p = \frac{1}{2}$; thus the general solution of Eq. (18) is

$$R(\rho) = \rho^{-1/2}[AJ_{1/2}(\lambda\rho) + BY_{1/2}(\lambda\rho)].$$

We know that near $\rho = 0$,

$$J_{1/2}(\lambda\rho) \sim \text{const} \times \rho^{1/2}$$

$$Y_{1/2}(\lambda\rho) \sim \text{const} \times \rho^{-1/2}.$$

Thus in order to satisfy Eq. (20), we must have $B = 0$. It is possible to show that

$$J_{1/2}(\lambda\rho) = \frac{2}{\pi} \frac{\sin \lambda\rho}{\sqrt{\lambda\rho}}, \qquad Y_{1/2}(\lambda\rho) = -\frac{2}{\pi} \frac{\cos \lambda\rho}{\sqrt{\lambda\rho}}.$$

Our solution to Eqs. (18) and (20) is therefore

$$R(\rho) = \frac{\sin \lambda\rho}{\rho} \qquad (21)$$

and Eq. (19) is satisfied if $\lambda_n^2 = (n\pi/a)^2$. The solution of the problem Eqs. (13)–(16) can be written in the form

$$u(\rho, t) = \sum_{n=1}^{\infty} \frac{\sin \lambda_n \rho}{\rho} [a_n \cos \lambda_n ct + b_n \sin \lambda_n ct]. \qquad (22)$$

The as and bs are, as usual, chosen so that the initial conditions Eqs. (15) and (16) are satisfied.

C Pressure in a bearing

The pressure in the lubricant inside a plane pad bearing satisfies the problem

$$\frac{\partial}{\partial x}\left(x^3 \frac{\partial p}{\partial x}\right) + x^3 \frac{\partial^2 p}{\partial y^2} = -1, \qquad a < x < b, \quad -c < y < c \qquad (23)$$

$$p(a, y) = 0, \qquad p(b, y) = 0, \qquad -c < y < c \qquad (24)$$

$$p(x, -c) = 0, \qquad p(x, c) = 0, \qquad a < x < b. \qquad (25)$$

(Here a and c are positive constants, and $b = a + 1$.) Equation (23) is elliptic and nonhomogeneous. To reduce this equation to a more familiar one, let $d(x, y) = v(x) + u(x, y)$, where $v(x)$ satisfies the problem

$$(x^3 v')' = -1, \qquad a < x < b \tag{26}$$

$$v(a) = 0, \qquad v(b) = 0. \tag{27}$$

Then, when v is found, u must be the solution of the problem

$$\frac{\partial}{\partial x}\left(x^3 \frac{\partial u}{\partial x}\right) + x^3 \frac{\partial^2 u}{\partial y^2} = 0, \qquad a < x < b, \qquad -c < y < c \tag{28}$$

$$u(a, y) = 0, \qquad u(b, y) = 0, \qquad -c < y < c \tag{29}$$

$$u(x, \pm c) = -v(x), \qquad a < x < b \tag{30}$$

If we now assume that $u(x, y) = X(x)Y(y)$, the variables can be separated:

$$(x^3 X')' + \lambda^2 x^3 X = 0, \qquad a < x < b \tag{31}$$

$$X(a) = 0, \qquad X(b) = 0 \tag{32}$$

$$Y'' - \lambda^2 Y = 0, \qquad -c < y < c. \tag{33}$$

Equation (31) may be put in the form

$$X'' + \frac{3}{x} X' + \lambda^2 X = 0, \qquad a < x < b.$$

By comparing to Eq. (1) we find that $\alpha = -1$, $\gamma = 1$, and $p = 1$, and that the general solution of Eq. (31) is

$$X(x) = \frac{1}{x}(AJ_1(\lambda x) + BY_1(\lambda x)).$$

Since the point $x = 0$ is not included in the interval $a < x < b$, there is no problem with boundedness. Instead we must satisfy the boundary conditions Eq. (32), which after some algebra have the form

$$AJ_1(\lambda a) + BY_1(\lambda a) = 0, \qquad AJ_1(\lambda b) + BY_1(\lambda b) = 0.$$

Not both A and B may be zero, so the determinant of these simultaneous equations must be zero:

$$J_1(\lambda a)Y_1(\lambda b) - J_1(\lambda b)Y_1(\lambda a) = 0.$$

Some solutions of the equation are tabulated for various values of b/a. For instance, if $b/a = 2.5$, the first three eigenvalues λ^2 are

$$\left(\frac{2.156}{a}\right)^2, \qquad \left(\frac{4.223}{a}\right)^2, \qquad \left(\frac{6.307}{a}\right)^2.$$

We now can take X_n to be

$$X_n(x) = \frac{1}{x} (Y_1(\lambda_n a) J_1(\lambda_n x) - J_1(\lambda_n a) Y_1(\lambda_n x)) \tag{34}$$

and the solution of Eqs. (28)–(30) has the form

$$u(x, y) = \sum_{n=1}^{\infty} a_n X_n(x) \frac{\cosh \lambda_n y}{\cosh \lambda_n c}. \tag{35}$$

The as are chosen to satisfy the boundary conditions Eq. (30), using the orthogonality principle

$$\int_a^b X_n(x) X_m(x) x^3 \, dx = 0, \qquad n \neq m.$$

Notice that Eqs. (31) and (32) make up a regular Sturm–Liouville problem.

Exercises

1. Find the general solution of the differential equation
    $$(x^n \phi')' + \lambda^2 x^n \phi = 0$$
    where $n = 0, 1, 2, \ldots$.

2. Find the solution of the equation in Exercise 1 which is bounded at $x = 0$.

3. Find the solutions of Eq. (9), including the case $\lambda^2 = 0$, and prove that Eq. (12) is a solution of Eqs. (2)–(4).

4. Show that any function of the form
    $$u(\rho, t) = \frac{1}{\rho} (\phi(\rho + ct) + \psi(\rho - ct))$$
    is a solution of Eq. (13) if ϕ and ψ have at least two derivatives.

5. Find functions ϕ and ψ, such that $u(\rho, t)$ as given in Exercise 4 satisfies Eqs. (14)–(16).

6. Give the formula for the as and bs in Eq. (12).

7. What is the orthogonality relation for the eigenfunctions of Eqs. (18)–(20)? Use it to find the as and bs in Eq. (22).

8. Sketch the first few eigenfunctions of Eqs. (18)–(20).

9. Find the function $v(x)$ which is the solution of Eqs. (26) and (27).

10. Use the technique of Example C to change the following problem into a potential problem:

$$\frac{\partial^2 u}{\partial x^2} + \frac{\partial^2 u}{\partial y^2} = -f(x), \qquad 0 < x < a, \quad 0 < y < b,$$

$u = 0$ on all boundaries.

11. Will the same technique work if $f(x)$ is replaced by $f(x, y)$?

12. Verify that Eqs. (31) and (32) is a regular Sturm–Liouville problem. Show the eigenfunctions' orthogonality by using the orthogonality of the Bessel functions.

13. Find a formula for the a_n of Eq. (35).

14. Verify that Eq. (34) is a solution of Eqs. (28)–(30).

9 Spherical coordinates. Legendre polynomials

After the Cartesian and cylindrical coordinate systems, the one most frequently encountered is the spherical system (Fig. 5.11), in which

$$x = \rho \cos \phi \cos \theta$$
$$y = \rho \cos \phi \sin \theta$$
$$z = \rho \sin \phi.$$

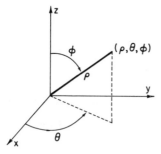

Figure 5.11 Spherical coordinates.

The variables are restricted by $0 \le \rho, 0 \le \theta < 2\pi, 0 \le \phi \le \pi$. In this coordinate system the Laplacian operator is

$$\nabla^2 u = \frac{1}{\rho^2} \left\{ \frac{\partial}{\partial \rho} \left(\rho^2 \frac{\partial u}{\partial \rho} \right) + \frac{1}{\sin \phi} \frac{\partial}{\partial \phi} \left(\sin \phi \frac{\partial u}{\partial \phi} \right) + \frac{1}{\sin^2 \phi} \frac{\partial^2 u}{\partial \theta^2} \right\}.$$

From what we have seen in other cases, we expect solvable problems in spherical coordinates to reduce to one of the following:

Problem 1 $\nabla^2 u = -\lambda^2 u$ in R, plus homogeneous boundary conditions.

Problem 2 $\nabla^2 u = 0$ in R, plus homogeneous boundary conditions on facing sides (where R is a generalized rectangle in spherical coordinates).

Problem 1 would come from a heat or wave equation after separating out the time variable. Problem 2 is a part of the potential problem.

The complete solution of either of these problems is very complicated, but a number of special cases are simple, important and not uncommon. We have already seen Problem 1 solved (Section 8) when u is a function of ρ only. A second important case is Problem 2, when u is independent of the variable θ. We shall state a complete boundary value problem and solve it by separation of variables.

$$\frac{1}{\rho^2}\left\{\frac{\partial}{\partial\rho}\left(\rho^2\frac{\partial u}{\partial\rho}\right) + \frac{1}{\sin\phi}\frac{\partial}{\partial\phi}\left(\sin\phi\frac{\partial u}{\partial\phi}\right)\right\} = 0, \qquad 0 < \rho < c, \quad 0 < \phi < \pi \quad (1)$$

$$u(c, \phi) = f(\phi), \qquad 0 < \phi < \pi. \qquad (2)$$

From the assumption $u(\rho, \phi) = R(\rho)\Phi(\phi)$, it follows that

$$\frac{(\rho^2 R')'}{R} + \frac{(\sin\phi\ \Phi')'}{\sin\phi\ \Phi} = 0.$$

Both terms are constant, and the second is negative, $-\mu^2$, because the boundary condition at $\rho = c$ will have to be satisfied by a linear combination of functions of ϕ. The separated equations are

$$(\rho^2 R')' - \mu^2 R = 0, \qquad 0 < \rho < c \qquad (3)$$

$$(\sin\phi\ \Phi')' + \mu^2 \sin\phi\ \Phi = 0, \qquad 0 < \phi < \pi. \qquad (4)$$

Neither equation has a boundary condition. But $\rho = 0$ is a singular point of the first equation, and both $\phi = 0$ and $\phi = \pi$ are singular points of the second equation. (At these points, the coefficient of the highest-order derivative is zero, while some other coefficient is nonzero.) At each of the singular points, we impose a boundedness condition:

$$R(0) \text{ bounded}, \qquad \Phi(0) \quad \text{and} \quad \Phi(\pi) \text{ bounded}.$$

The second equation can be simplified by the change of variables $x = \cos \phi$, $\Phi(\phi) = y(x)$. By the chain rule, the relevant derivatives are

$$\frac{d\Phi}{d\phi} = -\sin \phi \, \frac{dy}{dx}$$

$$\frac{d}{d\phi}\left(\sin \phi \, \frac{d\Phi}{d\phi}\right) = \sin^3 \phi \, \frac{d^2 y}{dx^2} - 2 \sin \phi \cos \phi \, \frac{dy}{dx}.$$

The differential equation becomes

$$\sin^2 \phi \, \frac{d^2 y}{dx^2} - 2 \cos \phi \, \frac{dy}{dx} + \mu^2 y = 0$$

or, in terms of x alone

$$(1 - x^2)y'' - 2xy' + \mu^2 y = 0, \qquad -1 < x < 1. \tag{5}$$

In addition, we require that $y(x)$ be bounded at $x = \pm 1$.

Solutions of the differential equation are usually found by the power series method. Assume that $y(x) = a_0 + a_1 x + \cdots + a_k x^k + \cdots$. The terms of the differential equation are then

$$y'' = 2a_2 + 3 \cdot 2a_3 x + 4 \cdot 3a_4 x^2 + \cdots + (k + 2)(k + 1)a_{k+2} x^k + \cdots$$
$$-x^2 y'' = \qquad\qquad\qquad -2a_2 x^2 - \cdots \qquad\qquad - k(k - 1)a_k x^k + \cdots$$
$$-2xy' = \qquad -2a_1 x \quad -2a_2 x^2 - \cdots \qquad\qquad -2ka_k x^k - \cdots$$
$$\mu^2 y = \mu^2 a_0 + \mu^2 a_1 x \quad + \mu^2 a_2 x^2 + \cdots \qquad\qquad\quad + \mu^2 a_k x^k + \cdots.$$

When this tableau is added vertically, the left-hand side is zero, according to the differential equation. The right-hand side adds up to a power series, each of whose coefficients must be zero. We therefore obtain the following relations

$$2a_2 + \mu^2 a_0 = 0$$
$$6a_3 + (\mu^2 - 2)a_1 = 0$$
$$(k + 2)(k + 1)a_{k+2} + [\mu^2 - k(k + 1)]a_k = 0.$$

The last formula actually includes the first two, apparently special, cases. We may write the general relation as

$$a_{k+2} = \frac{k(k + 1) - \mu^2}{(k + 2)(k + 1)} a_k$$

valid for $k = 0, 1, 2, \ldots$.

Suppose for the moment that μ^2 is given. A short calculation gives the first few coefficients.

$$a_2 = \frac{-\mu^2}{2} a_0, \qquad\qquad a_3 = \frac{2 - \mu^2}{6} a_1$$

$$a_4 = \frac{6 - \mu^2}{12} a_2, \qquad\qquad a_5 = \frac{12 - \mu^2}{20} a_3$$

$$= \frac{6 - \mu^2}{12} \frac{-\mu^2}{2} a_0, \qquad\qquad = \frac{12 - \mu^2}{20} \frac{2 - \mu^2}{6} a_1 .$$

It is clear that all the as with even index will be multiples of a_0 and those with odd index will be multiples of a_1. Thus $y(x)$ equals a_0 times an even function plus a_1 times an odd function, with both a_0 and a_1 arbitrary.

It is not difficult to prove that both the odd and even series produced by this process diverge at both $x = \pm 1$, for general μ^2. However, when μ^2 has one of the special values

$$\mu^2 = \mu_n{}^2 = n(n + 1), \qquad n = 0, 1, 2, \ldots$$

one of the two series turns out to have all zero coefficients after a_n. For instance, if $\mu^2 = 3 \cdot 4$, then $a_5 = 0$, and all subsequent coefficients with odd index are also zero. Hence, one of the solutions of

$$(1 - x^2)y'' - 2xy' + 12y = 0$$

is the polynomial $a_1(x - \frac{5}{3}x^3)$. The other solution is an even function unbounded at both $x = \pm 1$.

Now we see that the boundedness conditions can be satisfied only if μ^2 is one of the numbers $0, 2, 6, \ldots, n(n + 1), \ldots$. In such a case, one solution of the differential equation is a polynomial (naturally bounded at $x = \pm 1$). When normalized by the condition $y(1) = 1$, these are called Legendre polynomials, written $P_n(x)$. In Table 5.3 are the first five Legendre polynomials. Figure 5.12 shows their graphs.

Table 5.3

$$P_0(x) = 1$$
$$P_1(x) = x$$
$$P_2(x) = \tfrac{1}{2}(3x^2 - 1)$$
$$P_3(x) = \tfrac{1}{2}(5x^3 - 3x)$$
$$P_4(x) = \tfrac{1}{8}(35x^4 - 30x^2 + 3)$$

Since the differential equation (5) is easily put into self-adjoint form

$$((1 - x^2)y')' + \mu^2 y = 0, \qquad -1 < x < 1,$$

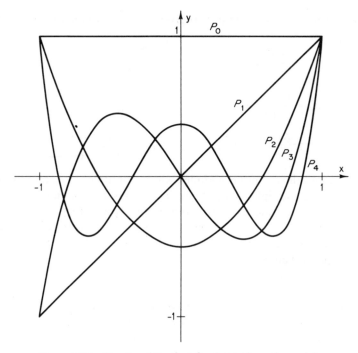

Figure 5.12 Graphs of the first five Legendre polynomials.

it is routine to show that the Legendre polynomials satisfy the orthogonality relation

$$\int_{-1}^{1} P_n(x)P_m(x)\, dx = 0, \qquad n \neq m.$$

By direct calculation, it can be shown that

$$\int_{-1}^{1} P_n^2(x)\, dx = \frac{2}{2n+1}. \tag{6}$$

A compact way of representing the Legendre polynomials is by means of Rodrigues' formula,

$$P_n(x) = \frac{1}{n!\, 2^n} \frac{d^n}{dx^n} [(x^2 - 1)^n]. \tag{7}$$

Elementary algebra and calculus show that the nth derivative of $(x^2 - 1)^n$ is a polynomial of degree n. Substituting this polynomial into the differential equation (5), with $\mu = n(n+1)$, shows that it is a solution—bounded, of course. Therefore, it is a multiple of the Legendre polynomial $P_n(x)$. Through

Rodrigues' formula or otherwise, it is possible to prove the following two formulas, which relate three consecutive Legendre polynomials:

$$(2n + 1)P_n(x) = P'_{n+1}(x) - P'_{n-1}(x) \tag{8}$$

$$(n + 1)P_{n+1}(x) + nP_{n-1}(x) = (2n + 1)xP_n(x). \tag{9}$$

We return now to the original problem. We have found the eigenfunctions of

$$(\sin \phi \Phi')' + \mu^2 \sin \phi \Phi = 0, \qquad 0 < \phi < \pi$$

to be $\Phi_n(\phi) = P_n(\cos \phi)$, corresponding to the eigenvalues $\mu_n^2 = n(n + 1)$. We must still solve Eq. (3) for R. After the differentiation has been carried out, the problem for R becomes

$$\rho^2 R_n'' + 2\rho R_n' - n(n + 1)R_n = 0, \qquad 0 < \rho < c$$
$$R_n \text{ bounded at } \rho = 0.$$

The equation is of the Euler–Cauchy type, solved by assuming $R = \rho^\alpha$ and determining α. Two solutions, ρ^n and $\rho^{-(n+1)}$, are found, of which the second is unbounded at $\rho = 0$. Hence $R_n = \rho^n$, and our product solutions of the potential equation have the form

$$u_n(\rho, \phi) = R_n(\rho)\Phi_n(\phi) = \rho^n P_n(\cos \phi).$$

The general solution of the partial differential equation which is bounded in the region $0 < \rho < c, 0 < \phi < \pi$ is thus the linear combination

$$u(\rho, \phi) = \sum_0^\infty b_n \rho^n P_n(\cos \phi).$$

At $\rho = c$, the boundary condition becomes

$$u(c, \phi) = \sum_0^\infty b_n c^n P_n(\cos \phi) = f(\phi), \qquad 0 < \phi < \pi.$$

The coefficients b_n are then found, by using the orthogonality relation for the functions $P_n(\cos \phi)$, to be

$$b_n = \frac{2n + 1}{2c^n} \int_0^\pi f(\phi)P_n(\cos \phi) \sin \phi \, d\phi. \tag{10}$$

For a number of simple functions f, these coefficients can be found explicitly.

The Legendre polynomials $P_n(\cos \phi)$ are often called zonal harmonics because their nodal lines (loci of solutions of $P_n(\cos \phi) = 0$) divide a sphere into zones, as shown in Fig. 5.13.

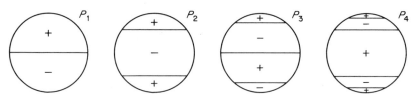

Figure 5.13 Nodal curves of the "zonal harmonics." Circles shown are solutions of $P_n(\cos \phi) = 0$ for $n = 1, 2, 3, 4$.

Exercises

1. Equation (4) may be solved by assuming

$$\Phi(\phi) = \tfrac{1}{2}a_0 + \sum_{1}^{\infty} a_k \cos k\phi.$$

Find the relations among the coefficients a_k by computing the terms of the equation in the form of series. Use

$$\sin \phi \sin k\phi = \tfrac{1}{2}[\cos(k - 1)\phi - \cos(k + 1)\phi].$$

Show that the coefficients are all zero after a_n if $\mu^2 = n(n + 1)$.

2. Derive the formula for the coefficients b_n, as shown in Eq. (10).

3. Find $P_5(x)$.

4. Verify the formula Eq. (6) for the first few Legendre polynomials.

5. We know that one of the solutions of $(1 - x^2)y'' - 2xy' = 0$ is $y(x) = 1(\mu^2 = 0)$. Find another independent solution of this differential equation.

6. Show that the orthogonality relation for the functions $\Phi_n(\phi) = P_n(\cos \phi)$ is

$$\int_0^\pi \Phi_n(\phi)\Phi_m(\phi) \sin \phi \, d\phi = 0, \qquad n \neq m.$$

7. Solve the following problem by separation of variables:

$$\frac{1}{\rho^2}\left[\frac{\partial}{\partial\rho}\left(\rho^2 \frac{\partial u}{\partial\rho}\right) + \frac{1}{\sin \phi}\frac{\partial}{\partial\phi}\left(\sin \phi \frac{\partial u}{\partial\phi}\right)\right] = -\lambda^2 u, \quad 0 < \rho < c, \quad 0 < \phi < \pi$$

$$u(c, \phi) = 0.$$

Assume $u(\rho, \phi) = R(\rho)\Phi(\phi)$. Separate Φ first. See Section 8, Eq. (1) to determine R.

8.   Let $F = (x^2 - 1)^n$. Show that F satisfies the differential equation

$$(x^2 - 1)F' = 2nxF.$$

9.   Differentiate both sides of the above equation $n + 1$ times to show that the nth derivative of F satisfies Legendre's equation (5). Use Leibnitz's rule.

10. Obtain Eq. (6) by these manipulations:

   a.   Multiply through Eq. (9) by P_{n+1}, integrate from -1 to 1, and use the orthogonality of P_{n+1} with P_{n-1}.

   b.   Replace $(2n + 1)P_n$ by means of Eq. (8).

   c. P_{n+1} is orthogonal to xP'_{n-1}, which is a polynomial of degree n.

   d.   Solve what remains for the desired integral.

11. Solve the problem

$$\nabla^2 u = 0, \qquad 0 < \rho < 1, \quad 0 < \phi < \pi$$

$$u(1, \phi) = \begin{cases} 1, & 0 < \phi < \pi/2 \\ 0, & \pi/2 < \phi < \pi. \end{cases}$$

10   References

We have seen just a few problems in two dimensions, but they are sufficient to illustrate the complications that may arise. A serious drawback to the solution by separation of variables is that double series tend to converge slowly, if at all. Thus, if a numerical solution to a two-dimensional problem is needed, it may be advisable to sidestep the analytical solution by using an approximate numerical technique from the beginning.

One advantage of using special coordinate systems is that some problems which are two-dimensional in cartesian coordinates may be one-dimensional in another system. This is the case, for instance, when distance from a point (r in polar or ρ in spherical coordinates) is the only significant space variable. Of course, nonrectangular systems may arise naturally from the geometry of a problem.

The eigenfunctions of the Laplacian operator ∇^2 are well known for a wide variety of coordinate systems. Extensive information may be found in *Methods of Theoretical Physics* by Morse and Feshbach [20], and in *Handbook of Mathematical Functions* by Abramowitz and Stegun [1].

Miscellaneous exercises

1. Solve the heat problem

$$\nabla^2 u = \frac{1}{k}\frac{\partial u}{\partial t}, \qquad\qquad 0 < x < a, \quad 0 < y < b, \quad 0 < t$$

$$\frac{\partial u}{\partial x}(0, y, t) = 0, \frac{\partial u}{\partial x}(a, y, t) = 0, \qquad 0 < y < b, \quad 0 < t$$

$$u(x, 0, t) = 0, \, u(x, b, t) = 0, \qquad 0 < x < a, \quad 0 < t$$

$$u(x, y, 0) = \frac{Tx}{a}, \qquad\qquad 0 < x < a, \quad 0 < y < b.$$

2. Same as Exercise 1, but the initial condition is

$$u(x, y, 0) = \frac{Ty}{b}, \qquad 0 < x < a, \quad 0 < y < b.$$

3. Let $u(x, y, t)$ be the solution of the heat equation in a rectangle as stated below. Find an expression for $u(a/2, b/2, t)$. Write out the first three nonzero terms for the case $a = b$.

$$\nabla^2 u = \frac{1}{k}\frac{\partial u}{\partial t}, \qquad 0 < x < a, \quad 0 < y < b, \quad 0 < t$$

$$u = 0 \text{ on all boundaries}$$

$$u(x, y, 0) = T, \qquad 0 < x < a, \quad 0 < y < b$$

4. Find the nodal lines of the square membrane. These are loci of points satisfying $\phi_{mn}(x, y) = 0$ where ϕ_{mn} satisfies $\nabla^2 \phi = -\lambda^2 \phi$ in the square and $\phi = 0$ on the boundary.

5. Find the solution of the boundary value problem

$$\frac{1}{r}\frac{d}{dr}\left(r\frac{du}{dr}\right) = -1, \qquad 0 < r < a$$

$$u(0) \text{ bounded}, \qquad u(a) = 0$$

both directly and by assuming that both $u(r)$ and the constant function 1 have Bessel series on the interval $0 < r < a$:

$$u(r) = \sum_1^\infty C_n J_0(\lambda_n r)$$

$$1 = \sum_1^\infty c_n J_0(\lambda_n r).$$

$$\left(\text{Hint:} \frac{1}{r}\frac{d}{dr}\left(r\frac{dJ_0(\lambda r)}{dr}\right) = -\lambda^2 J_0(\lambda r).\right)$$

6. Suppose that $w(x, t)$ and $v(y, t)$ are solutions of the partial differential equations

$$\frac{\partial^2 w}{\partial x^2} = \frac{1}{k}\frac{\partial w}{\partial t}, \qquad \frac{\partial^2 v}{\partial y^2} = \frac{1}{k}\frac{\partial v}{\partial t}.$$

Show that $u(x, y, t) = w(x, t) \cdot v(y, t)$ satisfies the two-dimensional heat equation

$$\frac{\partial^2 u}{\partial x^2} + \frac{\partial^2 u}{\partial y^2} = \frac{1}{k}\frac{\partial u}{\partial t}.$$

7. Use the idea of Exercise 6 to solve the problem stated in Exercise 1.

8. Let $w(x, y)$ and $v(z, t)$ satisfy the equations

$$\frac{\partial^2 w}{\partial x^2} + \frac{\partial^2 w}{\partial y^2} = 0, \qquad \frac{\partial^2 v}{\partial z^2} = \frac{1}{c^2}\frac{\partial^2 v}{\partial t^2}.$$

Show that the product $u(x, y, z, t) = w(x, y)v(z, t)$ satisfies the three-dimensional wave equation

$$\frac{\partial^2 u}{\partial x^2} + \frac{\partial^2 u}{\partial y^2} + \frac{\partial^2 u}{\partial z^2} = \frac{1}{c^2}\frac{\partial^2 u}{\partial t^2}.$$

9. Find product solutions of the equation

$$\frac{1}{r}\frac{\partial}{\partial r}\left(r\frac{\partial u}{\partial r}\right) = \frac{1}{k}\frac{\partial u}{\partial t}, \qquad 0 < r, \quad 0 < t$$

which are bounded as $r \to 0+$ and as $r \to \infty$.

10. Show that the boundary value problem

$$((1 - x^2)\phi')' = -f(x), \qquad -1 < x < 1$$

$$\phi(x) \qquad \text{bounded at } x = \pm 1$$

has as its solution

$$\phi(x) = \int_0^x \frac{1}{1 - y^2} \int_y^1 f(z) \, dz \, dy$$

provided that the function f satisfies

$$\int_{-1}^1 f(z) \, dz = 0.$$

11. Suppose that the functions $f(x)$ and $\phi(x)$ in the preceding exercise have expansions in terms of Legendre polynomials

$$f(x) = \sum_0^\infty b_k P_k(x), \qquad -1 < x < 1$$

$$\phi(x) = \sum_0^\infty B_k P_k(x), \qquad -1 < x < 1.$$

What is the relation between B_k and b_k?

12. By applying separation of variables to the problem

$$\nabla^2 u = 0, \qquad 0 < \rho < a, \quad 0 \leq \phi < \pi$$

with u bounded at $\phi = 0$, π and u periodic (2π) in θ, derive the following equation for the factor function $\Phi(\phi)$:

$$\sin \phi (\sin \phi \Phi')' - m^2 \Phi + \mu^2 \sin^2 \phi \Phi = 0,$$

where $m = 0, 1, 2, \ldots$ comes from the factor $\Theta(\theta)$.

13. Using the change of variables $x = \cos \phi$, $\Phi(\phi) = y(x)$ on the equation of Exercise 12, derive a differential equation for $y(x)$.

14. Solve the heat conduction problem

$$\frac{1}{r} \frac{\partial}{\partial r} \left(r \frac{\partial u}{\partial r} \right) = \frac{1}{k} \frac{\partial u}{\partial t}, \qquad 0 < r < a, \quad 0 < t$$

$$\frac{\partial u}{\partial r}(a, t) = 0, \qquad 0 < t$$

$$u(r, 0) = T_0 - \Delta \left(\frac{r}{a} \right)^2, \qquad 0 < r < a.$$

15. Solve the following potential problem in a cylinder:

$$\frac{1}{r}\frac{\partial}{\partial r}\left(r\frac{\partial u}{\partial r}\right) + \frac{\partial^2 u}{\partial z^2} = 0, \qquad\qquad 0 < r < a, \quad 0 < z < b$$

$$u(a, z) = 0, \qquad\qquad 0 < z < b$$

$$u(r, 0) = 0, \, u(r, b) = U_0, \qquad 0 < r < a.$$

16. Find the solution of the heat conduction problem

$$\frac{1}{r}\frac{\partial}{\partial r}\left(r\frac{\partial u}{\partial r}\right) = \frac{1}{k}\frac{\partial u}{\partial t}, \qquad 0 < r < a, \quad 0 < t$$

$$u(a, t) = T_0, \qquad 0 < t$$

$$u(r, 0) = T_1, \qquad 0 < r < a.$$

17. Find some frequencies of vibration of a cylinder by finding product solutions of the problem

$$\frac{1}{r}\frac{\partial}{\partial r}\left(r\frac{\partial u}{\partial r}\right) + \frac{\partial^2 u}{\partial z^2} = \frac{1}{c^2}\frac{\partial^2 u}{\partial t^2}, \qquad 0 < r < a, \quad 0 < z < b, \quad 0 < t$$

$$u(r, 0, t) = 0, \, u(r, b, t) = 0, \qquad 0 < r < a, \quad \alpha < t$$

$$u(a, z, t) = 0, \qquad\qquad\qquad 0 < z < b, \quad 0 < t.$$

18. Derive the given formula for the solution of the following potential equation in a spherical shell:

$$\nabla^2 u = 0, \qquad\qquad a < \rho < b, \quad 0 < \phi < \pi$$

$$u(a, \phi) = f(\cos \phi), \, u(b, \phi) = 0, \qquad 0 < \phi < \pi$$

$$u(\rho, \phi) = \sum A_n \frac{b^{2n+1} - \rho^{2n+1}}{b^{2n+1} - a^{2n+1}}\left(\frac{a}{\rho}\right)^{n+1} P_n(\cos \phi),$$

$$A_n = \frac{2n+1}{2}\int_{-1}^{1} f(x)P_n(x)\, dx.$$

chapter 6
Laplace Transform

1 Definition and elementary properties

The Laplace transform serves as a device for simplifying or mechanizing the solution of ordinary and partial differential equations. It associates a function $f(t)$ with a function of another variable, $F(s)$, from which the original function can be recovered.

Let $f(t)$ be sectionally continuous in every interval $0 < t < T$. The Laplace transform of f, written $\mathscr{L}(f)$ or $F(s)$, is defined by the integral

$$\mathscr{L}(f) = F(s) = \int_0^\infty e^{-st} f(t)\, dt. \tag{1}$$

We use the convention that a function of t is represented by a lower case letter and its transform by the corresponding capital letter. The variable s may be real or complex, but in the computation of transforms by the definition, s is usually assumed to be real. Two simple examples are

$$\mathscr{L}(1) = \int_0^\infty e^{-st} \cdot 1\, dt = \frac{1}{s}$$

$$\mathcal{L}(e^{at}) = \int_0^\infty e^{-st} e^{at}\, dt = \frac{-e^{-(s-a)t}}{s-a}\Big|_0^\infty = \frac{1}{s-a}.$$

Not every function of t has a Laplace transform, for the defining integral may fail to converge: the function $\exp t^2$ has no transform. However, if $f(t)$ does not grow too rapidly with t, the integral for the Laplace transform of f does converge, at least for large enough values of s. The basic requirement for the existence of the transform—besides sectional continuity—is that f be of "exponential order." That means that some constants, M and k, exist for which

$$|f(t)| \le Me^{kt}$$

when $t > 0$. Then the Laplace transform of f certainly exists if s, or just the real part of s, is greater than k. Existence is not much of a worry in our work.

The Laplace transform inherits two important properties from the integral used in its definition:

$$\mathcal{L}(cf(t)) = c\mathcal{L}(f(t)), \qquad c \text{ constant} \tag{2}$$

$$\mathcal{L}(f(t) + g(t)) = \mathcal{L}(f(t)) + \mathcal{L}(g(t)). \tag{3}$$

By exploiting these properties, we easily determine that

$$\mathcal{L}(\cosh at) = \mathcal{L}\left[\frac{1}{2}(e^{at} + e^{-at})\right]$$

$$= \frac{1}{2}\left(\frac{1}{s-a} + \frac{1}{s+a}\right) = \frac{s}{s^2 - a^2}$$

$$\mathcal{L}(\sin \omega t) = \mathcal{L}\left[\frac{1}{2i}(e^{i\omega t} - e^{-i\omega t})\right]$$

$$= \frac{1}{2i}\left(\frac{1}{s-i\omega} - \frac{1}{s+i\omega}\right) = \frac{\omega}{s^2 + \omega^2}.$$

Notice the use of linearity properties with complex constants and functions.

Because of the factor e^{-st} in the definition of the Laplace transform, exponential multipliers are easily handled by the "shifting theorem":

$$\mathcal{L}(e^{bt}f(t)) = \int_0^\infty e^{-st} e^{bt} f(t)\, dt$$

$$= \int_0^\infty e^{-(s-b)t} f(t)\, dt = F(s-b)$$

where $F(s) = \mathcal{L}(f(t))$. For instance,

$$\mathcal{L}(e^{bt} \sin \omega t) = \frac{\omega}{(s-b)^2 + \omega^2} = \frac{\omega}{s^2 - 2sb + b^2 + \omega^2}$$

since $\mathcal{L}(\sin \omega t) = \omega/(s^2 + \omega^2)$.

The real virtue of the Laplace transform is revealed by its effect on derivatives. Suppose $f(t)$ is continuous and has a sectionally continuous derivative $f'(t)$. Then by definition

$$\mathcal{L}(f'(t)) = \int_0^\infty e^{-st} f'(t)\, dt.$$

Integrating by parts, we get

$$\mathcal{L}(f'(t)) = e^{-st} f(t)\Big|_0^\infty - \int_0^\infty (-s) e^{-st} f(t)\, dt.$$

If $f(t)$ is of exponential order, $e^{-st} f(t)$ must tend to 0 as t tends to infinity (for large enough s) so that the equation above becomes

$$\mathcal{L}(f'(t)) = -f(0) + s \int_0^\infty e^{-st} f(t)\, dt$$

$$= -f(0) + s\mathcal{L}(f(t)).$$

(If $f(t)$ has a jump at $t = 0$, $f(0)$ is to be interpreted as $f(0+)$.)

Similarly, if f and f' are continuous, f'' is sectionally continuous, and if all three functions are of exponential order, then

$$\mathcal{L}(f''(t)) = -f'(0) + s\mathcal{L}(f'(t))$$

$$= -f'(0) - sf(0) + s^2 \mathcal{L}(f(t)).$$

An easy generalization extends this formula to the nth derivative:

$$\mathcal{L}[f^{(n)}(t)] = -f^{(n-1)}(0) - sf^{(n-2)}(0) - \cdots - s^{n-1} f(0) + s^n \mathcal{L}(f(t)) \qquad (4)$$

on the assumption that f and its first $n - 1$ derivatives are continuous, $f^{(n)}$ is sectionally continuous, and all are of exponential order.

We may apply Eq. (4) to the function $f(t) = t^k$, k a nonnegative integer. Here

$$f(0) = f'(0) = \cdots = f^{(k-1)}(0) = 0, \qquad f^{(k)}(0) = k!, \qquad f^{(k+1)}(t) = 0.$$

Thus, Formula (4) with $n = k + 1$ yields

$$0 = -k! + s^{k+1} \mathcal{L}(t^k)$$

$$\mathcal{L}(t^k) = \frac{k!}{s^{k+1}}.$$

A different application of the derivative rule is used to transform integrals. If $f(t)$ is sectionally continuous, then $\int_0^t f(t')\, dt'$ is a continuous function, equal to zero at $t = 0$, and has derivative $f(t)$. Hence

$$\mathcal{L}(f(t)) = s\mathcal{L}\left[\int_0^t f(t')\, dt'\right]$$

or

$$\mathcal{L}\left[\int_0^t f(t')\, dt'\right] = \frac{1}{s}\mathcal{L}(f(t)) \qquad (5)$$

Differentiation and integration with respect to s may produce transforms of previously inaccessible functions. We need the two formulas

$$-\frac{de^{-st}}{ds} = te^{-st}, \qquad \int_s^\infty e^{-s't}\, ds' = \frac{1}{t} e^{-st}$$

to derive the results

$$\mathcal{L}(tf(t)) = -\frac{dF(s)}{ds}, \qquad \mathcal{L}\left(\frac{1}{t}f(t)\right) = \int_s^\infty F(s')\, ds'. \tag{6}$$

(Note that, unless $f(0) = 0$, the transform of $f(t)/t$ will not exist.) Examples of the use of these formulas are

$$\mathcal{L}(t\sin\omega t) = -\frac{d}{ds}\left(\frac{\omega}{s^2 + \omega^2}\right) = \frac{2s\omega}{(s^2 + \omega^2)^2}$$

$$\mathcal{L}\left(\frac{\sin t}{t}\right) = \int_s^\infty \frac{ds'}{s'^2 + 1} = \frac{\pi}{2} - \tan^{-1} s = \tan^{-1}\left(\frac{1}{s}\right).$$

When a problem is solved by use of Laplace transforms, a prime difficulty is computation of the corresponding function of t. Methods for computing the "inverse transform," $f(t) = \mathcal{L}^{-1}(F(s))$, include integration in the complex plane, convolution, partial fractions (discussed in Section 2), and tables of transforms. The last method, which involves the least work, is the most popular. Following this section is a short table of transforms, all calculated from the definition or by use of formulas in this section.

Exercises

1. By using linearity and the transform of e^{at}, compute the transform of each of the following functions:

   a. $\sinh at$ b. $\cos\omega t$
   c. $\cos^2\omega t$ d. $\sin(\omega t - \phi)$
   e. $e^{2(t+1)}$ f. $\sin^2\omega t$.

2. Use differentiation with respect to t to find the transforms of

   a. te^{at} from $\mathcal{L}(e^{at})$
   b. $\sin\omega t$ from $\mathcal{L}(\cos\omega t)$
   c. $\cosh at$ from $\mathcal{L}(\sinh at)$.

3. Compute the transform of each of the following directly from the definition:

   a. $f(t) = \begin{cases} 0, & 0 < t < a \\ 1, & a < t \end{cases}$

b. $f(t) = \begin{cases} 0, & 0 < t < a \\ 1, & a < t < b \\ 0, & b < t \end{cases}$

c. $f(t) = \begin{cases} t, & 0 < t < a \\ a, & a < t. \end{cases}$

4. The " Heaviside step function " is defined by the formula

$$H_a(t) = \begin{cases} 1, & t > a \\ 0, & t < a. \end{cases}$$

Assuming $a \geq 0$, show that the Laplace transform of H_a is

$$\mathscr{L}(H_a(t)) = e^{-as}/s.$$

5. Use completion of squares and the shifting theorem to find the inverse transform of

a. $1/(s^2 + 2s)$ **b.** $s/(s^2 + 2s)$
c. $1/(s^2 + 2as + b^2)$, $b > a.$

6. Find the Laplace transform of the square-wave function

$$f(t) = \begin{cases} 1, & 0 < x < a \\ 0, & a < x < 2a \end{cases}, \qquad f(x + 2a) = f(x).$$

(Hint: Break up the integral as shown below, evaluate the integrals, and add up a geometric series

$$F(s) = \sum_{n=0}^{\infty} \int_{2na}^{2(n+1)a} f(t)e^{-st}\, dt.)$$

7. Use any method to find the inverse transform of:

a. $1/(s - a)(s - b)$ **b.** $s/(s^2 - a^2)^2$
c. $s^2/(s^2 + \omega^2)^2$ **d.** $1/(s - a)^3$
e. $(1 - e^{-s})/s.$

8. Use any theorem or formula to find the transform of

a. $\dfrac{1 - \cos \omega t}{t}$ **b.** $\displaystyle\int_0^t \dfrac{\sin at'}{t'}\, dt'$
c. $t^2 e^{-at}$ **d.** $t \cos \omega t$
e. $\sinh at \sin \omega t.$

Short table of Laplace transforms

$f(t)$	$F(s)$
0	0
1	$1/s$
e^{at}	$1/(s-a)$
$\cosh at$	$s/(s^2 - a^2)$
$\sinh at$	$a/(s^2 - a^2)$
$\cos \omega t$	$s/(s^2 + \omega^2)$
$\sin \omega t$	$\omega/(s^2 + \omega^2)$
t	$1/s^2$
t^k	$k!/s^{k+1}$
$e^{bt} \sin \omega t$	$\omega/((s-b)^2 + \omega^2)$
$e^{bt} \cos \omega t$	$s/((s-b)^2 + \omega^2)$
$e^{bt} t^k$	$k!/(s-b)^{k+1}$
$1 - e^{at}$	$-a/s(s-a)$
$t \sin \omega t$	$2s\omega/(s^2 + \omega^2)^2$
$t \cos \omega t$	$(s^2 - \omega^2)/(s^2 + \omega^2)^2$

Properties of Laplace transform

$$\mathcal{L}(f) = F(s) = \int_0^\infty e^{-st} f(t)\, dt$$

$$\mathcal{L}(cf(t)) = c\mathcal{L}(f(t))$$

$$\mathcal{L}(f(t) + g(t)) = \mathcal{L}(f(t)) + \mathcal{L}(g(t))$$

$$\mathcal{L}(f'(t)) = -f(0) + sF(s)$$

$$\mathcal{L}(f''(t)) = -f'(0) - sf(0) + s^2 F(s)$$

$$\mathcal{L}(f^{(n)}(t)) = -f^{(n-1)}(0) - sf^{(n-2)}(0) - \cdots - s^{n-1}f(0) + s^n F(s)$$

$$\mathcal{L}(e^{bt} f(t)) = F(s-b)$$

$$\mathcal{L}\left(\int_0^t f(t')\, dt' \right) = \frac{1}{s} F(s)$$

$$\mathcal{L}(tf(t)) = -\frac{dF}{ds}$$

$$\mathcal{L}\left(\frac{1}{t} f(t) \right) = \int_s^\infty F(s')\, ds'$$

2 Elementary applications. Partial fractions and convolutions

Because of the formula for the transform of derivatives, the Laplace transform finds important application to linear differential equations with constant coefficients, subject to initial conditions. In order to solve the simple problem

$$u' + au = 0, \qquad u(0) = 1$$

we transform the entire equation, obtaining

$$\mathscr{L}(u') + a\mathscr{L}(u) = 0$$

or

$$sU - 1 + aU = 0$$

where $U = \mathscr{L}(u)$. The derivative has been "transformed out," and U is determined by simple algebra to be

$$U(s) = \frac{1}{(s+a)}.$$

By consulting the table we find that $u(t) = e^{-at}$.

Equations of higher order can be solved the same way. When transformed, the problem

$$u'' + \omega^2 u = 0, \qquad u(0) = 1, \qquad u'(0) = 0$$

becomes

$$s^2 U - s \cdot 1 - 0 + \omega^2 U = 0,$$

and U is found to be

$$U(s) = \frac{s}{(s^2 + \omega^2)}$$

the transform of cos ωt.

In general we may outline our procedure in the figure below:

Original problem   Solution of original problem

$\mathscr{L}$ $\mathscr{L}^{-1}$

Transformed problem ⟶ Solution of transformed problem

The step symbolized by $\mathscr{L}^{-1}$ is usually the sticky one.

A simple mass–spring–damper system leads to the initial value problem

$$u'' + au' + \omega^2 u = 0, \qquad u(0) = u_0, \qquad u'(0) = u_1$$

whose transform is

$$s^2 U - s u_0 - u_1 + a(sU - u_0) + \omega^2 U = 0.$$

Determination of U gives it as the ratio of two polynomials

$$U(s) = \frac{s u_0 + (u_1 + a u_0)}{s^2 + as + \omega^2}.$$

Although this expression is not in our table, it can be worked around to a function of $s + \tfrac{1}{2}a$ whose inverse transform is available. The shift theorem then gives $u(t)$. But there is a better way.

The inversion of a rational function of s (that is, the ratio of two polynomials) can be accomplished by the technique of "partial fractions." Suppose we wish to compute the inverse transform of

$$U(s) = \frac{cs + d}{s^2 + as + b}.$$

The denominator has two roots r_1 and r_2, which we assume for the moment to be distinct. Thus

$$s^2 + as + b = (s - r_1)(s - r_2)$$

$$U(s) = \frac{cs + d}{(s - r_1)(s - r_2)}.$$

The rules of elementary algebra suggest that U can be written as a sum,

$$\frac{cs + d}{(s - r_1)(s - r_2)} = \frac{A_1}{s - r_1} + \frac{A_2}{s - r_2} \tag{1}$$

for some choice of A_1 and A_2. Indeed, by finding the common-denominator form for the right-hand side and matching powers of s in the numerator, we obtain

$$\frac{cs + d}{(s - r_1)(s - r_2)} = \frac{A_1(s - r_2) + A_2(s - r_1)}{(s - r_1)(s - r_2)}$$

$$c = A_1 + A_2, \qquad d = -A_1 r_2 - A_2 r_1.$$

When A_1 and A_2 are determined, the inverse transform of the right-hand side of Eq. (1) is easily found:

$$\mathscr{L}^{-1}\!\left(\frac{A_1}{s - r_1} + \frac{A_2}{s - r_2} \right) = A_1 \exp r_1 t + A_2 \exp r_2 t.$$

For a specific example, suppose that

$$U(s) = \frac{s + 4}{s^2 + 3s + 2}.$$

The roots of the denominator are $r_1 = -1$ and $r_2 = -2$. Thus

$$\frac{s + 4}{s^2 + 3s + 2} = \frac{A_1}{s + 1} + \frac{A_2}{s + 2} = \frac{(A_1 + A_2)s + (2A_1 + A_2)}{(s + 1)(s + 2)}.$$

We find $A_1 = 3$, $A_2 = -2$, hence

$$\mathcal{L}^{-1}\left(\frac{s + 4}{s^2 + 3s + 2}\right) = \mathcal{L}^{-1}\left(\frac{3}{s + 1} - \frac{2}{s + 2}\right) = 3e^{-t} - 2e^{-2t}.$$

A little calculus takes us much further. Suppose that U has the form:

$$U(s) = \frac{q(s)}{p(s)}$$

where p and q are polynomials, and the degree of q is less than the degree of p. Assume that p has distinct roots $r_1, \ldots, r_k$:

$$p(s) = (s - r_1)(s - r_2) \cdots (s - r_k).$$

We try to write U in the fraction form

$$U(s) = \frac{A_1}{s - r_1} + \frac{A_2}{s - r_2} + \cdots + \frac{A_k}{s - r_k} = \frac{q(s)}{p(s)}.$$

The algebraic determination of the As is very tedious, but notice that

$$\frac{(s - r_1)q(s)}{p(s)} = A_1 + A_2 \frac{s - r_1}{s - r_2} + \cdots + A_k \frac{s - r_1}{s - r_k}.$$

If s is set equal to r_1, the right-hand side is just A_1. The left-hand side becomes $0/0$, but L'Hôpital's rule gives

$$\lim_{s \to r_1} \frac{(s - r_1)q(s)}{p(s)} = \lim_{s \to r_1} \frac{(s - r_1)q'(s) + q(s)}{p'(s)} = \frac{q(r_1)}{p'(r_1)}.$$

Therefore A_1 and all the other As are given by

$$A_i = \frac{q(r_i)}{p'(r_i)}.$$

Consequently, our rational function takes the form

$$\frac{q(s)}{p(s)} = \frac{q(r_1)}{p'(r_1)} \frac{1}{s - r_1} + \cdots + \frac{q(r_k)}{p'(r_k)} \frac{1}{s - r_k}.$$

From this form we readily compute the inverse transform:

$$\mathcal{L}^{-1}\left(\frac{q(s)}{p(s)}\right) = \frac{q(r_1)}{p'(r_1)} \exp r_1 t + \cdots + \frac{q(r_k)}{p'(r_k)} \exp r_k t. \tag{2}$$

(Eq. (2) is known as the Heaviside formula.) In our numerical example, $q(s) = s + 4$, $p(s) = s^2 + 3s + 2$, $p'(s) = 2s + 3$. Thus

$$U(s) = \frac{-1+4}{2(-1)+3} \frac{1}{s+1} + \frac{-2+4}{2(-2)+3} \frac{1}{s+2}$$

$$= \frac{3}{s+1} - \frac{2}{s+2}.$$

The special case of multiple roots will be dealt with later. We should note that these decompositions remain valid when some or all roots of the denominator are complex.

In nonhomogeneous problems also, the Laplace transform is a useful tool. To solve the problem:

$$u' + au = f(t), \qquad u(0) = u_0$$

we again transform the entire equation, obtaining

$$sU - u_0 + aU = F(s)$$

$$U(s) = \frac{u_0}{s+a} + \frac{1}{s+a} F(s).$$

The first term in this expression is recognized as the transform of $u_0 e^{-at}$. If $F(s)$ is a rational function, partial fractions may be used to invert the second term. However, we can identify that term by solving the problem another way.

$$e^{at}(u' + au) = e^{at}(f(t))$$

$$(ue^{at})' = e^{at}f(t)$$

$$ue^{at} = \int_0^t e^{at'}f(t')\, dt' + c$$

$$u(t) = \int_0^t e^{-a(t-t')} f(t')\, dt' + ce^{-at}.$$

The initial condition requires that $c = u_0$. On comparing the two results, we see that

$$\mathcal{L}\left[\int_0^t e^{-a(t-t')}f(t')\, dt'\right] = \frac{1}{s-a} F(s).$$

Thus the transform of the combination of e^{-at} and $f(t)$ on the left is the product of the transforms of e^{-at} and $f(t)$. This simple result can be generalized in the following way:

If $g(t)$ and $f(t)$ have Laplace transforms $G(s)$ and $F(s)$ respectively, then

$$\mathscr{L}\left[\int_0^t g(t-t')f(t')\,dt'\right] = G(s)F(s). \tag{3}$$

(This is known as the convolution theorem.) The integral on the left is called the *convolution* of g and f, written

$$g(t) * f(t) = \int_0^t g(t-t')f(t')\,dt'.$$

It can be shown that the convolution follows these rules:

$$g * f = f * g \tag{4a}$$

$$f * (g * h) = (f * g) * h \tag{4b}$$

$$f * (g + h) = f * g + f * h. \tag{4c}$$

The convolution theorem provides an important device for inverting Laplace transforms, which we shall apply to find the general solution of the nonhomogeneous problem

$$u'' - au = f(t), \qquad u(0) = u_0, \, u'(0) = u_1.$$

The transformed equation is readily solved, yielding

$$U(s) = \frac{su_0 + u_1}{s^2 - a} + \frac{1}{s^2 - a} F(s).$$

Since $1/(s^2 - a)$ is the transform of $\sinh\sqrt{a}\,t/\sqrt{a}$, we easily determine that u is

$$u(t) = u_0 \cosh\sqrt{a}\,t + \frac{u_1}{\sqrt{a}}\sinh\sqrt{a}\,t + \int_0^t \frac{\sinh\sqrt{a}(t-t')}{\sqrt{a}} f(t')\,dt'. \tag{5}$$

A slightly different problem occurs if the mass in a spring–mass system is struck while the system is in motion. The mathematical model of the system might be

$$u'' + \omega^2 u = f(t), \qquad u(0) = u_0, \qquad u'(0) = u_1$$

where $f(t) = F_0$ for $t_0 < t < t_1$ and $f(t) = 0$ for other values. The transform of U is

$$U(s) = \frac{su_0 + u_1}{s^2 + \omega^2} + \frac{1}{s^2 + \omega^2} F(s).$$

The inverse transform is then

$$u(t) = u_0 \cos \omega t + u_1 \frac{\sin \omega t}{\omega} + \int_0^t \frac{\sin \omega(t - t')}{\omega} f(t')\, dt'.$$

The convolution in this case is easy to calculate:

$$\int_0^t \frac{\sin \omega(t - t')}{\omega} f(t')\, dt' = \begin{cases} 0, & t < t_0 \\[2mm] F_0 \dfrac{1 - \cos \omega(t - t_0)}{\omega^2}, & t_0 < t < t_1 \\[2mm] F_0 \dfrac{\cos \omega(t - t_1) - \cos \omega(t - t_0)}{\omega^2}, & t_1 < t. \end{cases}$$

Exercises

1. Solve the initial value problems

    a. $u' - 2u = 0$, $u(0) = 1$
    b. $u' + 2u = 0$, $u(0) = 1$
    c. $u'' + 4u' + 3u = 0$, $u(0) = 1$, $u'(0) = 0$
    d. $u'' + 9u = 0$, $u(0) = 0$, $u'(0) = 1$.

2. Solve the initial value problem

    $$u'' + 2au' + u = 0, \qquad u(0) = u_0, \qquad u'(0) = u_1$$

    in the three cases: $0 < a < 1$, $a = 1$, $a > 1$.

3. Solve the inhomogeneous problems with zero initial conditions:

    a. $u' + au = 1$ b. $u'' + u = t$
    c. $u'' + 4u = \sin t$ d. $u'' + 4u = \sin 2t$
    e. $u'' + 2u' = 1 - e^{-t}$ f. $u'' - u = 1$.

4. Complete the square in the denominator and use the shift theorem $[F(s - a) = \mathscr{L}(e^{at}f(t))]$ to invert

    $$U(s) = \frac{su_0 + (u_1 + 2au_0)}{s^2 + 2as + \omega^2}.$$

    There are three cases corresponding to

    $$\omega^2 - a^2 > 0, \ = 0, \ < 0.$$

5. Use partial fractions to invert the following transforms:

    a. $1/(s^2 - 4)$ b. $1/(s^2 + 4)$
    c. $(s + 3)/s(s^2 + 2)$ d. $4/s(s + 1)$.

6. Prove properties (4a) and (4c) of the convolution.

7. Compute the convolution $f * g$ for
   a. $f(t) = 1$, $g(t) = \sin t$
   b. $f(t) = e^t$, $g(t) = \cos \omega t$
   c. $f(t) = t$, $g(t) = \sin t$.

8. Demonstrate the following properties of convolution either directly or by using Laplace transform:
   a. $1 * f'(t) = f(t) - f(0)$
   b. $(t * f(t))'' = f(t)$
   c. $(f * g)' = f' * g = f * g'$, if $f(0) = g(0) = 0$.

3 Applications to partial differential equations

A function of two (or more) variables may also have a Laplace transform:

$$\mathcal{L}(u(x, t)) = \int_0^\infty e^{-st} u(x, t) \, dt = U(x, s).$$

For instance, we easily find

$$\mathcal{L}(e^{-at} \sin \pi x) = \frac{1}{s + a} \sin \pi x$$

$$\mathcal{L}(\sin(x + t)) = \frac{s \sin x + \cos x}{s^2 + 1}.$$

The transform U naturally is a function not only of s but also of the "untransformed" variable x. We assume that derivatives or integrals with respect to the untransformed variable pass through the transform:

$$\mathcal{L}\left(\frac{\partial u}{\partial x}\right) = \int_0^\infty \frac{\partial u(x, t)}{\partial x} e^{-st} \, dt$$

$$= \frac{\partial}{\partial x} \int_0^\infty u(x, t) e^{-st} \, dt = \frac{\partial}{\partial x} (U(x, s))$$

Often the ordinary derivative is used, as in

$$\mathcal{L}\left(\frac{\partial u}{\partial x}\right) = \frac{dU}{dx}.$$

The rule for transforming a derivative with respect to t can be found, as before, with integration by parts:

$$\mathcal{L}\left(\frac{\partial u}{\partial t}\right) = s\mathcal{L}(u(x, t)) - u(x, 0).$$

If the Laplace transform is applied to a boundary value–initial value problem in x and t, all time derivatives disappear, leaving an ordinary differential equation in x. We shall illustrate this technique with some trivial examples. Incidentally, we assume from here on that dimensional analysis has been used to eliminate as many parameters as possible. There are enough complexities already in the problems.

Example 1

$$\frac{\partial^2 u}{\partial x^2} = \frac{\partial u}{\partial t}, \qquad\qquad 0 < x < 1, \quad 0 < t$$

$$u(0, t) = 1, \qquad u(1, t) = 1, \qquad 0 < t$$

$$u(x, 0) = 1 + \sin \pi x, \qquad\qquad 0 < x < 1.$$

The partial differential equation *and the boundary conditions* (that is, everything that is valid for $t > 0$) are transformed, while the initial condition is incorporated by the transform.

$$\frac{d^2 U}{dx^2} = sU - (1 + \sin \pi x), \qquad 0 < x < 1$$

$$U(0, s) = \frac{1}{s}, \qquad U(1, s) = \frac{1}{s}.$$

This boundary value problem is solved to obtain

$$U(x, s) = \frac{1}{s} + \frac{\sin \pi x}{s + \pi^2}.$$

We direct our attention now to U as a function of s. Since $\sin \pi x$ is a constant with respect to s, tables may be used to find

$$u(x, t) = 1 + \sin \pi x \exp(- \pi^2 t).$$

Example 2

$$\frac{\partial^2 u}{\partial x^2} = \frac{\partial^2 u}{\partial t^2}, \qquad\qquad 0 < x < 1, \quad 0 < t$$

$$u(0, t) = 0, \qquad u(1, t) = 0, \qquad 0 < t$$

$$u(x, 0) = \sin \pi x, \qquad\qquad 0 < x < 1$$

$$\frac{\partial u}{\partial t}(x, 0) = - \sin \pi x, \qquad\qquad 0 < x < 1.$$

Under transformation the problem becomes

$$\frac{d^2U}{dx^2} = s^2U - s \sin \pi x + \sin \pi x, \qquad 0 < x < 1$$

$$U(0, s) = 0, \qquad U(1, s) = 0.$$

The function U is found to be

$$U(x, s) = \frac{s-1}{s^2 + \pi^2} \sin \pi x$$

from which we find the solution,

$$u(x, t) = (\sin \pi x)\left(\cos \pi t - \frac{1}{\pi} \sin \pi t\right).$$

Example 3 Now we consider a problem that we know to have a more complicated solution:

$$\frac{\partial^2 u}{\partial x^2} = \frac{\partial u}{\partial t}, \qquad\qquad 0 < x < 1, \quad 0 < t$$

$$u(0, t) = 1, \qquad u(1, t) = 1, \qquad 0 < t$$

$$u(x, 0) = 0, \qquad\qquad\qquad 0 < x < 1.$$

The transformed problem is

$$\frac{d^2U}{dx^2} = sU, \qquad 0 < x < 1$$

$$U(0, s) = \frac{1}{s}, \qquad U(1, s) = \frac{1}{s}.$$

The general solution of the differential equation is well known to be a combination of $\sinh \sqrt{s}x$ and $\cosh \sqrt{s}x$. Application of the boundary conditions yields

$$U(x, s) = \frac{1}{s} \cosh \sqrt{s}x + \frac{(1 - \cosh \sqrt{s}) \sinh \sqrt{s}x}{s \sinh \sqrt{s}}$$

$$= \frac{\sinh \sqrt{s}x + \sinh \sqrt{s}(1 - x)}{s \sinh \sqrt{s}}.$$

This function rarely appears in a table of transforms. However, by extending the Heaviside formula, we can compute an inverse transform.

When U is the ratio of two transcendental functions (not polynomials) of s, we wish to write

$$U(x, s) = \sum A_k(x) \frac{1}{s - r_k}.$$

In this formula, the numbers r_k are values of s for which the "denominator" of U is zero or, rather, for which $|U(x, s)|$ becomes infinite; and the A_k are functions of x but not s. From this form we expect to determine

$$u(x, t) = \sum A_k(x) \exp r_k t.$$

This solution should be checked for convergence.

The hyperbolic sine (also the cosh, cos, sin, and exponential functions) is not infinite for any finite value of its argument. Thus $U(x, s)$ becomes infinite only where s or $\sinh \sqrt{s}$ is zero. Since $\sinh \sqrt{s} = 0$ has no real root besides zero, we seek complex roots by setting $\sqrt{s} = \xi + i\eta$ (ξ and η real).

The addition rules for hyperbolic and trigonometric functions remain valid for complex arguments. Furthermore, we know that

$$\cosh iA = \cos A, \qquad \sinh iA = i \sin A.$$

By combining the addition rule and these identities we find

$$\sinh(\xi + i\eta) = \sinh \xi \cos \eta + i \cosh \xi \sin \eta.$$

This function is zero only if both the real and imaginary parts are zero. Thus ξ and η must be chosen to satisfy simultaneously

$$\sinh \xi \cos \eta = 0, \qquad \cosh \xi \sin \eta = 0.$$

Of the four possible combinations, only $\sinh \xi = 0$ and $\sin \eta = 0$ produce solutions. Therefore $\xi = 0$ and $\eta = \pm n\pi$ ($n = 0, 1, 2, \ldots$); whence

$$\sqrt{s} = \pm in\pi, \qquad s = -n^2\pi^2.$$

Recall that only the value of s, not the value of $\sqrt{s}$, is significant.

Finally then, we have located $r_0 = 0$, and $r_n = -n^2\pi^2$ ($n = 1, 2, \ldots$). We proceed to find the $A_k(x)$ by the same method used in Section 2. The computations are done piecemeal and then the solution is assembled.

Part a ($r_0 = 0$). In order to find A_0 we multiply both sides of our proposed partial fractions development

$$U(x, s) = \sum_{k=0}^{\infty} A_k(x) \frac{1}{s - r_k}$$

by $s - r_0 = s$, and take the limit as s approaches r_0. The right-hand side goes to A_0. On the left we have

$$\lim_{s \to 0} s \, \frac{\sinh \sqrt{s} x + \sinh \sqrt{s}(1 - x)}{s \sinh \sqrt{s}} = x + 1 - x = 1 = A_0(x).$$

Thus the part of $u(x, t)$ corresponding to $s = 0$ is $1 \cdot e^{0t} = 1$, easily recognized as the steady-state solution.

Part b ($r_n = -n^2\pi^2$, $n = 1, 2, \ldots$). For these cases, we find

$$A_n = \frac{q(r_n)}{p'(r_n)}$$

where q and p are the obvious choices. We take $\sqrt{r_n} = +in\pi$ in all calculations:

$$p'(s) = \sinh \sqrt{s} + s \, \frac{1}{2\sqrt{s}} \cosh \sqrt{s}$$

$$p'(r_n) = \tfrac{1}{2} in\pi \cosh in\pi = \tfrac{1}{2} in\pi \cos n\pi$$

$$q(r_n) = \sinh in\pi x + \sinh in\pi(1 - x)$$

$$= i[\sin n\pi x + \sin n\pi(1 - x)].$$

Hence the portion of $u(x, t)$ which arises from each r_n is

$$A_n(x)\exp r_n t = 2 \, \frac{\sin n\pi x + \sin n\pi(1 - x)}{n\pi \cos n\pi} \exp(-n^2\pi^2 t).$$

Part c On assembling the various pieces of the solution, we get

$$u(x, t) = 1 + \frac{2}{\pi} \sum_{1}^{\infty} \frac{\sin n\pi x + \sin n\pi(1 - x)}{n \cos n\pi} \exp(-n^2\pi^2 t).$$

The same solution would be found by separation of variables.

Example 4 Now consider the wave problem

$$\frac{\partial^2 u}{\partial x^2} = \frac{\partial^2 u}{\partial t^2}, \qquad\qquad 0 < x < 1, \quad 0 < t$$

$$u(0, t) = 0, \qquad \frac{\partial u(1, t)}{\partial x} = 0, \qquad 0 < t$$

$$u(x, 0) = 0, \qquad \frac{\partial u(x, 0)}{\partial t} = x, \qquad 0 < x < 1.$$

The transformed problem is

$$\frac{d^2 U}{dx^2} = s^2 U - x, \qquad 0 < x < 1$$

$$U(0, s) = 0, \qquad U'(1, s) = 0$$

and its solution (by undetermined coefficients or otherwise) gives

$$U(x, s) = \frac{sx \cosh s - \sinh sx}{s^3 \cosh s}.$$

The numerator of this function is never infinite. The denominator is zero at $s = 0$ and $s = \pm i(2n - 1)\pi/2$ $(n = 1, 2, \ldots)$. We shall again use the Heaviside formula to determine the inverse transform of U.

Part a $(r_0 = 0)$.   The limit as s approaches zero of $sU(x, s)$ may be found by L'Hôpital's rule or by using the Taylor series for sinh and cosh. From the latter,

$$sU(x, s) = \frac{sx\left(1 + \dfrac{s^2}{2} + \cdots\right) - \left(sx + \dfrac{s^3 x^3}{6} + \cdots\right)}{s^2\left(1 + \dfrac{s^2}{2} + \cdots\right)}$$

$$= \frac{s^3\left(\dfrac{x}{2} - \dfrac{x^3}{6} - \cdots\right)}{s^2\left(1 + \dfrac{s^2}{2} + \cdots\right)} \to 0.$$

Thus, in spite of the formidable appearance of s^3 in the denominator, $s = 0$ is not really a significant value and contributes nothing to $u(x, t)$.

Part b   It is convenient to take the remaining roots in pairs. We label $\pm i(2n - 1)\pi/2 = \pm i\rho_n$.

The derivative of the denominator is

$$p'(s) = 3s^2 \cosh s + s^3 \sinh s$$

$$p'(\pm i\rho_n) = \pm i^3 \rho_n^3 \sinh(\pm i\rho_n)$$

$$= \rho_n^3 \sin \rho_n$$

since sinh $i\rho = i \sin \rho$, and $(\pm i)^4 = 1$. The contribution of these two roots together may be calculated using the exponential definition of sine:

$$\frac{q(i\rho_n)}{p'(i\rho_n)} \exp i\rho_n t + \frac{q(-i\rho_n)}{p'(-i\rho_n)} \exp - i\rho_n t$$

$$= \frac{-\sinh(i\rho_n x) \exp i\rho_n t + \sinh(i\rho_n x) \exp - i\rho_n t}{\rho_n{}^3 \sin \rho_n}$$

$$= \frac{\sin \rho_n x}{\rho_n{}^3 \sin \rho_n} i(- \exp i\rho_n t + \exp - i\rho_n t)$$

$$= \frac{2 \sin \rho_n x \sin \rho_n t}{\rho_n{}^3 \sin \rho_n}.$$

Part c The final form of $u(x, t)$, found by adding up all the contributions from Part b, is the same as would be found by separation of variables:

$$u(x, t) = 2 \sum_1^\infty \frac{\sin \rho_n x \sin \rho_n t}{\rho_n{}^3 \sin \rho_n}.$$

Exercises

1. Find all values of s, real and complex, for which the following functions are zero:

    a. $\cosh \sqrt{s}$
    b. $\cosh s$
    c. $\sinh s$
    d. $\cosh s - s \sinh s$
    e. $\cosh s + s \sinh s$.

2. Find the inverse transforms of the following functions in terms of an infinite series:

    a. $\dfrac{1}{s} \tanh s$
    b. $\dfrac{\sinh sx}{s \cosh s}$.

3. Find the transform $U(x, s)$ of the solution of each of the following problems:

    a.
    $$\frac{\partial^2 u}{\partial x^2} = \frac{\partial u}{\partial t}, \qquad 0 < x < 1, \quad 0 < t$$

    $$u(0, t) = 0, \qquad u(1, t) = t, \qquad 0 < t$$

    $$u(x, 0) = 0$$

b.
$$\frac{\partial^2 u}{\partial x^2} = \frac{\partial u}{\partial t},$$
$$0 < x < 1, \quad 0 < t$$

$$u(0, t) = u(1, t) = e^{-t}, \qquad 0 < t$$

$$u(x, 0) = 1, \qquad 0 < x < 1.$$

4. Solve each of the problems in Exercise 3, inverting the transform by means of the extended Heaviside formula.

5. Solve each of the following problems by Laplace transform methods:

a.
$$\frac{\partial^2 u}{\partial x^2} = \frac{\partial u}{\partial t},$$
$$0 < x < 1, \qquad 0 < t$$

$$u(0, t) = 0, \qquad u(1, t) = 1, \qquad 0 < t$$

$$u(x, 0) = 0, \qquad 0 < x < 1$$

b.
$$\frac{\partial^2 u}{\partial x^2} = \frac{\partial u}{\partial t},$$
$$0 < x < 1, \quad 0 < t$$

$$u(0, t) = 0, \qquad u(1, t) = 0 \qquad 0 < t$$

$$u(x, 0) = 1, \qquad 0 < x < 1.$$

4   More difficult examples

The technique of separation of variables, once mastered, seems more straightforward than the Laplace transform. However, when time-dependent boundary conditions or inhomogeneities are present, the Laplace transform offers a distinct advantage. Following are some examples which display the power of transform methods.

Example 1   A uniform insulated rod is attached at one end to an insulated container of fluid. The fluid is circulated so well that its temperature is uniform and equal to that at the end of the rod. The other end of the rod is maintained at a constant temperature. A dimensionless initial value-boundary value problem which describes the temperature in the rod is

$$\frac{\partial^2 u}{\partial x^2} = \frac{\partial u}{\partial t},$$
$$0 < x < 1, \quad 0 < t$$

$$\frac{\partial u(0, t)}{\partial x} = \gamma \frac{\partial u(0, t)}{\partial t}, \qquad u(1, t) = 1, \qquad 0 < t$$

$$u(x, 0) = 0.$$

The transformed problem and its solution are

$$\frac{d^2 U}{dx^2} = sU, \qquad 0 < x < 1$$

$$\frac{dU}{dx}(0, s) = s\gamma U(0, s), \qquad U(1, s) = \frac{1}{s}$$

$$U(x, s) = \frac{\cosh \sqrt{s}x + \sqrt{s}\gamma \sinh \sqrt{s}x}{s(\cosh \sqrt{s} + \sqrt{s}\gamma \sinh \sqrt{s})} = \frac{q(s)}{p(s)}.$$

Aside from $s = 0$, the denominator has no real roots. Thus we again search for complex roots by employing $\sqrt{s} = \xi + i\eta$. The real and imaginary parts of the denominator are to be computed by using the addition formulas for cosh and sinh. The requirement that both real and imaginary parts be zero leads to the equations

$$(\cosh \xi + \xi\gamma \sinh \xi) \cos \eta - \eta\gamma \cosh \xi \sin \eta = 0 \qquad (1)$$

$$\eta\gamma \sinh \xi \cos \eta + (\sinh \xi + \xi\gamma \cosh \xi) \sin \eta = 0. \qquad (2)$$

We may think of these as simultaneous equations in $\sin \eta$ and $\cos \eta$. Since $\sin^2\eta + \cos^2\eta = 1$, the system has a solution only when its determinant is zero. Thus after some algebra, we arrive at the condition

$$(1 + \xi^2\gamma^2 + \eta^2\gamma^2) \sinh \xi \cosh \xi + \xi\gamma(\sinh^2\xi + \cosh^2\xi) = 0.$$

The only solution occurs when $\xi = 0$, for otherwise both terms of this equation have the same sign.

After setting $\xi = 0$, Eq. (1) reduces to

$$\tan \eta = \frac{1}{\eta\gamma}$$

for which there are an infinite number of solutions. We shall number the positive solutions $\eta_1, \eta_2, \ldots$. Now we have found that the roots of $p(s)$ are $r_0 = 0$ and $r_k = (i\eta_k)^2 = -\eta_k^2$. The computation of the inverse transform follows.

Part a The limit of $sU(x, s)$ as s tends to zero is easily found to be 1. Thus this root contributes $1 \cdot e^{0t} = 1$ to $u(x, t)$.

Part b $(r_k = -\eta_k^2)$. First we compute

$$p'(s) = \cosh \sqrt{s} + \sqrt{s}\gamma \sinh \sqrt{s} + \tfrac{1}{2}\sqrt{s}(1 + \gamma) \sinh \sqrt{s} + \tfrac{1}{2}\gamma s \cosh \sqrt{s}.$$

Using the fact that $\cosh \sqrt{r_k} + \sqrt{r_k}\gamma \sinh \sqrt{r_k} = 0$ we may reduce the above to

$$p'(r_k) = \frac{-1}{2\gamma}(1 + \gamma + \eta_k^2\gamma^2)\cos \eta_k.$$

Hence the contribution to $u(x, t)$ of r_k is

$$\frac{q(r_k)}{p'(r_k)}\exp r_k t = -2\gamma \frac{\cos \eta_k x - \eta_k\gamma \sin \eta_k x}{(1 + \gamma + \eta_k^2\gamma^2)\cos \eta_k}\exp(-\eta_k^2 t).$$

Part c The construction of the final solution is left to the reader. We note that an attempt to solve this problem by separation of variables would find difficulties, for the eigenfunctions are not orthogonal.

Example 2 Sometimes one is interested not in the complete solution of a problem, but only in part of it. For example, in the problem of heat conduction in a semi-infinite solid with time-varying boundary conditions, we may seek that part of the solution which persists after a long time. (This may or may not be a steady-state solution.) Any initial condition that is bounded in x gives rise only to transient temperatures; these being of no interest, we assume a zero initial condition. Thus, the problem to be studied is

$$\frac{\partial^2 u}{\partial x^2} = \frac{\partial u}{\partial t}, \qquad 0 < x, \quad 0 < t$$

$$u(0, t) = f(t), \qquad 0 < t$$

$$u(x, 0) = 0, \qquad 0 < x.$$

The transformed equation and its general solution are

$$\frac{d^2 U}{dx^2} = sU, \qquad 0 < x, \qquad U(0, s) = F(s)$$

$$U(x, s) = A \exp -\sqrt{sx} + B \exp \sqrt{sx}.$$

We make two further assumptions about the solution. First, that $u(x, t)$ is bounded as x tends to infinity; second, that $\sqrt{s}$ means the square root of s which has a nonnegative real part. Under these two assumptions, we must choose $B = 0$ and $A = F(s)$, making

$$U(x, s) = F(s) \exp -\sqrt{sx}.$$

In order to find the persistent part of $u(x, t)$, we apply the Heaviside inversion formula to those values of s having nonnegative real parts. For, a value of s with *negative* real part corresponds to a function containing a decaying exponential—a transient.

For instance, let

$$f(t) = 1 - e^{-\beta t} + \alpha \sin \omega t$$

$$F(s) = \frac{1}{s} - \frac{1}{s+\beta} + \frac{\alpha\omega}{s^2 + \omega^2}.$$

The values of s for which $U(x, s) = F(s) \exp -\sqrt{s}x$ becomes infinite are 0, $\pm i\omega$, $-\beta$. The last value is discarded, since it is negative. Thus, the persistent part of the solution is given by

$$A_0 e^{0t} + A_1 e^{i\omega t} + A_2 e^{-i\omega t}$$

where

$$A_0 = \lim_{s \to 0} [sF(s) \exp -\sqrt{s}x] = 1$$

$$A_1 = \lim_{s \to i\omega} [(s - i\omega)F(s) \exp -\sqrt{s}x] = \frac{\alpha}{2i} \exp -\sqrt{i\omega}x$$

$$A_2 = \lim_{s \to -i\omega} [(s + i\omega)F(s) \exp -\sqrt{s}x] = -\frac{\alpha}{2i} \exp -\sqrt{-i\omega}x.$$

We also need to know that the roots of $\pm i$ with positive real part are

$$\sqrt{i} = \frac{1}{\sqrt{2}}(1 + i), \qquad \sqrt{-i} = \frac{1}{\sqrt{2}}(1 - i).$$

Thus the function we seek is

$$1 + \frac{\alpha}{2i} \exp\left[i\omega t - \sqrt{\frac{\omega}{2}}(1 + i)x\right] - \frac{\alpha}{2i} \exp\left[-i\omega t - \sqrt{\frac{\omega}{2}}(1 - i)x\right]$$

$$= 1 + \alpha \exp\left(-\sqrt{\frac{\omega}{2}}x\right)\sin\left(\omega t - \sqrt{\frac{\omega}{2}}x\right).$$

Example 3 If a steel wire is exposed to a sinusoidal magnetic field, the boundary value–initial value problem which describes its displacement is

$$\frac{\partial^2 u}{\partial x^2} = \frac{\partial^2 u}{\partial t^2} - \sin \omega t, \qquad 0 < x < 1, \quad 0 < t$$

$$u(0, t) = 0, \qquad u(1, t) = 0, \qquad 0 < t$$

$$u(x, 0) = 0, \qquad \frac{\partial u}{\partial t}(x, 0) = 0, \qquad 0 < x < 1.$$

The nonhomogeneity in the partial differential equation represents the effect of the force due to the field. The transformed equation and its solution are

$$\frac{d^2 U}{dx^2} = s^2 U - \frac{\omega}{s^2 + \omega^2}, \qquad 0 < x < 1$$

$$U(0, s) = 0, \qquad U(1, s) = 0.$$

$$U(x, s) = \frac{\omega}{s^2(s^2 + \omega^2)} \frac{\cosh \tfrac{1}{2}s - \cosh s(\tfrac{1}{2} - x)}{\cosh \tfrac{1}{2}s}.$$

Several methods are available for the inverse transformation of U. An obvious one would be to compute

$$v(x, t) = \mathscr{L}^{-1}\left(\frac{\cosh \tfrac{1}{2}s - \cosh s(\tfrac{1}{2} - x)}{s^2 \cosh \tfrac{1}{2}s}\right)$$

and write $u(x, t)$ as a convolution:

$$u(x, t) = \int_0^t \sin \omega(t - t')v(x, t') \, dt'.$$

The details of this development are left as an exercise.

We could also use the Heaviside formula. The application is now routine, except in the interesting case where $\cosh \tfrac{1}{2}i\omega = 0$: that is, where $\omega = (2n - 1)\pi$, one of the natural frequencies of the wire.

Let us suppose $\omega = \pi$, so that

$$U(x, s) = \frac{\pi}{s^2(s^2 + \pi^2)} \frac{\cosh \tfrac{1}{2}s - \cosh s(\tfrac{1}{2} - x)}{\cosh \tfrac{1}{2}s}.$$

At

$$s = 0, \qquad s = \pm i\pi, \qquad s = \pm(2n - 1)i\pi, \qquad n = 2, 3, \ldots$$

U becomes undefined. The computation of the parts of the inverse transform corresponding to the points other than $\pm i\pi$ is easily carried out. However, at these two troublesome points, our usual procedure will not work. Instead of expecting a partial-fraction decomposition containing

$$\frac{A_{-1}}{s + i\pi} + \frac{A_1}{s - i\pi}$$

and other terms of the same sort, we must seek terms like

$$\frac{A_{-1}(s + i\pi) + B_{-1}}{(s + i\pi)^2} + \frac{A_1(s - i\pi) + B_1}{(s - i\pi)^2}$$

whose contribution to the inverse transform of U would be

$$A_{-1}e^{-i\pi t} + B_{-1}te^{-i\pi t} + A_1 e^{i\pi t} + B_1 te^{i\pi t}.$$

One can compute A_1 and B_1, for example, by noting that

$$B_1 = \lim_{s \to i\pi} [(s - i\pi)^2 U(x, s)]$$

$$A_1 = \lim_{s \to i\pi} \left\{ (s - i\pi) \left[U(x, s) - \frac{B_1}{(s - i\pi)^2} \right] \right\}$$

and similarly for A_{-1} and B_{-1}. The limit for B_1 is not too difficult:

$$B_1 = \lim_{s \to i\pi} \left\{ \frac{\pi}{s^2(s + i\pi)} \frac{\cosh \frac{1}{2}s - \cosh s(\frac{1}{2} - x)}{(\cosh \frac{1}{2}s)/(s - i\pi)} \right\}$$

$$= \frac{\pi}{-\pi^2(2i\pi)} \frac{\cosh \frac{1}{2}i\pi - \cosh i\pi(\frac{1}{2} - x)}{\frac{1}{2}\sinh \frac{1}{2}i\pi}$$

$$= \frac{-1}{\pi^2} \cos \pi(\tfrac{1}{2} - x) = \frac{-1}{\pi^2} \sin \pi x.$$

The limit for A_1 is rather more complicated but may be computed by L'Hôpital's rule. Nevertheless, since $B_{-1} = B_1$, we already see that $u(x, t)$ contains the term

$$B_1 te^{i\pi t} + B_{-1} te^{-i\pi t} = -\frac{2t}{\pi^2} \sin \pi x \cos \pi t$$

whose amplitude increases with time. This, of course, is the expected resonance phenomenon.

Exercises

1. Find the persistent part of the solution of the heat problem

$$\frac{\partial^2 u}{\partial x^2} = \frac{\partial u}{\partial t}, \qquad\qquad 0 < x < 1, \quad 0 < t$$

$$\frac{\partial u}{\partial x}(0, t) = 0, \qquad \frac{\partial u}{\partial x}(1, t) = 1, \qquad 0 < t$$

$$u(x, 0) = 0, \qquad\qquad 0 < x < 1.$$

2. Verify that the persistent part of the solution to Example 2 actually satisfies the heat equation. What boundary condition does it satisfy?

3. Find the function $v(x, t)$ whose transform is

$$\frac{\cosh \frac{1}{2}s - \cosh s(\frac{1}{2} - x)}{s^2 \cosh \frac{1}{2}s}.$$

What boundary value–initial value problem does $v(x, t)$ satisfy?

4.   Solve

$$\frac{\partial^2 u}{\partial x^2} = \frac{\partial^2 u}{\partial t^2},$$ $0 < x < 1, \quad 0 < t$

$u(0, t) = 0,$ $u(1, t) = 0,$ $0 < t$

$u(x, 0) = 0,$ $\dfrac{\partial u}{\partial t}(x, 0) = 1,$ $0 < x < 1.$

5.   a.   Solve

$$\frac{\partial^2 u}{\partial x^2} = \frac{\partial^2 u}{\partial t^2} - \sin \pi x \sin \omega t,$$ $0 < x < 1, \quad 0 < t$

$u(0, t) = 0,$ $u(1, t) = 0,$ $0 < t$

$u(x, 0) = 0,$ $\dfrac{\partial u}{\partial t}(x, 0) = 0,$ $0 < x < 1.$

b.   Examine the special case $\omega = \pi$.

6.   Obtain the complete solution of Example 1 and verify that it satisfies the boundary conditions and the heat equation.

7.   a.   Solve

$$\frac{\partial^2 u}{\partial x^2} = \frac{\partial u}{\partial t},$$ $0 < x < 1, \quad 0 < t$

$u(0, t) = 0,$ $u(1, t) = 1 - e^{-at},$ $0 < t$

$u(x, 0) = 0,$ $0 < x < 1.$

b.   Examine the special case where $a = n^2 \pi^2$ for some integer n.

5   Comments and references

Laplace had virtually nothing to do with the Laplace transform, although a method of his for solving certain differential equations can be interpreted as an example of its use. The real development began in the late 19th century, when Oliver Heaviside invented a powerful, but unjustified, symbolic method for studying the ordinary and partial differential equations of mathematical physics. By the 1920's, Heaviside's method had been legitimatized and recast as the Laplace transform which we now use. Later generalizations are Schwartz's theory of distributions (1940's) and Mikusinski's operational calculus (1950's). The former seems to be the more general. (See *Differential*

Equations of Applied Mathematics by Duff and Naylor [12] for a brief intro-duction.) Both theories give an interpretation of $F(s) = 1$, which is not the Laplace transform of any function, in the sense we use.

There are a number of other transforms, under the names of Fourier, Mellin, Haenkel, and others, similar in intent to the Laplace transform, in which some other function replaces e^{-st} in the defining integral. *The Laplace Transform* by Rainville [25] and *Operational Mathematics* by Churchill [8] have more information about the applications of transforms. Extensive tables of transforms will be found in *Tables of Integral Transforms* by Erdelyi et al. [13].

Miscellaneous exercises

1. Solve the heat conduction problem

$$\frac{\partial^2 u}{\partial x^2} - \gamma^2(u - T) = \frac{\partial u}{\partial t}, \qquad\qquad 0 < x < 1, \quad 0 < t$$

$$\frac{\partial u}{\partial x}(0, t) = 0, \quad \frac{\partial u}{\partial x}(1, t) = 0, \qquad 0 < t$$

$$u(x, 0) = T_0, \qquad\qquad 0 < x < 1.$$

2. Find the "persistent part" of the solution of

$$\frac{\partial^2 u}{\partial x^2} = \frac{\partial u}{\partial t}, \qquad\qquad 0 < x < 1, \quad 0 < t$$

$$\frac{\partial u}{\partial x}(0, t) = 0, \quad u(1, t) = t, \qquad 0 < t$$

$$u(x, 0) = 0, \qquad\qquad 0 < x < 1.$$

3. Find the complete solution of the problem in Exercise 2.

4. A solid object and a surrounding fluid exchange heat by convection. The temperatures u_1 and u_2 are governed by the equations below. Solve them by means of Laplace transforms.

$$\frac{du_1}{dt} = -\beta_1(u_1 - u_2)$$

$$\frac{du_2}{dt} = -\beta_2(u_2 - u_1)$$

$$u_1(0) = 1, \qquad u_2(0) = 0$$

5. Solve the following nonhomogeneous problem with transforms:

$$\frac{\partial^2 u}{\partial x^2} = \frac{\partial u}{\partial t} - 1, \qquad 0 < x < 1, \quad 0 < t$$

$$u(0, t) = 0, \quad u(1, t) = 0, \qquad 0 < t$$

$$u(x, 0) = 0, \qquad 0 < x < 1.$$

6. Find the transform of the solution of the problem

$$\frac{\partial^2 u}{\partial x^2} = \frac{\partial u}{\partial t}, \qquad 0 < x < 1, \quad 0 < t$$

$$u(0, t) = 0, \quad u(1, t) = 1, \qquad 0 < t$$

$$u(x, 0) = 0, \qquad 0 < x < 1.$$

7. Find the solution of the problem in Exercise 6 by using the extended Heaviside formula.

8. Solve the heat problem

$$\frac{\partial^2 u}{\partial x^2} = \frac{\partial u}{\partial t}, \qquad 0 < x, \quad 0 < t$$

$$u(0, t) = 0, \qquad 0 < t$$

$$u(x, 0) = \sin x, \qquad 0 < x.$$

9. Find the transform of the solution of

$$\frac{\partial^2 u}{\partial x^2} = \frac{\partial u}{\partial t}, \qquad 0 < x, \quad 0 < t$$

$$u(0, t) = 0, \qquad 0 < t$$

$$u(x, 0) = 1, \qquad 0 < x$$

$$u(x, t) \qquad \text{bounded as } x \to \infty.$$

10. In Section 11, Chapter 2, Exercise 8, the problem above was solved by other means. Use this fact to identify

$$\frac{1}{s}(1 - e^{-\sqrt{s}x}) = \mathscr{L}\left[\text{erf}\left(\frac{x}{\sqrt{4t}}\right)\right]$$

and

$$\frac{1}{s}e^{-\sqrt{s}x} = \mathscr{L}\left[1 - \text{erf}\left(\frac{x}{\sqrt{4t}}\right)\right]$$

(The latter function is called the complementary error function, erfc(q) ≡ 1 − erf(q).)

11. Find the function of t whose Laplace transform is

$$F(s) = e^{-x\sqrt{s}}.$$

12. Using the definition of sinh in terms of exponentials and a geometric series, show that

$$\frac{\sinh\sqrt{s}x}{\sinh\sqrt{s}} = \sum_{n=0}^{\infty} (e^{-\sqrt{s}(2n+1-x)} - e^{-\sqrt{s}(2n+1+x)}).$$

13. Use the series in Exercise 12 to find a solution of the problem in Exercise 6 in terms of complementary error functions.

14. Show the following relation by using Exercise 11 and differentiating with respect to s:

$$\mathcal{L}\left[\frac{1}{\sqrt{\pi t}}\exp\left(\frac{-k^2}{4t}\right)\right] = \frac{1}{\sqrt{s}}e^{-k\sqrt{s}}.$$

15. Find the Laplace transform of the odd periodic extension of the function

$$f(t) = \pi - t, \qquad 0 < t < \pi$$

by transforming its Fourier series term-by-term.

16. Suppose that the function $f(t)$ is periodic with period $2a$. Show that the Laplace transform of f is given by the formula

$$F(s) = \frac{G(s)}{1 - e^{-2as}},$$

where

$$G(s) = \int_0^{2a} f(t)e^{-st}\,dt.$$

(Hint: See Section 1, Exercise 6.)

17. Apply the extended Heaviside method to the inversion of a transform with the form

$$F(s) = \frac{G(s)}{1 - e^{-2as}}$$

where $G(s)$ does not become infinite for any value of s.

18. Show that for a periodic function $f(t)$ the quantities

$$c_n = \frac{1}{2a} G\left(\frac{in\pi}{a}\right)$$

($G(s)$ is defined in Exercise 16) are the complex Fourier coefficients.

19. How is it possible to determine that a Laplace transform $F(s)$ corresponds to a periodic $f(t)$?

20. Is this function the transform of a periodic function?

$$F(s) = \frac{1}{s^2 + a^2}$$

21. Use the method of Exercise 16 to find the transform of the periodic extension of

$$f(t) = \begin{cases} 1, & 0 < t < \pi \\ -1, & \pi < t < 2\pi. \end{cases}$$

22. Same as Exercise 21, but use the function of Exercise 15.

23. Use the method of Exercise 16 to find the transform of

$$f(t) = |\sin t|.$$

24. Find the transform of the solution of the problem

$$\frac{\partial^2 u}{\partial x^2} = \frac{\partial^2 u}{\partial t^2}, \qquad\qquad 0 < x, \quad 0 < t$$

$$u(0, t) = h(t), \qquad\qquad 0 < t$$

$$u(x, 0) = 0, \quad \frac{\partial u}{\partial t}(x, 0) = 0, \qquad 0 < x$$

$$u(x, t) \qquad \text{bounded as } x \to \infty.$$

Use the solution of the same problem as found in Chapter 3, Section 6, to verify the rule

$$\mathscr{L}^{-1}(e^{-sx}H(s)) = \begin{cases} h(t - x), & t > x \\ 0, & t < x. \end{cases}$$

chapter 7
Numerical Methods

1 Boundary value problems

More often than not, significant practical problems in partial—and even ordinary—differential equations cannot be solved by analytical methods. Difficulties may arise from variable coefficients, irregular regions, unsuitable boundary conditions, interfaces, or just overwhelming detail. Now that machine computation is cheap and easily accessible, numerical methods provide reliable answers to formerly difficult problems. In this chapter we examine a few methods which are simple and equally adaptable to machine or hand computation.

The definition of the derivative of a function

$$\frac{du}{dx} = \lim_{\Delta x \to 0} \frac{u(x + \Delta x) - u(x)}{\Delta x}$$

implies that the derivative can be approximated by a difference quotient, which is the central idea of the numerical methods we use. Three different types of approximation are common:

276

Forward difference: $\dfrac{du}{dx} \simeq \dfrac{u(x + \Delta x) - u(x)}{\Delta x}$

Backward difference: $\dfrac{du}{dx} \simeq \dfrac{u(x) - u(x - \Delta x)}{\Delta x}$

Central difference: $\dfrac{du}{dx} \simeq \dfrac{u(x + \frac{1}{2}\Delta x) - u(x - \frac{1}{2}\Delta x)}{\Delta x}$.

The order of magnitude of the error caused by such an approximation is easily found if the function u has a Taylor series. For instance

$$\frac{u(x + \Delta x) - u(x)}{\Delta x} = \frac{u(x) + \Delta x u'(x) + \frac{1}{2}\Delta x^2 u''(x) + \cdots - u(x)}{\Delta x}$$

$$= u'(x) + \tfrac{1}{2}\Delta x u''(x) + \cdots.$$

Thus the error due to replacing a derivative by a forward difference is about $\frac{1}{2}\Delta x u''(x)$. (Terms containing higher powers of Δx are neglected; this is justified when Δx is small.) The backward difference also has an error of "order Δx." The central difference, however, deviates from the derivative by about $\frac{1}{24}(\Delta x)^2 u^{(3)}(x)$, and for this reason is usually superior to forward and backward differences when Δx is small.

As an example, let $u(x) = \sin x$ at $x = 1$ rad. Table 7.1 contains approximations to $u'(1) = \cos 1 = 0.540302$.

Table 7.1

	$\Delta x = 0.1$	$\Delta x = 0.01$
Forward	0.4974	0.536086
Backward	0.5814	0.544501
Central	0.5400	0.540300

The difference analog of the second derivative can be formed as a difference of differences. For instance, using forward differences twice, we find

$$\frac{d^2 u}{dx^2} \simeq \frac{1}{\Delta x}\left(\frac{du}{dx}(x + \Delta x) - \frac{du}{dx}(x)\right)$$

$$\simeq \frac{1}{\Delta x}\left(\frac{u(x + 2\Delta x) - u(x + \Delta x)}{\Delta x} - \frac{u(x + \Delta x) - u(x)}{\Delta x}\right).$$

Typical replacements for second derivatives are:

Forward: $\dfrac{d^2u}{dx^2} \to \dfrac{u(x + 2\,\Delta x) - 2u(x + \Delta x) + u(x)}{(\Delta x)^2}$

Backward: $\dfrac{d^2u}{dx^2} \to \dfrac{u(x) - 2u(x - \Delta x) + u(x - 2\,\Delta x)}{(\Delta x)^2}$

Central: $\dfrac{d^2u}{dx^2} \to \dfrac{u(x + \Delta x) - 2u(x) + u(x - \Delta x)}{(\Delta x)^2}$

(It is also possible to compound forward with central, backward with forward, and so forth.) An error analysis with the Taylor series of u shows the errors of the forward and backward formulas to be $\Delta x u^{(3)}(x)$, and of the central difference formula to be $\frac{1}{12}(\Delta x)^2 u^{(4)}(x)$.

Now consider the simple boundary value problem,

$$\frac{d^2u}{dx^2} = f(x), \qquad 0 < x < 1$$
$$u(0) = 0, \qquad u(1) = 1$$

(I)

which might represent a steady-state temperature distribution. In order to solve this problem numerically, we first choose a set (mesh) of points on the interval $0 \le x \le 1$:

$$0 = x_0, x_1, \ldots, x_n = 1.$$

(We assume here that the points are equally spaced, so $x_{i+1} - x_i = \Delta x = 1/n$. Equal spacing usually provides the best accuracy.) The function u certainly satisfies

$$u(x_0) = 0, \qquad u(x_n) = 1$$
$$\frac{d^2u(x_i)}{dx^2} = f(x_i), \qquad i = 1, 2, \ldots, n - 1.$$

If a set of numbers $u_0, u_1, \ldots, u_n$ is found which satisfy the difference equation

$$\frac{u_{i+1} - 2u_i + u_{i-1}}{(\Delta x)^2} = f(x_i), \qquad i = 1, 2, \ldots, n - 1$$

with $u_0 = 0$, $u_n = 1$, then we expect the numbers u_i to approximate $u(x_i)$. In general, the u_i are not equal to the values of u at points x_i, nor does the true solution $u(x)$ of the boundary value problem satisfy the difference equation. Nevertheless, when Δx is small, u_i approaches $u(x_i)$.

For a specific example, we take $f(x) = -1$ and $n = 4$, so $\Delta x = \frac{1}{4}$. The equations that replace the differential equation

$$\frac{d^2u}{dx^2} = -1, \qquad 0 < x < 1$$

are

$$16(u_0 - 2u_1 + u_2) = -1$$
$$16(u_1 - 2u_2 + u_3) = -1$$
$$16(u_2 - 2u_3 + u_4) = -1.$$

After replacing u_0 and u_4 by their known values, 0 and 1, and doing some algebra, we have the following equations to solve:

$$-2u_1 + u_2 \qquad\qquad = -\tfrac{1}{16} \tag{1}$$

$$u_1 - 2u_2 + u_3 \ = -\tfrac{1}{16} \tag{2}$$

$$u_2 - 2u_3 = -\tfrac{17}{16}. \tag{3}$$

(Notice that the boundary conditions have been incorporated into the equations for the unknowns.)

These three equations in three unknowns are easily solved by elimination, as illustrated in Table 7.2. The solution procedure is simple: use the first

Table 7.2

Equation	Number	Operation
$u_1 - \tfrac{1}{2}u_2 = \tfrac{1}{32}$	(1′)	$-\tfrac{1}{2} \times$ Eq. (1)
$-\tfrac{3}{2}u_2 + u_3 = -\tfrac{3}{32}$	(2′)	Eq. (2) $-$ Eq. (1′)
$u_2 - \tfrac{2}{3}u_3 = \tfrac{1}{16}$	(2″)	$-\tfrac{2}{3} \times$ Eq. (2′)
$-\tfrac{4}{3}u_3 = -\tfrac{18}{16}$	(3′)	Eq. (3) $-$ Eq. (2″)
$u_3 = \tfrac{27}{32}$	(3″)	$-\tfrac{3}{4} \times$ Eq. (3′)
$u_2 = \tfrac{1}{16} + \tfrac{2}{3} \cdot \tfrac{27}{32} = \tfrac{5}{8}$		
$u_1 = \tfrac{1}{32} + \tfrac{1}{2} \cdot \tfrac{5}{8} = \tfrac{11}{32}$		

equation to eliminate u_1 from the second and third equations; use the modified second equation to eliminate u_2 from the third equation; solve for u_3; solve for u_2; solve for u_1. Of course, u_1 does not actually appear in the third equation, so one operation is saved. In fact, it is typical of difference approximations to boundary value problems that each unknown appears in just a few equations. This process will be recognized by some as Gaussian elimination.

A slightly more complicated example is the problem

$$\frac{d^2u}{dx^2} - u = -2x, \qquad 0 < x < 1$$

$$u(0) = 0, \qquad u(1) = 1.$$

(II)

Again we use a mesh size $\Delta x = \frac{1}{4}$. The replacement equations are

$$16(u_0 - 2u_1 + u_2) - u_1 = -\frac{1}{2}$$
$$16(u_1 - 2u_2 + u_3) - u_2 = -1$$
$$16(u_2 - 2u_3 + u_4) - u_3 = -\frac{3}{2}.$$

In simplified form they become

$$-\tfrac{33}{16} u_1 + \quad u_2 \qquad\qquad = -\tfrac{1}{32}$$
$$u_1 - \tfrac{33}{16} u_2 + \quad u_3 = -\tfrac{1}{16}$$
$$u_2 - \tfrac{33}{16} u_3 = -\tfrac{35}{32}.$$

These equations were solved by elimination on a calculator. The results are compared in Table 7.3 to the exact solution, $u(x) = 2x - (\sinh x/\sinh 1)$.

Table 7.3

x_i	u_i	$u(x_i)$	Error
$\frac{1}{4}$	0.284885	0.285048	0.000163
$\frac{1}{2}$	0.556326	0.556591	0.000265
$\frac{3}{4}$	0.800037	0.800276	0.000239

In the boundary value problem

$$\frac{d^2u}{dx^2} = f(x), \qquad 0 < x < 1$$

$$u(0) = 0, \qquad \frac{du}{dx}(1) = -1$$

(III)

the value of u at the right endpoint is unknown. We assume that the differential equation continues to hold there. Thus the substitute equations are

$$\frac{u_{i+1} - 2u_i + u_{i-1}}{(\Delta x^2)} = f(x_i)$$

supposed valid for $i = 1, 2, \ldots, n$. The last equation

$$\frac{u_{n+1} - 2u_n + u_{n-1}}{(\Delta x)^2} = f(x_n)$$

involves u_{n+1}, which would correspond to a value of u beyond the right end of the interval $0 < x < 1$. However, when the boundary condition at $x = 1$ is replaced by a central difference

$$\frac{u_{n+1} - u_{n-1}}{2\,\Delta x} = -1$$

this extraneous unknown may be eliminated. After this procedure, the last substitute equation becomes

$$\frac{2u_{n-1} - 2u_n - 2\,\Delta x}{(\Delta x)^2} = f(x_n).$$

For instance, with $f(x) = -1$ and $\Delta x = \frac{1}{5}$ $(n = 5)$, the substitute equations are

$$
\begin{aligned}
-2u_1 + u_2 &= -\tfrac{1}{25} \\
u_1 - 2u_2 + u_3 &= -\tfrac{1}{25} \\
u_2 - 2u_3 + u_4 &= -\tfrac{1}{25} \\
u_3 - 2u_4 + u_5 &= -\tfrac{1}{25} \\
2u_4 - 2u_5 &= -\tfrac{1}{25} + \tfrac{2}{5}.
\end{aligned}
\tag{4}
$$

The boundary condition involving both u and its derivative may occur also. As in the preceding case, u itself at the boundary is unknown, and a difference equation for that value of u must be included with the substitute equations. Then any extraneous value of u is eliminated by using the approximation to the boundary condition. For example, in

$$\frac{d^2u}{dx^2} = f(x), \qquad 0 < x < 1 \tag{IV}$$

$$u(0) - \gamma u'(0) = c, \qquad u(1) = 0$$

the first substitute equation and the boundary condition at $x = 0$ are

$$\frac{u_{-1} - 2u_0 + u_1}{(\Delta x)^2} = f(0) \tag{5}$$

$$u_0 - \gamma\,\frac{u_1 - u_{-1}}{2\,\Delta x} = c \tag{6}$$

One finds $u_{-1} = (2\,\Delta x/\gamma)(c - u_0) + u_1$ and replaces the u_{-1} in Eq. (6) by this expression.

Warning: Not all boundary value problems have solutions; some have solutions that become very large or oscillate violently. The problem $u'' + ku = 0$, $u(0) = 0$, $u(1) = -1$, has no solution when $k = \pi^2$. When $k = 10$, the solution exists, but has a maximum of about 50. One qualitative check on numerical results is to sketch a graph from the numerical solution. If the graph bends sharply or suddenly the higher-order derivatives of u are probably large. In such cases a very small Δx is necessary for good accuracy.

Exercises

1. Determine the exact solution of the first example problem and compare $u(x_i)$ to u_i ($i = 1, 2, 3$). Why is there no error?

2. Eliminate $u'(x)$ from the following Taylor series to find the central difference approximation to $u''(x)$:

$$u(x + \Delta x) = u(x) + \Delta x u'(x) + \tfrac{1}{2}(\Delta x)^2 u''(x) + \cdots$$
$$u(x - \Delta x) = u(x) - \Delta x u'(x) + \tfrac{1}{2}(\Delta x)^2 u''(x) - \cdots.$$

3. Same as Exercise 2, but take $u(x + \Delta x_1)$ and $u(x - \Delta x_2)$, where Δx_1 and Δx_2 may be different.

4. Use the Taylor series to find the order of magnitude of the error due to replacement of a second derivative with the difference expressions of Exercises 2 and 3.

5. Find a central difference approximation to the fourth derivative $u^{(4)}(x)$, using values of u at x, $x \pm \Delta x$, $x \pm 2\,\Delta x$.

6. Set up and solve substitute equations for problem (IV), taking $n = 4$. Remember that u_0 is unknown, u_4 is known. Take $\gamma = 1$, $c = 1$, $f(x) = x$.

7. Solve by elimination Eq. (4).

8. Set up and solve the substitute equations for $u'' + 10u = 0$, $u(0) = 0$, $u(1) = -1$, taking $n = 3$ and $n = 4$. Sketch your results.

9. Set up and solve replacement equations for this problem, using $\Delta x = \tfrac{1}{4}$.

$$\frac{d^2u}{dx^2} - 32xu = 0$$

$$u(0) = 0, \qquad u(1) = 1.$$

10. Use Taylor series to find the errors in the forward, backward and central difference approximations to the second derivative.

2   The heat equation

In the numerical treatment of a heat conduction problem such as*

$$\frac{\partial^2 u}{\partial x^2} = \frac{\partial u}{\partial t}, \qquad\qquad 0 < x < 1, \quad 0 < t$$

$$u(0, t) = 0, \qquad u(1, t) = 0, \qquad 0 < t$$

$$u(x, 0) = f(x), \qquad\qquad 0 < x < 1$$

(I)

we replace both the space and time derivatives by difference approximations. We shall take a uniform space mesh $\Delta x = 1/n$ and a time step Δt. The partial derivative with respect to x is replaced by a central difference and the partial with respect to t by a forward difference. We use the notation $u_i(m)$ to represent the approximation to $u(x_i, t_m)$, where $x_i = i \, \Delta x$ and $t_m = m \, \Delta t$. The replacement equations for the problem above are

$$\frac{u_{i-1}(m) - 2u_i(m) + u_{i+1}(m)}{(\Delta x)^2} = \frac{u_i(m+1) - u_i(m)}{\Delta t}$$

supposed valid for $i = 1, 2, \ldots, n-1$ and $m = 0, 1, 2, \ldots$.

The point of using a forward difference for the time derivative is that these equations may be solved for $u_i(m+1)$:

$$u_i(m+1) = r u_{i-1}(m) + (1 - 2r)u_i(m) + r u_{i+1}(m)$$

where $r = \Delta t/(\Delta x)^2$. Thus each $u_i(m+1)$ is calculated from values of u at the preceding time. Since the initial condition gives each $u_i(0)$, the values of the us at time 1 can be computed, then the values of the us at time 2 are gotten from these, and so on into the future.

As an example, we take $\Delta x = \frac{1}{4}$ and $r = \frac{1}{2}$ ($\Delta t = \frac{1}{32}$). The equations giving the us at $m + 1$ are

$$u_1(m+1) = \tfrac{1}{2}(u_0(m) + u_2(m))$$
$$u_2(m+1) = \tfrac{1}{2}(u_1(m) + u_3(m)) \qquad\qquad (1)$$
$$u_3(m+1) = \tfrac{1}{2}(u_2(m) + u_4(m)).$$

Recall that the boundary conditions specify $u_0(m) = 0$ and $u_4(m) = 0$, if $m = 1, 2, 3, \ldots$, and the initial condition gives $u_i(0) = f(x_i)$, for $i = 0, 1,$

---

* We assume throughout that dimensional analysis has been applied to reduce the number of parameters to a minimum.

2, 3, 4. In Table 7.4 are calculated values of $u_i(m)$ for the initial condition $f(x) = x$. Numbers in italics represent *given* information.

Table 7.4

m \ i	0	1	2	3	4
0	*0*	*0.25*	*0.5*	*0.75*	*1*
1	*0*	0.25	0.5	0.75	*0*
2	*0*	0.25	0.5	0.25	*0*
3	*0*	0.25	0.25	0.25	*0*
4	*0*	0.125	0.25	0.125	*0*
5	*0*	0.125	0.125	0.125	*0*

The initial condition, $u(x, 0) = x$, $0 < x < 1$, suggests that $u(1, 0)$ should be 1, while the boundary condition suggests that it should be 0. In fact, neither condition specifies $u(1, 0)$, nor is there a hard and fast rule telling what to do in case of conflict. Fortunately, it doesn't matter much, either. (See Exercise 1.)

Equation (1) says that each entry in the unitalicized portion of the table is simply the average of the entries to the left and right on the preceding line. There is also a simple graphical interpretation. Suppose the initial temperature distribution $u(x, 0)$ is graphed along with the vertical lines $x = 0$, $x = \Delta x$, $x = 2\,\Delta x, \ldots, x = 1$. Then $u_i(1)$ is located at the intersection of the line $x = i\,\Delta x$ with the line joining $u_{i-1}(0)$ and $u_{i+1}(0)$. When the $u_i(1)$ are found for $i = 1, 2, \ldots, n - 1$, the $u_i(2)$ are located by the same proceedure. In Fig. 7.1 this process is illustrated for the initial condition $u(x, 0) = 4x(1 - x)$.

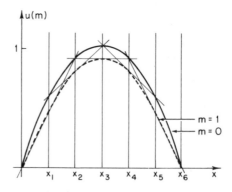

Figure 7.1 Graphical solution of the heat equation with temperature zero at the boundaries and initial condition $u(x, 0) = 4x(1 - x)$.

The choice we made of $r = \frac{1}{2}$ seems natural, perhaps, because it simplifies the computation. It might also seem desirable to take a larger value of r (signifying a larger time step) to get into the future more rapidly. For example, with $r = 1(\Delta t = \frac{1}{16})$ the replacement equations take the form

$$u_i(m + 1) = u_{i-1}(m) - u_i(m) + u_{i+1}(m).$$

In Table 7.5 are values of $u_i(m)$ computed from this formula.

Table 7.5

i \ m	0	1	2	3	4
0	0	0.25	0.50	0.75	1.0
1	0	0.25	0.50	0.75	0
2	0	0.25	0.50	−0.25	0
3	0	0.25	−0.50	0.75	0
4	0	−0.75	1.50	−1.25	0
5	0	2.25	−3.50	2.75	0

No one can believe that these wildly fluctuating values approximate the solution to the heat problem in any sense. Indeed, they suffer from *instability* due to using a time step too long relative to the mesh size. The analysis of instability requires familiarity with matrix theory, but there are some simple rules of thumb that guarantee stability.

First, write out the equations for each $u_i(m + 1)$ in terms of values of u at the preceding time level. The coefficients of these equations must satisfy two conditions:

(a)   No coefficient of a $u(m)$ is negative;
(b)   The sum of the coefficients of the $u(m)$ does not exceed 1.

In the example, the replacement equations were

$$u_1(m + 1) = ru_0(m) + (1 - 2r)u_1(m) + ru_2(m)$$
$$u_2(m + 1) = ru_1(m) + (1 - 2r)u_2(m) + ru_3(m)$$
$$u_3(m + 1) = ru_2(m) + (1 - 2r)u_3(m) + ru_4(m).$$

The second requirement is satisfied automatically, since $r + (1 - 2r) + r = 1$. But the first condition is satisfied only for $r \leq \frac{1}{2}$. Thus the first choice of $r = \frac{1}{2}$ corresponded to the longest stable time step.

Different problems give different maximum values for r. For the heat conduction problem

$$\frac{\partial^2 u}{\partial x^2} = \frac{\partial u}{\partial t}, \qquad 0 < x < 1, \quad 0 < t$$

$$u(0, t) = 1, \qquad \frac{\partial u}{\partial x}(1, t) + \gamma u(1, t) = 0, \qquad 0 < t \tag{II}$$

$$u(x, 0) = 0, \qquad 0 < x < 1$$

the substitute equations are found to be (for $n = 4$)

$$u_1(m + 1) = r u_0(m) + (1 - 2r)u_1(m) + r u_2(m)$$
$$u_2(m + 1) = r u_1(m) + (1 - 2r)u_2(m) + r u_3(m)$$
$$u_3(m + 1) = r u_2(m) + (1 - 2r)u_3(m) + r u_4(m)$$
$$u_4(m + 1) = 2 r u_3(m) + (1 - 2r - \tfrac{1}{2}r\gamma)u_4(m).$$

(Remember that $u(1, t)$, corresponding to u_4, is an unknown. The boundary condition has been incorporated into the equation for $u_4(m + 1)$.) Again, the second stability requirement is satisfied automatically; but the first rule requires that

$$1 - 2r - \tfrac{1}{2}r\gamma \geq 0 \qquad \text{or} \qquad r \leq \frac{1}{(2 + \tfrac{1}{2}\gamma)}.$$

Exercises

1. Compute an approximate solution to the first example problem with $f(x) = x$. Take $u_4(0) = 0$, $r = \tfrac{1}{2}$ and compare your results with Table 7.4.

2. Compute a numerical solution to the first example problem using first $r = \tfrac{1}{2}$ then $r = \tfrac{1}{4}$, and compare the results at corresponding times.

3. For the second example, find the longest stable time step when $\gamma = 1$, and compute the numerical solution using the corresponding value of r.

4. Compute a graphical solution to the first example.

For each problem below, set up the replacement equations for $n = 4$, compute the longest stable time step, and calculate the numerical solution for a few values of m.

5. $\dfrac{\partial^2 u}{\partial x^2} = \dfrac{\partial u}{\partial t}$, $\qquad u(0, t) = u(1, t) = t$, $\qquad u(x, 0) = 0$

6. $\dfrac{\partial^2 u}{\partial x^2} - u = \dfrac{\partial u}{\partial t}$, $u(0, t) = u(1, t) = 1$, $u(x, 0) = 0$

7. $\dfrac{\partial^2 u}{\partial x^2} = \dfrac{\partial u}{\partial t} - 1$, $u(0, t) = u(1, t) = 0$, $u(x, 0) = 0$

8. $\dfrac{\partial^2 u}{\partial x^2} = \dfrac{\partial u}{\partial t}$, $u(0, t) = 0$, $\dfrac{\partial u}{\partial x}(1, t) + u(1, t) = 1$, $u(x, 0) = 0$

9. $\dfrac{\partial^2 u}{\partial x^2} = \dfrac{\partial u}{\partial t}$, $\dfrac{\partial u}{\partial x}(0, t) = 0$, $u(1, t) = 1$, $u(x, 0) = x$

3   Two-dimensional heat equation

The variety of two-dimensional problems solvable by separation of variables is severely limited by the kinds of regions and boundary conditions that can be handled. Even problems that can be solved suffer from considerable difficulty, as we noted before. However, simple numerical techniques work quite well on many two-dimensional problems.

In order to see how the numerical attack works, consider the problem:

$$\frac{\partial^2 u}{\partial x^2} + \frac{\partial^2 u}{\partial y^2} = \frac{\partial u}{\partial t}, \qquad 0 < x < 1.25, \quad 0 < y < 1$$

$$u(0, y, t) = u(1.25, y, t) = u(x, 0, t) = u(x, 1, t) = 0 \tag{I}$$

$$u(x, y, 0) = 1.$$

First, we make a rectangular mesh or grid of points with spacings Δx and Δy (Fig. 7.2). (Variable spacing is, of course, possible.) The Laplacian operator

Figure 7.2 Five points for the discretized Laplacian operator.

in the heat equation is to be replaced by a sum of central differences (five-point formula):

$$\frac{\partial^2 u}{\partial x^2} + \frac{\partial^2 u}{\partial y^2} \to \frac{u(x + \Delta x, y, t) - 2u(x, y, t) + u(x - \Delta x, y, t)}{(\Delta x)^2}$$

$$+ \frac{u(x, y + \Delta y, t) - 2u(x, y, t) + u(x, y - \Delta y, t)}{(\Delta y)^2}. \quad (1)$$

The time derivative is to be replaced by a forward difference, as before:

$$\frac{\partial u}{\partial t} \to \frac{u(x, y, t + \Delta t) - u(x, y, t)}{\Delta t}. \quad (2)$$

In the sample problem we choose $\Delta x = \Delta y = \frac{1}{4}$ and number each mesh point with a single integer as shown in Fig. 7.3. (Sometimes it is more con-

Figure 7.3

venient to use two integers corresponding to x and y distances, as shown in Fig. 7.4.) The natural extension of our previous notation makes the correspondence

$$u(\tfrac{1}{4}, \tfrac{1}{2}, m\,\Delta t) \to u_5(m)$$

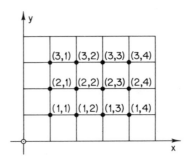

Figure 7.4 Scheme for numbering points with two indices.

and so forth. It must be borne in mind that some points (in this example, all but 6 and 7) are adjacent to boundary points where the temperature is known —and may vary with time.

The typical replacement equation is

$$\frac{u_i(m+1) - u_i(m)}{\Delta t} = 16[u_N(m) + u_S(m) + u_E(m) + u_W(m) - 4u_i(m)]$$

where the subscripts N, S, E, W stand for the numbers associated with points adjacent to point i. When this equation is solved for $u_i(m+1)$, it becomes

$$u_i(m+1) = r[u_N(m) + u_S(m) + u_E(m) + u_W(m)] + (1 - 4r)u_i(m) \qquad (3)$$

in which

$$r = \frac{\Delta t}{\Delta x^2} = \frac{\Delta t}{\Delta y^2} = 16 \, \Delta t.$$

The stability considerations of Section 2 are still important and the rules of thumb still valid. We must limit r by the requirement that $1 - 4r \geq 0$, or, in this case, $\Delta t \leq \frac{1}{64}$. We shall take the longest acceptable time step, $\Delta t = \frac{1}{64}$, $r = \frac{1}{4}$, which makes the equations a little simpler.

At $m = 0$, all temperatures are given as 1. For $m \geq 1$, all the boundary temperatures are zero and the $u_i(m)$ are all found to equal 1. For $m = 2$, we calculate

$$u_1(2) = \tfrac{1}{4}(u_2(1) + u_5(1) + 0 + 0) = \tfrac{1}{2}$$
$$u_2(2) = \tfrac{1}{4}(u_1(1) + u_3(1) + u_6(1) + 0) = \tfrac{3}{4}$$
$$\vdots$$
$$u_5(2) = \tfrac{1}{4}(u_1(1) + u_6(1) + u_9(1) + 0) = \tfrac{3}{4}$$
$$u_6(2) = \tfrac{1}{4}(u_2(1) + u_5(1) + u_7(1) + u_{10}(1)) = 1.$$

The zeros in these equations stand for boundary temperatures.

An alert calculator will notice that only the unknowns u_1, u_2, u_5, u_6 need be calculated, since, in this example, the others will be given at each time step by symmetry:

$$u_1(m) = u_4(m) = u_9(m) = u_{12}(m), \quad u_5(m) = u_8(m),$$
$$u_6(m) = u_7(m), \quad u_2(m) = u_3(m) = u_{10}(m) = u_{11}(m).$$

In Table 7.6 are computed values of the significant us at a few times.

Now we look at a problem not solvable by separation. A long L-shaped bar is initially at temperature zero. The temperature on the boundary is increased steadily until a certain temperature is reached, then is held at that temperature. In Fig. 7.5 is a representation of the L-shaped region we consider, on which a mesh with spacing $\Delta x = \Delta y = \frac{1}{5}$ has been set up.

Table 7.6

i $\diagup$ m	1	2	5	6
0	1	1	1	1
1	$\frac{1}{2}$	$\frac{3}{4}$	$\frac{3}{4}$	1
2	$\frac{3}{8}$	$\frac{9}{16}$	$\frac{1}{2}$	$\frac{13}{16}$
3	$\frac{17}{64}$	$\frac{7}{16}$	$\frac{25}{64}$	$\frac{39}{64}$

The replacement equations are found, as in the first example, to be

$$u(x, y, t + \Delta t) = r[u(x + \Delta x, y, t) + u(x - \Delta x, y, t)$$
$$+ u(x, y + \Delta y, t) + u(x, y - \Delta y, t)]$$
$$+ (1 - 4r)u(x, y, t)$$

where

$$r = \frac{\Delta t}{\Delta x^2} = \frac{\Delta t}{\Delta y^2} = 25\,\Delta t.$$

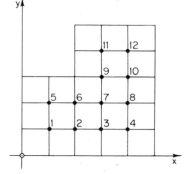

Figure 7.5

The longest stable time step is $\Delta t = 1/100$, corresponding to $r = \frac{1}{4}$. Symmetry about the line through points 4 and 7 reduces the problem to determining u_1 through u_7. Some of the equations used are:

$$u_1(m + 1) = \tfrac{1}{4}(u_2(m) + u_5(m) + 2f(m))$$
$$u_2(m + 1) = \tfrac{1}{4}(u_1(m) + u_3(m) + u_6(m) + f(m))$$
$$u_4(m + 1) = \tfrac{1}{2}(u_3(m) + f(m))$$
$$u_7(m + 1) = \tfrac{1}{2}(u_3(m) + u_6(m)).$$

In these equations, $f(m)$ represents the boundary temperature

$$f(m) = \begin{cases} m\,\Delta t, & 0 \le m \le M \\ M\,\Delta t, & M < m. \end{cases}$$

Table 7.7 contains calculated values of u for the first four time levels.

	i	1	2	3	4	5	6	7	f(m)
m									
0		0	0	0	0	0	0	0	0
1		0	0	0	0	0	0	0	1.0
2		0.5	0.25	0.25	0.5	0.5	0.25	0	2.0
3		1.22	0.75	0.69	0.75	1.22	0.69	0.25	3.0
4		1.99	1.40	1.19	1.84	1.98	1.30	0.69	4.0

Table 7.7   Entries are $100 \times u_i(m)$

Exercises

1. Find the complete set of replacement equations (using the symmetry conditions) for the first sample problem, and extend Table 7.6 for several more steps.

2. Same as Exercise 1 for the second example.

3. Set up replacement equations for the heat conduction problem in a square, when the conductivity is different in the x and y directions:

$$\frac{\partial^2 u}{\partial x^2} + \gamma \frac{\partial^2 u}{\partial y^2} = \frac{\partial u}{\partial t}, \qquad 0 < x < 1, \quad 0 < y < 1, \quad 0 < t$$

with initial conditions zero and boundary condition $u = 1$. How is the longest stable time step related to Δx, Δy, and γ?

4   Potential equation

In this section we shall be concerned with the approximate solution of the potential and other elliptic equations. The simple problem

$$\frac{\partial^2 u}{\partial x^2} + \frac{\partial^2 u}{\partial y^2} = 0, \qquad\qquad 0 < x < 1, \quad 0 < y < 1 \quad (1)$$

$$u(0, y) = 0, \qquad u(1, y) = 0, \qquad\qquad 0 < y < 1 \qquad\qquad (2)$$

$$u(x, 0) = u(x, 1) = f(x) = \begin{cases} 2x, & 0 < x < \frac{1}{2} \\ 2 - 2x, & \frac{1}{2} \le x < 1 \end{cases} \qquad (3)$$

which was solved in Chapter 4, will serve as a model. We first choose a rectangular mesh in the region considered and replace the derivatives in the partial differential equation by central differences. The potential equation (1) then becomes

$$\frac{u(x + \Delta x, y) - 2u(x, y) + u(x - \Delta x, y)}{(\Delta x)^2}$$

$$+ \frac{u(x, y + \Delta y) - 2u(x, y) + u(x, y - \Delta y)}{(\Delta y)^2} \simeq 0.$$

Obviously we should choose $\Delta x = \Delta y$ in this case; then we may multiply through the replacement equation by $(\Delta x)^2$ to obtain

$$u(x + \Delta x, y) + u(x - \Delta x, y) + u(x, y + \Delta y) + u(x, y - \Delta y) - 4u(x, y) \simeq 0.$$

(Incidentally, when solved for $u(x, y)$, this equation expresses the finite-difference analog of the mean-value property mentioned in Chapter 4. In words, $u(x, y)$ equals the average of its four neighbors.)

Taking a mesh size of $\Delta x = \Delta y = \frac{1}{4}$, and numbering the mesh points as shown in Fig. 7.6 we will find the following equations to replace Eqs. (1)–(3)

Figure 7.6

(the entries on the right come from the boundary conditions):

$$-4u_1 + u_2 + u_4 = -\tfrac{1}{2}$$
$$u_1 - 4u_2 + u_3 + u_5 = -1$$
$$u_2 - 4u_3 + u_6 = -\tfrac{1}{2}$$
$$u_1 - 4u_4 + u_5 + u_7 = 0$$
$$u_2 + u_4 - 4u_5 + u_6 + u_8 = 0$$

and so forth. Symmetry in the problem shows that only u_1, u_2, u_4, u_5 need be found. Elimination was used to get the solution shown in Fig. 7.7.

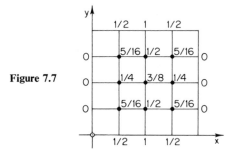

Figure 7.7

The mesh size we have used is laughably coarse, however, and the approximation to the solution of the analytical problem Eqs. (1)–(3) is correspondingly crude. The more realistic choice of $\Delta x = \Delta y = \frac{1}{10}$ would lead to 81 equations (reduced to 25 by symmetry). Less regular boundary conditions could easily force a still finer mesh. In fact, problems involving tens of thousands of unknowns occur daily in industry.

Even for a digital computer, the solution of large systems of algebraic equations by elimination may not be practicable because of (a) storage requirements and (b) accumulation of roundoff errors after many operations. Thus we turn our attention to the *iterative* solution of algebraic equations arising from discretization of elliptic equations.

The simplest iterative scheme (a form of Jacobi iteration) is based on this idea: The original problem Eqs. (1)–(3) can be interpreted as the limiting case $(t \rightarrow \infty)$ of the heat-conduction problem

$$\frac{\partial^2 u}{\partial x^2} + \frac{\partial^2 u}{\partial y^2} = \frac{\partial u}{\partial t}, \qquad\qquad 0 < x < 1, \quad 0 < y < 1, \quad 0 < t \quad (4)$$

$$u(0, y, t) = 0, \qquad u(1, y, t) = 0, \qquad 0 < y < 1, \quad 0 < t \qquad\qquad (5)$$

$$u(x, 0, t) = u(x, 1, t) = f(x), \qquad 0 < x < 1, \quad 0 < t \qquad\qquad (6)$$

under *any* initial condition. Thus we may solve the original discretized problem by setting up the heat problem Eqs. (4)–(6) in discretized form (using the same x–y mesh, of course) with some initial condition, then compute the approximations $u_i(m)$ until the differences between successive time steps are insignificant. It is natural to use the longest stable time step; indeed, the whole object of stability analysis is to guarantee that the solution of the discretized heat problem actually tends to the solution of the discretized steady-state problem.

If this procedure is carried out for the example problem, the equations we work with are

$$u_1(m + 1) = \tfrac{1}{4}(u_2(m) + u_4(m) + \tfrac{1}{2})$$
$$u_2(m + 1) = \tfrac{1}{4}(2u_1(m) + u_5(m) + 1)$$
$$u_4(m + 1) = \tfrac{1}{4}(2u_1(m) + u_5(m))$$
$$u_5(m + 1) = \tfrac{1}{4}(2u_2(m) + 2u_4(m)).$$

(The symmetry conditions $u_1 = u_3 = u_7 = u_9$, $u_2 = u_8$, $u_4 = u_6$ have been used.)

To get started, we must choose $u_1(0)$, $u_2(0)$, $u_4(0)$, $u_5(0)$; the choice is arbitrary, since the steady-state is independent of the initial conditions. One may simply take all $u_i(0) = 0$, but iterative methods share the property that *a good initial guess is rewarded by shortened computation*. Indeed, if we could guess the solution to the equations, one iteration would be excessive. Table 7.8 illustrates the effect of a good initial guess in shortening work, or rather in giving improved accuracy with the same amount of work.

Table 7.8   Jacobi iterations[a]

m \ i	1	2	4	5	1	2	4	5
0	0	0	0	0	16	48	16	32
1	8	16	0	0	24	32	16	32
2	12	20	4	8	20	36	20	24
3	14	24	8	12	22	32	16	28
4	16	26	10	16	20	34	18	24
5	17	28	12	18	21	32	16	26
6	18	29	13	20	20	33	17	24
⋮								
∞	20	32	16	24	20	32	16	24

[a] Entries are $u_i(m) \times 64$.

It is clear that iterative methods cannot be expected to give the exact solution of the difference equations. However, since these are approximations to the boundary value problem (which itself is an approximate representation of a physical problem), the exact solution of the difference equations may not be a worthwhile goal.

A distinct advantage of iterative methods over elimination is that they are *self correcting*, in the sense that an arithmetic error at any stage is eventually suppressed. In fact, in hand computation one often makes deliberate altera-tions to hurry things along. This feature is very important in machine com-putation, for it diminishes the effect of roundoff error.

The method of Jacobi can be improved upon by always using the best—most recently calculated—value of each variable. This method, called Gauss–Seidel iteration, usually converges about twice as fast as Jacobi iteration for these problems and has the additional advantage of saving some storage space in a computer. The advantages are so substantial that Jacobi iteration probably should never be used.

To effect Gauss–Seidel iteration, the u_is are calculated in a fixed sequence at every stage. In our simple example, using the natural sequence 1, 2, 4, 5, the equations employed are

$$u_1(m+1) = \tfrac{1}{4}(u_2(m) + u_4(m) + \tfrac{1}{2})$$
$$u_2(m+1) = \tfrac{1}{4}(2u_1(m+1) + u_5(m) + 1)$$
$$u_4(m+1) = \tfrac{1}{4}(2u_1(m+1) + u_5(m))$$
$$u_5(m+1) = \tfrac{1}{4}(2u_2(m+1) + 2u_4(m+1)).$$

In Table 7.9 are the first few iterates of the Jacobi and Gauss–Seidel methods, both starting from zero.

Table 7.9[a]

	Jacobi				Gauss–Seidel			
i m	1	2	4	5	1	2	4	5
0	0	0	0	0	0	0	0	0
1	8	16	0	0	8	20	4	12
2	12	20	4	8	14	26	10	18
3	14	24	8	12	17	29	13	21
⋮								
∞	20	32	16	24	20	32	16	24

[a] Entries are $u_i(m) \times 64$.

The Gauss–Seidel method is easy to understand and to program. It may be used occasionally for small problems. However, methods are available that are far superior to this when large problems are to be solved.

Exercises

Set up and solve replacement equations for each of the following problems. Use symmetry to reduce the number of unknowns.

1. $\nabla^2 u = -1$, $0 < x < 1$, $0 < y < 1$, $u = 0$ on the boundary. $\Delta x = \Delta y = \tfrac{1}{4}$.

2. Same as Exercise 1 with $\Delta x = \Delta y = \tfrac{1}{8}$. Compare the solutions.

3. $\nabla^2 u = 0$, $0 < x < 1$, $0 < y < 1$, $u(0, y) = 0$, $u(x, 0) = 0$, $u(1, y) = y$, $u(x, 1) = x$. $\Delta x = \Delta y = \frac{1}{4}$.

4. Same as Exercise 3 with $\Delta x = \Delta y = \frac{1}{8}$.

5. The region R is a square of side 1 from the center of which a similar square of side $\frac{1}{7}$ has been removed. $\nabla^2 u = 0$ in R, $u = 0$ on the outside boundary, and $u = 1$ on the inside boundary. $\Delta x = \Delta y = \frac{1}{7}$.

6. Same as Exercise 5, but the partial differential equation is $\nabla^2 u = -1$, and the boundary condition is $u = 0$ on all boundaries.

7. The L-shaped region R is a square of side 1 from which is removed a similar square of side $\frac{1}{4}$ in the upper left corner. $\nabla^2 u = -1$ in R and $u = 0$ on the boundary. $\Delta x = \Delta y = \frac{1}{4}$.

5 Wave equation

The simple vibrating string problem that we studied in Chapter 3,

$$\frac{\partial^2 u}{\partial x^2} = \frac{\partial^2 u}{\partial t^2}, \qquad\qquad 0 < x < 1, \quad 0 < t \qquad (1)$$

$$u(0, t) = 0, \qquad u(1, t) = 0, \qquad\qquad 0 < t \qquad (2)$$

$$u(x, 0) = f(x), \qquad \frac{\partial u}{\partial t}(x, 0) = g(x), \qquad 0 < x < 1, \qquad (3)$$

rarely needs treatment by numerical methods, since the D'Alembert solution provides a simple and direct means of calculating the solution $u(x, t)$ for arbitrary x and t. Even if the partial differential equation is nonhomogeneous or if the boundary conditions are of more complicated types, the D'Alembert solution can often be adapted to provide a solution. Indeed, the so called "method of characteristics," of which D'Alembert's solution is the simplest case, provides a technique for solving even nonlinear wave equations.

In two dimensions, however, D'Alembert's idea doesn't work, and the analytical solutions are unsatisfactory. There, simple numerical techniques are quite rewarding. As a guide for ourselves in the two-dimensional case, we first develop a numerical technique for the approximate solution of the vibrating string problem.

In order to convert the wave equation (1) to difference form, we replace both the space and time derivatives with their central difference approximations:

$$\frac{\partial^2 u}{\partial x^2} \rightarrow \frac{u(x + \Delta x, t) - 2u(x, t) + u(x - \Delta x, t)}{(\Delta x)^2}$$

$$\frac{\partial^2 u}{\partial t^2} \rightarrow \frac{u(x, t + \Delta t) - 2u(x, t) + u(x, t - \Delta t)}{(\Delta t)^2}.$$

The replacement equations are then found to be

$$u_i(m + 1) - 2u_i(m) + u_i(m - 1) = \rho^2(u_{i+1}(m) - 2u_i(m) + u_{i-1}(m))$$

where $\rho = \Delta t / \Delta x$ and, as in the case of the heat equation, the subscript i refers to point $x_i = i \, \Delta x$ and the m in parentheses refers to time $t_m = m \, \Delta t$. (It is assumed that $\Delta x = 1/n$ for some integer n.)

The replacement equations may be solved for the unknowns $u_i(m + 1)$, yielding the equation

$$u_i(m + 1) = \rho^2 u_{i-1}(m) + 2(1 - \rho^2)u_i(m) + \rho^2 u_{i+1}(m) - u_i(m - 1), \quad (4)$$

valid for $i = 1, 2, \ldots, n - 1$. Naturally, the boundary conditions, Eq. (2), carry over as $u_0(m) = 0$, $u_n(m) = 0$. It is obvious that Eq. (4) requires us to know the approximate solution at time levels m and $m - 1$ in order to find it at time level $m + 1$. In other words, to get $u_i(1)$ we need $u_{i-1}(0)$, $u_i(0)$, $u_{i+1}(0)$ —which are available from the initial condition—and also $u_i(-1)$! Of course, we have not yet applied the second initial condition,

$$\frac{\partial u}{\partial t}(x, 0) = g(x), \qquad 0 < x < 1.$$

If we replace the time derivative by a central difference approximation, this equation translates into

$$\frac{u_i(1) - u_i(-1)}{2 \, \Delta t} = g(x_i) \qquad (5)$$

for $i = 1, 2, \ldots, n - 1$. Equation (5), together with a slightly modified version of Eq. (4) (with $m = 0$ and $u_i(0) = f(x_i)$), yields the system

$$u_i(1) + u_i(-1) = \rho^2 f(x_{i-1}) + 2(1 - \rho^2)f(x_i) + \rho^2 f(x_{i+1})$$
$$u_i(1) - u_i(-1) = 2 \, \Delta t g(x_i), \qquad\qquad\qquad\qquad (6)$$

which we can easily solve for the us at the first time level:

$$u_i(1) = \tfrac{1}{2}\rho^2 f(x_{i-1}) + (1 - \rho^2)f(x_i) + \tfrac{1}{2}\rho^2 f(x_{i+1}) + \Delta t g(x_i). \qquad (7)$$

Let us now attempt to solve a simple problem using Eq. (7) to start up and Eq. (4) to run. Suppose that $g(x) \equiv 0$ for $0 < x < 1$ and that $f(x)$ is given by

$$f(x) = \begin{cases} 2x, & 0 < x < \tfrac{1}{2} \\ 2(1-x), & \tfrac{1}{2} < x < 1. \end{cases}$$

Also, we shall choose $n = 4$ and $\rho = 1$ for convenience. (That is, $\Delta t = \Delta x = \tfrac{1}{4}$). Our rule for calculation, Eq. (4), is then

$$u_i(m+1) = u_{i-1}(m) + u_{i+1}(m) - u_i(m-1).$$

In Table 7.10 are the calculated values of $u_i(m)$. Entries in italics are given data. It is easy to check that this numerical solution is identical with the

Table 7.10

m \ i	0	1	2	3	4
0	*0*	*1/2*	*1*	*1/2*	*0*
1	*0*	$\tfrac{1}{2}$	$\tfrac{1}{2}$	$\tfrac{1}{2}$	*0*
2	*0*	0	0	0	*0*
3	*0*	$-\tfrac{1}{2}$	$-\tfrac{1}{2}$	$-\tfrac{1}{2}$	*0*
4	*0*	$-\tfrac{1}{2}$	-1	$-\tfrac{1}{2}$	*0*
5	*0*	$-\tfrac{1}{2}$	$-\tfrac{1}{2}$	$-\tfrac{1}{2}$	*0*
6	*0*	0	0	0	*0*

D'Alembert solution of this particular problem. (See Exercise 6.) However, if the initial velocity were not identically zero, the numerical solution would in general be only an approximation to the true solution.

In our study of the heat equation, Section 2, we saw that the choice of Δx and Δt was not free. The same is true for the wave equation. Suppose we attempt to solve the same problem as above, but with $\rho^2 = (\Delta t / \Delta x)^2$ chosen to be 2. Then Eq. (4) becomes

$$u_i(m+1) = 2(u_{i-1}(m) - u_i(m) + u_{i+1}(m)) - u_i(m-1)$$

and the "solution" corresponding to this rule of calculation is shown in Table 7.11. Of course, the results bear no resemblance to the solution of the wave equation. They suffer from the same sort of instability as that observed in Section 2. There is a rule of thumb, similar to the one to be found there, applicable to the wave equation.

First, write out the equations for each $u_i(m+1)$ in terms of values of u at time levels m and $m-1$. The coefficients of these equations must satisfy the two conditions:

Table 7.11

i \ m	0	1	2	3	4
0	0	$1/2$	1	$1/2$	0
1	0	$\frac{1}{2}$	0	$\frac{1}{2}$	0
2	0	$-\frac{3}{2}$	1	$-\frac{3}{2}$	0
3	0	$\frac{9}{2}$	-8	$\frac{9}{2}$	0
4	0	$-\frac{47}{2}$	33	$-\frac{47}{2}$	0

(a)   No coefficient of a $u(m)$ is negative;

(b)   the sum of the coefficients of the $u(m)$ does not exceed 2.

Of course, $u_i(m-1)$ appears with a coefficient of -1; nothing can be done about that, nor does it enter into the rules above.

In Eq. (4) we see that both conditions are met when $\rho = \Delta t/\Delta x$ is less than or equal to one: in other words, the time step must not exceed the space step.

We have by now built up enough experience to deal with two-dimensional wave problems fairly easily. In order to illustrate the techniques, let us consider the vibrations of a square membrane, as described by this problem:

$$\frac{\partial^2 u}{\partial x^2} + \frac{\partial^2 u}{\partial y^2} = \frac{\partial^2 u}{\partial t^2}, \qquad\qquad 0 < x < 1, \quad 0 < y < 1, \quad 0 < t \qquad (8)$$

$$u(x, 0, t) = 0, \, u(x, 1, t) = 0, \qquad 0 < x < 1, \quad 0 < t \qquad\qquad (9)$$

$$u(0, y, t) = 0, \, u(1, y, t) = 0, \qquad 0 < y < 1, \quad 0 < t \qquad\qquad (10)$$

$$u(x, y, 0) = f(x, y), \qquad\qquad 0 < x < 1, \quad 0 < y < 1 \qquad\qquad (11)$$

$$\frac{\partial u}{\partial t}(x, y, 0) = g(x, y), \qquad\qquad 0 < x < 1, \quad 0 < y < 1 \qquad\qquad (12)$$

All the second derivatives appearing in Eq. (8) are to be replaced by central difference approximations:

$$\frac{\partial^2 u}{\partial x^2} + \frac{\partial^2 u}{\partial y^2} \rightarrow \frac{u(x + \Delta x, y, t) - 2u(x, y, t) + u(x - \Delta x, y, t)}{(\Delta x)^2}$$

$$+ \frac{u(x, y + \Delta y, t) - 2u(x, y, t) + u(x, y - \Delta y, t)}{(\Delta y)^2}$$

$$\frac{\partial^2 u}{\partial t^2} \rightarrow \frac{u(x, y, t + \Delta t) - 2u(x, y, t) + u(x, y, t - \Delta t)}{(\Delta t)^2}.$$

Following our previous practice, we shall use a (single or double) subscript to refer to position and a letter or number in parentheses to refer to time level. Then a typical replacement equation is

$$\frac{u_i(m+1) - 2u_i(m) + u_i(m-1)}{(\Delta t)^2} = \frac{u_E(m) - 2u_i(m) + u_W(m)}{(\Delta x)^2}$$
$$+ \frac{u_N(m) - 2u_i(m) + u_S(m)}{(\Delta y)^2}. \quad (13)$$

Here the subscripts E and W refer to points left and right of point i and N and S refer to points above and below.

If we take $\Delta x = \Delta y = 1/n$—a reasonable choice in a square region—and designate $\rho = \Delta t/\Delta x = \Delta t/\Delta y$, then our replacement equation simplifies to

$$u_i(m+1)$$
$$= \rho^2[u_E(m) + u_W(m) + u_N(m) + u_S(m)] + (2 - 4\rho^2)u_i(m) - u_i(m-1). \quad (14)$$

The stability rules given above still apply. Thus we must choose $\rho^2 \le \frac{1}{2}$ in order to get a sensible solution.

Let us now be specific. We shall take $\Delta x = \Delta y = \frac{1}{4}$, $\rho^2 = \frac{1}{2}$ (that is, $\Delta t = \frac{1}{4}\sqrt{2}$), and suppose that the initial data from Eqs. (11) and (12) are

$$f(x, y) = \begin{cases} 1 & \text{near } x = \frac{1}{4}, \ y = \frac{1}{4}, \\ 0 & \text{elsewhere}, \end{cases} \qquad g(x, y) \equiv 0.$$

The equation that we use for marching forward with the solution is Eq. (14):

$$u_i(m+1) = \frac{1}{2}[u_E(m) + u_W(m) + u_N(m) + u_S(m)] - u_i(m-1). \quad (15)$$

To get started, we solve simultaneously Eq. (15) with $m = 0$ and the discretized version of the initial velocity condition, Eq. (12):

$$u_i(1) + u_i(-1) = \frac{1}{2}[u_E(0) + u_W(0) + u_N(0) + u_S(0)]$$
$$u_i(1) - u_i(-1) = 2 \Delta t g_i.$$

Since $g(x, y) = 0$ in this instance, we find

$$u_i(1) = \frac{1}{4}[u_E(0) + u_W(0) + u_N(0) + u_S(0)]$$

as the starting equation; the right-hand side contains known values of u only. In Fig. 7.8 are representations of the numerical solution at various time levels.

The simple numerical technique that we have developed can be adapted

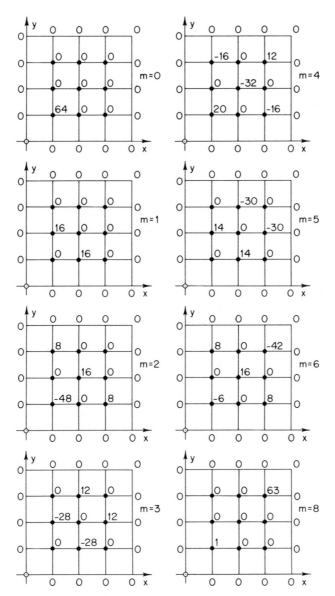

Figure 7.8   Displacements of the square membrane. Numbers shown are $u_i(m) \times 64$.

easily to treat nonhomogeneities, boundary conditions involving derivatives of u, or time-varying boundary conditions. Even nonrectangular regions can be handled, provided that they fit neatly on a rectangular grid. Several exercises illustrate these points.

Exercises

1. Obtain an approximate solution of Eqs. (1), (2), and (3) with $f(x) \equiv 0$ and $g(x) \equiv 1$. Take $\Delta x = \frac{1}{4}$, $\rho = 1$.

2. Compare the results of Exercise 1 with the D'Alembert solution.

3. Obtain an approximate solution of Eqs. (1), (2), and (3) with $f(x) \equiv 0$ and $g(x) = \sin \pi x$. Take $\Delta x = \frac{1}{4}$, $\rho = 1$.

4. Compare the results of Exercise 3 with the exact solution, $u(x, t) = (1/\pi) \sin \pi x \sin \pi t$.

5. Compare the entries of Table 7.10 with an approximate solution obtained using $\rho^2 = \frac{1}{2}$. Be sure to compare at corresponding times.

6. Compare the entries of Table 7.10 with the D'Alembert solution.

7. Obtain an approximate solution of Eq. (1) with zero initial conditions, and boundary conditions $u(0, t) = 0$, $u(1, t) = f(t)$, a square wave of amplitude 1 and period 2 ($f(0) = f(1) = 0$, $f(t) = 1$, $0 < t < 1$, $f(t) = -1$, $1 < t < 2$). Take $\Delta x = \Delta t = \frac{1}{4}$.

8. Obtain an approximate solution of Eqs. (8)–(12) with $f(x, y) \equiv 0$ and $g(x, y) = 4\sqrt{2}$. Take $\Delta x = \Delta y = \frac{1}{4}$ and $\rho^2 = \frac{1}{2}$.

9. Same as Exercise 8, but $f(x, y) \equiv 1$ and $g(x, y) \equiv 0$ in the square.

10. Obtain an approximate numerical solution to the wave equation on an L-shaped region (a 1×1 square with a $\frac{1}{4} \times \frac{1}{4}$ square removed from the upper right corner). Assume initial displacement $= 1$ in the lower right corner, initial velocity equal to 0 and zero displacement on the boundary. Take $\Delta x = \Delta y = \frac{1}{4}$ and $\rho^2 = \frac{1}{2}$.

11. Approximate the solution of the wave equation in a semi-infinite strip 3 units wide. Assume $u = 0$ on all boundaries, zero initial velocity, and an initial value for u that is 1 in a corner and 0 elsewhere. Take $\Delta x = \Delta y = 1$ and $\rho^2 = \frac{1}{2}$.

6 References

The numerical methods we have introduced are usually called finite-difference methods because they involve replacing derivatives with difference quotients. The biggest question is, "What then?" We have used only the simplest explicit methods for the heat and wave equations. Many alternatives are available that are superior in stability, speed, or accuracy. For elliptic equations, we have employed some elementary iterative methods. Newer techniques are far faster; they are practical for use on problems involving many thousands of mesh points.

Of the dozens of books available on the topic of numerical solution of partial differential equations, two that combine readability and accuracy are: *Numerical Solution of Partial Differential Equations*, by W. F. Ames [2], and *Numerical Solution of Partial Differential Equations*, by G. D. Smith [29]. A hallowed reference in the field is *Matrix Iterative Analysis* by Richard Varga [31].

As the title of the last reference suggests, the symbolism and theory of matrices plays an important role in the study of numerical techniques. Two excellent references for background in matrix theory are *Applied Linear Algebra* by Noble [21] and *Matrices and Linear Transformation* by Cullen [10].

Bibliography

1. M. Abramowitz, and I. Stegun (eds.), *Handbook of Mathematical Functions*, Dover, New York, 1965.

2. W. F. Ames, *Numerical Solution of Partial Differential Equations*, Academic Press, New York, 1977.

3. P. W. Berg and J. L. McGregor, *Elementary Partial Differential Equations*, Holden-Day, San Francisco, 1966.

4. B. Bertman and D. J. Sandiford, "'Second Sound' in Solid Helium," *Scientific American*, May 1970, p. 92.

5. H. S. Carslaw and J. C. Jaeger, *Conduction of Heat in Solids* (2nd ed.), Oxford, New York and London, 1959.

6. R. V. Churchill, *Complex Variables and Applications*, McGraw-Hill, New York, 1960.

7. R. V. Churchill, *Fourier Series and Boundary Value Problems* (2nd ed.), McGraw-Hill, New York, 1963.

8. R. V. Churchill, *Operational Mathematics* (3rd ed.), McGraw-Hill, New York, 1972.

9. J. Crank, *The Mathematics of Diffusion* (2nd ed.), Oxford, New York and London, 1975.

10. C. G. Cullen, *Matrices and Linear Transformations* (2nd ed.), Addison-Wesley, Reading, Massachusetts, 1972.

11. R. Dennemeyer, *Introduction to Partial Differential Equations and Boundary Value Problems*, McGraw-Hill, New York, 1968.

12. G. F. D. Duff and D. Naylor, *Differential Equations of Applied Mathematics*, Wiley, New York, 1966.

13. A. Erdelyi, W. Magnus, F. Oberhettinger, and F. Tricomi, *Tables of Integral Transforms*, Vols. 1 and 2, McGraw-Hill, New York, 1954.

14. W. Feller, *Introduction to Probability Theory and Its Applications* (3rd ed.), Vol. 1, Wiley, New York, 1968.

15. J. Fourier, *The Analytical Theory of Heat*, Dover, New York, 1878.

16. R. W. Hamming, *Numerical Methods for Scientists and Engineers* (2nd ed.), McGraw-Hill, New York, 1973.

17. R. Hersh and R. J. Griego, "Brownian Motion and Potential Theory," *Scientific American*, March 1969, p. 66.

18. E. Kreyszig, *Advanced Engineering Mathematics* (3rd ed.), Wiley, New York, 1972.

19. P. Moon and D. Spencer, *Partial Differential Equations*, D. C. Heath, Boston, Massachusetts, 1969.

20. P. M. Morse and H. Feshbach, *Methods of Theoretical Physics*, McGraw-Hill, New York, 1953.

21. B. Noble, *Applied Linear Algebra*, Prentice-Hall, Englewood Cliffs, New Jersey, 1969.

22. C. S. Peskin, *Partial Differential Equations in Biology*, Courant Institute of Mathematical Sciences, New York, 1975.

23. M. H. Protter and H. F. Weinberger, *Maximum Principles in Differential Equations*, Prentice-Hall, Englewood Cliffs, New Jersey, 1967.

24. N. Rashevsky, *Mathematical Biophysics*, Dover, New York, 1960.

25. E. D. Rainville, *The Laplace Transform*, Macmillan, New York, 1963.

26. F. M. Reza, *An Introduction to Information Theory*, McGraw-Hill, New York, 1961.

27. H. Sagan, *Boundary and Eigenvalue Problems in Mathematical Physics*, Wiley, New York, 1966.

28. H. P. Schwan, (ed.), *Biological Engineering*, McGraw-Hill, New York, 1969.

29. G. D. Smith, *Numerical Solution of Partial Differential Equations*, Oxford, New York, 1965.

30. G. P. Tolstov, *Fourier Series*, Dover, New York, 1976.

31. R. S. Varga, *Matrix Iterative Analysis*, Prentice-Hall, Englewood Cliffs, New Jersey, 1962.

32. H. F. Weinberger, *First Course in Partial Differential Equations*, Wiley, New York, 1965.

33. D. V. Widder, *The Heat Equation*, Academic Press, New York, 1975.

Answers
to
Odd-Numbered
Exercises

Chapter 0

Section 1, p. 5

1. **a.** $v(x) = \dfrac{A}{p} \ln|b + px| + B$

   b. $\phi(x) = e^{-x/2}(A \cos \mu x + B \sin \mu x)$ where $\mu = \sqrt{\lambda^2 - \tfrac{1}{4}}$

   c. $R(\rho) = (A \cos \lambda\rho + B \sin \lambda\rho)/\rho$

   d. $w(r) = c_1 r^\lambda + c_2 r^{-\lambda}$

   e. $R(\rho) = c_1 \rho^n + c_2 \rho^{-(n+1)}$

3. **a.** $u(x) = \dfrac{bx^2}{2} + Ax + B$

   b. $u(x) = \dfrac{-1}{k^2} + A \cos kx + B \sin kx$

5. **a.** $u(x) = Ae^{\lambda x} + Be^{-\lambda x} + C \cos \lambda x + D \sin \lambda x$

   b. $u(x) = e^{\mu x}(A \cos \mu x + B \sin \mu x) + e^{-\mu x}(C \cos \mu x + D \sin \mu x)$ where $\mu = \lambda/\sqrt{2}$

7. **a.** $u(r) = A \cos (\lambda \ln r) + B \sin(\lambda \ln r)$

   b. $u(r) = A \ln r + B$

Section 2, p. 10

1. **a.** $u_1(x) = e^{-x}$, $u_2(x) = (c_1 + c_2 x)e^{-x}$

   b. $u_1(x) = e^{ax}$, $u_2(x) = (c_1 + c_2 x)e^{ax}$

   c. $u_1(x) = t^m$,   where $m = (1 - b)/2$,

   $u_2(x) = (c_1 t^{-(1+b)} + c_2)t^m$ if $b + 1 \neq 0$ or

   $u_2(x) = (c_1 \ln t + c_2)t^m$ if $b + 1 = 0$

3. **a.** $u(t) = T + ce^{-at}$

   b. $u(t) = ce^{-at} + e^{at}/2a$

   c. $u(t) = (c + t)e^{-at}$

5. $u_2(x) = \sec x$

Section 3, p. 19

1. **a.** $u(x) = B \sin x$, B arbitrary

   b. $u(x) = 1 - \cos x - \dfrac{1 - \cos 1}{\sin 1} \sin x$ (unique)

   c. No solution exists.

3. **a. and b.** $\lambda = (2n - 1)\pi/2a$

   c. $\lambda = n\pi$, including $\lambda = 0$

5. $c = -a/2$, $c' = h - \dfrac{1}{\mu} \cosh \left(\dfrac{\mu a}{2}\right)$

7. $u(x) = T + c_1 \cosh \gamma x + c_2 \sinh \gamma x$, where $\gamma = \sqrt{hC/\kappa A}$ and

   $c_1 = T_0 - T$, $c_2 = -\dfrac{\gamma \cosh \gamma a + \sinh \gamma a}{\gamma \sinh \gamma a + \cosh \gamma a} c_1$

9. $u(x) = T + T^* \left(1 - \cosh \gamma x - \dfrac{1 - \cosh \gamma a}{\sinh \gamma a} \sinh \gamma x \right)$

where $T^* = \theta I^2 R / hC$ and $\gamma = \sqrt{hC/\kappa A}$

Section 4, p. 24

1. a. $u'' + \dfrac{1}{r} u' - u = 0, \qquad r = 0$

    b. $u'' - \dfrac{2x}{1 - x^2} u' = 0, \qquad x = \pm 1$

    c. $u'' + \cot \phi u' - u = 0, \qquad \phi = 0, \pi$

    d. $u'' + \dfrac{2}{\rho} u' + \lambda^2 u = 0, \qquad \rho = 0$

3. $u(0)$ bounded

    $u(\rho) = \dfrac{H}{6\kappa} (c^2 - \rho^2) + \dfrac{Hc}{3h} + T$

5. $u(\rho) = \dfrac{1}{\rho} (A \cos \mu\rho + B \sin \mu\rho)$. $u(\rho) \equiv 0$ unless $\mu a = \pi, 2\pi, \ldots$. The critical radius is $a = \pi/\mu$.

Miscellaneous exercises, p. 26

1. $u(x) = T_0 \cosh \gamma x + (T_1 - T_0 \cosh \gamma a) \dfrac{\sinh \gamma x}{\sinh \gamma a}$

3. $u(x) = T_0$

5. $u(r) = p(a^2 - r^2)/4$

7. $u(\rho) = H(a^2 - \rho^2)/6 + T_0$

9. $u(x) = T + (T_1 - T) \cosh \gamma x / \cosh \gamma a$

11. $u(x) = T_0 + (T - T_0)e^{-\gamma x}$

13. $h(x) = \sqrt{ex(a - x) + h_0^2 + (h_1^2 - h_0^2)(x/a)}$

15. $u(x) = w(1 - e^{-\gamma x} \cos \gamma x)EI/k$ where $\gamma = (k/4EI)^{1/4}$

17. $u(x) = \begin{cases} T_0 + Ax, & 0 < x < \alpha a \\ T_1 - B(a - x), & \alpha a < x < a \end{cases}$

$A = \dfrac{\kappa_2}{\kappa_1(1 - \alpha) + \kappa_2 \alpha} \cdot \dfrac{T_1 - T_0}{a}, \qquad B = \dfrac{\kappa_1}{\kappa_2} A$

19. $u(x) = \dfrac{1}{2}(1 - e^{-2x}) - \dfrac{1}{2}(1 - e^{-2a})\dfrac{1 - e^{-x}}{1 - e^{-a}}$

21. **a.** $u(x) = \sinh px / \sinh pa$

   b. $u(x) = \cosh px - \dfrac{\cosh pa}{\sinh pa} \sinh px$

   c. $u(x) = \cosh px / \cosh pa$

   d. $u(x) = \cosh p(a - x) / \cosh pa$

   e. $u(x) = \cosh p(a - x) / p \sinh pa$

   f. $u(x) = \cosh px / p \sinh pa$

23. **a.** $u(x) = \dfrac{x}{2} \ln \left| \dfrac{1 + x}{1 - x} \right| - 1$

   b. $u(x) = 1 + (1 - x) \ln \left| \dfrac{x}{1 - x} \right|$

Chapter 1

Section 1, p. 34

1. **a.** $2\left(\sin x - \dfrac{1}{2} \sin 2x + \dfrac{1}{3} \sin 3x - + \cdots \right)$

   b. $\dfrac{\pi}{2} - \dfrac{4}{\pi}\left(\cos x + \dfrac{1}{9} \cos 3x + \dfrac{1}{25} \cos 5x + \cdots \right)$

   c. $\dfrac{1}{2} + \dfrac{2}{\pi}\left(\sin x + \dfrac{1}{3} \sin 3x + \dfrac{1}{5} \sin 5x + \cdots \right)$

   d. $\dfrac{2}{\pi} - \dfrac{4}{\pi}\left(\dfrac{1}{3} \cos 2x + \dfrac{1}{15} \cos 4x + \dfrac{1}{35} \cos 6x + \cdots \right).$

3. $f(x + p) = 1 = f(x)$ for any p and all x.

5. If c is a multiple of p, the graph of $f(x)$ between c and $c + p$ is the same as between 0 and p. Otherwise, let k be the integer such that kp lies

between c and $c + p$:

$$\int_c^{c+p} f(x)\,dx = \int_c^{kp} f(x)\,dx + \int_{kp}^{c+p} f(x)\,dx = \int_{c*}^{p} f(x)\,dx + \int_0^{c*} f(x)\,dx$$

where $c* = c - (k-1)p$.

7. **a.** $\cos^2 x = \dfrac{1}{2} + \dfrac{1}{2}\cos 2x$

   b. $\sin\left(x - \dfrac{\pi}{6}\right) = \cos\dfrac{\pi}{6}\sin x - \sin\dfrac{\pi}{6}\cos x$

   c. $\sin x \cos 2x = \dfrac{1}{2}\sin x + \dfrac{1}{2}\sin 3x$

Section 2, p. 41

1. **a.** $\dfrac{1}{2} - \dfrac{4}{\pi^2}\left[\cos \pi x + \dfrac{1}{9}\cos 3\pi x + \dfrac{1}{25}\cos 5\pi x + \cdots\right]$

   b. $\dfrac{4}{\pi}\left[\sin\dfrac{\pi x}{2} + \dfrac{1}{3}\sin\dfrac{3\pi x}{2} + \dfrac{1}{5}\sin\dfrac{5\pi x}{2} + \cdots\right]$

   c. $\dfrac{1}{12} - \dfrac{1}{\pi^2}\left[\cos 2\pi x - \dfrac{1}{4}\cos 4\pi x + \dfrac{1}{9}\cos 6\pi x - + \cdots\right].$

3. $\bar{f}(x) = f(x - 2na), \qquad 2na < x < 2(n+1)a$

   $$f(x) \sim a_0 + \sum_1^\infty a_n \cos(n\pi x/a) + b_n \sin(n\pi x/a)$$

   $$a_0 = \dfrac{1}{2a}\int_0^{2a} f(x)\,dx, \qquad a_n = \dfrac{1}{a}\int_0^{2a} f(x)\cos(n\pi x/a)\,dx$$

   $$b_n = \dfrac{1}{a}\int_0^{2a} f(x)\sin(n\pi x/a)\,dx$$

5. Odd: (a), (d), (e); even: (b), (c).

7. **a.** $\dfrac{2}{\pi}\left(\sin \pi x - \dfrac{1}{2}\sin 2\pi x + - \cdots\right)$

   b. This function is its own Fourier series.

   c. $\dfrac{4}{\pi^2}\left(\sin \pi x - \dfrac{1}{9}\sin 3\pi x + \dfrac{1}{25}\sin 5\pi x - + \cdots\right)$

9. If $f(-x) = -f(x)$ and $f(x) = f(a - x)$ for $0 < x < a$, sine coefficients with even indices are zero. Example: square wave.

11. **a.** $f(x) = 1 = \dfrac{2}{\pi} \sum_1^\infty \dfrac{1 - \cos n\pi}{n} \sin \dfrac{n\pi x}{a}$

    b. $f(x) = \dfrac{a}{2} - \dfrac{2a}{\pi^2} \sum_1^\infty \dfrac{1 - \cos n\pi}{n^2} \cos \dfrac{n\pi x}{a}$

    $= \dfrac{2a}{\pi} \sum_1^\infty \dfrac{-\cos n\pi}{n} \sin \dfrac{n\pi x}{a}$

    d. $f(x) = \dfrac{2}{\pi} \left[1 - \sum_2^\infty \dfrac{1 + \cos n\pi}{n^2 - 1} \cos nx \right]$

    $= \sin x$

13. Even, yes. Odd, yes only if $f(0) = f(a) = 0$.

Section 3, p. 46

1. **a.** Use $x = 0$;   **b.** $x = \dfrac{1}{2}$;   **c.** $x = 0$.

3. To $f(x)$ everywhere.

Section 4, p. 52

1. (c), (d), (f), (g) have uniformly convergent Fourier series.
3. All of the cosine series converge uniformly. The sine series converges uniformly only in case (b).
5. (a), (c)

Section 5, p. 56

1. $\displaystyle\sum_{n=1}^\infty \dfrac{1}{n^2} = \dfrac{\pi^2}{6}$

3. $f'(x) = 1, 0 < x < \pi$. The sine series cannot be differentiated, because the odd periodic extension of f is not continuous. But the cosine series can be differentiated.

5. For the sine series: $f(0+) = 0$ and $f(a-) = 0$. For the cosine series no additional condition is necessary.

7. No. The function $\ln \left| 2 \cos \dfrac{x}{2} \right|$ is not even sectionally continuous.

Section 6, p. 60

1. $\dfrac{1}{\pi}\displaystyle\int_{-\pi}^{\pi}\left(\ln\left|2\cos\dfrac{x}{2}\right|\right)^2 dx = \sum\dfrac{1}{n^2} = \dfrac{\pi^2}{6}$

3. **a.** Coefficients tend to zero.
  b. Coefficients tend to zero, although $\int_{-1}^{1}|x|^{-1}\,dx$ is infinite.
5. The integral must be infinite, because $\sum a_n^2 + b_n^2 = \infty$.

Section 7, p. 65

1. $\hat{a}_6 = -0.00701$, $a_6 = -0.00569$
3. $\hat{a}_0 = 1.357 \qquad \hat{b}_1 = -0.043$
  $\hat{a}_1 = -0.844 \qquad \hat{b}_2 = -0.115$
  $\hat{a}_2 = 0.208 \qquad \hat{b}_3 = -0.050$
  $\hat{a}_3 = 0.050 \qquad \hat{b}_4 = 0.00$
  $\hat{a}_4 = -0.042 \qquad \hat{b}_5 = 0.043$
  $\hat{a}_5 = -0.0064$
  $\hat{a}_6 = 0.0167$

Section 8, p. 70

1. Each function has the representations (for $x > 0$)

$$f(x) = \int_0^\infty A(\lambda)\cos \lambda x\, d\lambda = \int_0^\infty B(\lambda)\sin \lambda x\, d\lambda.$$

  a. $A(\lambda) = 2/\pi(1 + \lambda^2)$, $\qquad$ $B(\lambda) = 2\lambda/\pi(1 + \lambda^2)$
  b. $A(\lambda) = 2\sin\lambda/\pi\lambda$, $\qquad$ $B(\lambda) = 2(1 - \cos\lambda)/\pi\lambda$
  c. $A(\lambda) = 2(1 - \cos\lambda\pi)/\lambda^2\pi$, $\qquad$ $B(\lambda) = 2(\pi\lambda - \sin\lambda\pi)/\pi\lambda^2$

3. **a.** $\dfrac{1}{1 + x^2} = \displaystyle\int_0^\infty e^{-\lambda}\cos\lambda x\, d\lambda$

  b. $\dfrac{\sin x}{x} = \displaystyle\int_0^\infty A(\lambda)\cos\lambda x\, d\lambda$

  where $A(\lambda) = \begin{cases} 1, & 0 < x < 1 \\ 0, & 1 < x \end{cases}$

5. The cosine and sine coefficient functions of $f'(x)$ are, respectively, $\lambda B(\lambda)$ and $-\lambda A(\lambda)$.

7. Change the variable of integration from x to λx.

9. **a.** $A(\lambda) \equiv 0$, $\qquad B(\lambda) = \dfrac{2 \sin \lambda \pi}{\pi(1 - \lambda^2)}$

b. $A(\lambda) = \dfrac{1 + \cos \lambda \pi}{\pi(1 - \lambda^2)}$, $\qquad B(\lambda) = \dfrac{\sin \lambda \pi}{\pi(1 - \lambda^2)}$

c. $A(\lambda) = \dfrac{2(1 + \cos \lambda \pi)}{\pi(1 - \lambda^2)}$, $\qquad B(\lambda) \equiv 0$

Section 9, p. 73

1. **a.** $1 + \sum r^n \cos nx = \operatorname{Re} \sum_{0}^{\infty} (re^{ix})^n = \operatorname{Re} \dfrac{1}{1 - re^{ix}}$

b. $\sum_{n=1}^{\infty} \dfrac{\sin nx}{n!} = \operatorname{Im} \sum_{n=0}^{\infty} \dfrac{e^{inx}}{n!} = \operatorname{Im} \exp(e^{ix})$

3. $e^{ax} = 2 \dfrac{\sinh \alpha \pi}{\pi} \left(\dfrac{1}{2\alpha} + \sum \dfrac{(-1)^n}{\alpha^2 + n^2} (\alpha \cos nx - n \sin nx) \right)$

5. $f(x) = \displaystyle\int_{-\infty}^{\infty} C(\lambda) e^{i\lambda x} \, d\lambda$

a. $C(\lambda) = \dfrac{1}{2\pi(1 + i\lambda)}$

b. $C(\lambda) = \dfrac{1 + e^{-i\lambda \pi}}{2\pi(1 - \lambda^2)}$

Miscellaneous exercises, p. 79

1. $f(x) = \sum b_n \sin nx$

$$b_n = \begin{cases} 0, & n \text{ even} \\[2mm] \dfrac{4 \sin n\alpha}{\pi \alpha n^2}, & n \text{ odd} \end{cases}$$

3. Yes. As $\alpha \to 0$, $(\sin n\alpha)/n\alpha \to 1$.

5. $f(x) = \sum b_n \sin(n\pi x/a)$

$$b_n = \dfrac{2h \sin n\pi\alpha}{\pi^2 \, n^2} \left(\dfrac{1}{\alpha} + \dfrac{1}{1 - \alpha} \right)$$

7. **a.** $b_n = 0$, $\qquad a_n = 0$, $\qquad a_0 = 1$

b. $\sum b_n \sin(n\pi x/a)$, $\qquad b_n = \dfrac{2(1 - \cos n\pi)}{n\pi}$

c. and d. same as (a)

e. same as (b)

f. $a_0 + \sum a_n \cos(n\pi x/a) + b_n \sin(n\pi x/a)$

$$a_0 = \frac{1}{2}, \qquad a_n = 0, \qquad b_n = \frac{1 - \cos n\pi}{n\pi}$$

9. $f(x) = a_0 + \sum a_n \cos(n\pi x/a) + b_n \sin(n\pi x/a)$

$$a_0 = \tfrac{1}{2}a, \qquad a_n = -\frac{2a(1 - \cos n\pi)}{n^2\pi^2}, \qquad b_n = -\frac{2a\cos n\pi}{n\pi}$$

$x = -a,$	$-a/2,$	$0,$	$a,$	$2a$
sum $= a,$	$0,$	$0,$	$a,$	0

11. $f(x) = a_0 + \sum a_n \cos nx$

$$a_0 = \frac{3}{4}, \qquad a_n = \frac{\sin(n\pi/2)}{n\pi}$$

$x = 0,$	$\pi/2,$	$\pi,$	$3\pi/2,$	2π
sum $= 1,$	$\tfrac{3}{4},$	$\tfrac{1}{2},$	$\tfrac{3}{4},$	1

13. $f(x) = \sum b_n \sin n\pi x$, $\qquad b_n = 2(1 + \cos n\pi)/n\pi$

15. $f(x) = \sum b_n \sin nx$

$$b_2 = \frac{1}{2}, \qquad \text{other } b_n = \frac{4\sin(n\pi/2)}{\pi(4 - n^2)}$$

17. $\displaystyle\sum_1^N \cos nx = \operatorname{Re} \sum_1^N e^{inx} = \operatorname{Re}\frac{e^{ix} - e^{iNx}}{1 - e^{ix}} = \operatorname{Re}\frac{e^{ix/2} - e^{i(2N-1)x/2}}{e^{-ix/2} - e^{ix/2}}$

The denominator is now $-2i\sin(x/2)$.

19. $f(x) = \sum b_n \sin nx$, $\qquad b_n = \dfrac{2a\sin(na + \pi)}{n^2a^2 - \pi^2}$

21. $f(x) = \displaystyle\int_0^\infty \left(\frac{\sin\lambda a}{\lambda\pi}\cos\lambda x + \frac{1 - \cos\lambda a}{\lambda\pi}\sin\lambda x\right)d\lambda$

23. $f(x) = \displaystyle\int_0^\infty \frac{2\sin\lambda\pi}{\pi(1 - \lambda^2)}\sin\lambda x\, d\lambda \qquad (x > 0)$

29. Use $\int_0^\infty \dfrac{\sin \lambda t}{\lambda}\, d\lambda = \dfrac{\pi}{2}$

31. These answers are not unique.

    a. $\sum b_n \sin n\pi x, \qquad b_n = 2/n\pi$

    b. $a_0 + \sum a_n \cos n\pi x, \qquad a_0 = \tfrac{1}{2}, \qquad a_n = 2(1 - \cos n\pi)/n^2\pi^2$

    c. $\int_0^\infty B(\lambda) \sin \lambda x\, d\lambda, \qquad B(\lambda) = 2(\lambda - \sin \lambda)/(\pi\lambda^2)$

    d. $\int_0^\infty A(\lambda) \cos \lambda x\, d\lambda, \qquad A(\lambda) = 2(1 - \cos \lambda)/(\pi\lambda^2)$

    The integrals of parts (c) and (d) converge to 0 for $x > 1$.

33. $\hat{a}_0 = 0.78424 \qquad\quad \hat{a}_4 = -0.00924$
$\hat{a}_1 = 0.22846 \qquad\quad \hat{a}_5 = 0.00744$
$\hat{a}_2 = -0.02153 \qquad \hat{a}_6 = -0.00347$
$\hat{a}_3 = 0.01410$

35. $a_0 = a/6, \qquad a_n = 2a\left(\cos \dfrac{2n\pi}{3} - \cos \dfrac{n\pi}{3}\right) \Big/ n^2\pi^2$

37. $a_0 = \tfrac{5}{8}, \; a_n = 2\left(3 \cos \dfrac{n\pi}{2} - 2 - \cos n\pi\right) \Big/ n^2\pi^2$

39. $a_0 = \tfrac{1}{2}, \; a_n = 2(1 - \cos n\pi)/n^2\pi^2$

41. $a_0 = a^2/6, \; a_n = -2a^2(1 + \cos n\pi)/n^2\pi^2$

43. $a_0 = \tfrac{1}{2}, \; a_n = -\left(2 \sin \dfrac{n\pi}{2}\right) \Big/ n\pi$

45. $b_n = \left(1 + \cos \dfrac{n\pi}{2} - 2 \cos n\pi\right) \Big/ n\pi$

47. $b_n = a\left(\dfrac{2 \sin(n\pi/2)}{n^2\pi^2} - \dfrac{\cos n\pi}{n\pi}\right)$

49. $b_n = 2\left(\cos \dfrac{n\pi}{4} - \cos \dfrac{3n\pi}{4}\right) \Big/ n\pi$

51. $b_n = 2n\pi(1 - e^{ka} \cos n\pi)/(a^2k^2 + n^2\pi^2)$

53. $A(\lambda) = \dfrac{2}{\pi(1 + \lambda^2)}$

55. $A(\lambda) = 2(\sin \lambda b)/(\pi\lambda)$

57. $A(\lambda) = 2(1 - \cos \lambda)/(\pi\lambda^2)$

59. $B(\lambda) = \dfrac{2\lambda}{\pi(1 + \lambda^2)}$

61. $B(\lambda) = 2(1 - \cos \lambda b)/(\lambda\pi)$

63. $B(\lambda) = 2(\lambda - \sin \lambda)/(\lambda^2\pi)$

Chapter 2

Section 1, p. 91

1. If $\dfrac{\partial x}{\partial u}(0, t)$ is positive, then heat is flowing to the left, so $u(0, t)$ is greater than $T(t)$.

3. $-\kappa \dfrac{\partial u}{\partial x}(a, t) = \sigma(u^4(a, t) - T^4)$

5. One possibility: $u(x, t)$ is the temperature in a rod of length a whose lateral surface is insulated. The temperature at the left end is held constant at T_0. The right end is exposed to a medium at temperature T_1. Initially the temperature is $f(x)$.

Section 2, p. 96

1. $v'' - \gamma^2(v - T) = 0, \qquad 0 < x < a$

$\qquad v(0) = T_0, \qquad v(a) = T_1$

$\qquad v(x) = T + A \cosh \gamma x + B \sinh \gamma x,$

$\qquad A = T_0 - T, \qquad B = \dfrac{(T_1 - T) - (T_0 - T)\cosh \gamma a}{\sinh \gamma a}$

3. $v(x) = T$. Heat is being generated at a rate proportional to $u - T$. If $\gamma = \pi/a$, the steady-state problem does not have a unique solution.

5. $v(x) = A \ln(b + dx) + B, \qquad A = (T_1 - T_0)/\ln(1 + ad/b),$

$\qquad\qquad\qquad\qquad\qquad\qquad\qquad\qquad\qquad B = T_0 - A \ln b$

7. **a.** $v(x) = T_0 + r(2a - x)x/2$

$\qquad$ **b.** $v(x) = A + B \sinh \beta x + C \sinh \beta(a - x)$

$\qquad\quad A = \alpha/\beta^2, \qquad B = (T_1 - \alpha/\beta^2)/\sinh \beta a$

$\qquad\quad C = (T_0 - \alpha/\beta^2)/\sinh \beta a.$

Section 3, p. 103

1. $w(x, t) = -\dfrac{2}{a}(T_0 + T_1) \sin \dfrac{\pi x}{a} \exp\left(-\dfrac{\pi^2 kt}{a^2}\right)$

$\qquad -\dfrac{2}{\pi}(T_0 - T_1) \sin \dfrac{2\pi x}{a} \exp\left(-\dfrac{4\pi^2 kt}{a^2}\right)$

$\qquad - \cdots$

3. Let $\xi = x/a$, $\tau = kt/a^2$, and $u(x, t) = U(\xi, \tau)$. Then

$$\frac{\partial^2 U}{\partial \xi^2} = \frac{\partial U}{\partial \tau}, \qquad 0 < \xi < 1, \quad 0 < \tau$$

is the form taken by the partial differential equation.

5. $w(x, t) = \displaystyle\sum_{0}^{\infty} b_n \sin \dfrac{n\pi x}{a} \exp(-n^2\pi^2 kt/a)$

    a. $b_n = T_0 \dfrac{2}{\pi} \dfrac{1 - \cos n\pi}{n}$

    b. $b_n = -\dfrac{2\beta a}{\pi} \dfrac{\cos n\pi}{n}$

    c. $b_n = \dfrac{2\beta a}{\pi} \cdot \dfrac{1}{n}$

Section 4, p. 107

1. The series $\sum |A_n(t_1)|$ converges.
3. $a_0 = T_1/2$, $a_n = 2T_1(\cos n\pi - 1)/(n\pi)^2$
5. $a_0 = T_0 + T_1/3$, $a_n = 4T_1(-1)^n/n^2\pi^2$

Section 5, p. 112

1. $v(x, t) = T_0$

3. The graph of G in the interval $0 < x < 2a$ is made by reflecting the graph of g in the line $x = a$. (Like an even extension.)

5. $g(x) \sim \sum b_n \sin \dfrac{(2n - 1)\pi x}{2a}$, $\qquad 0 < x < a$

    a. $b_n = \dfrac{8a(-1)^{n+1}}{\pi^2(2n - 1)^2}$

b. $b_n = \dfrac{4T}{\pi(2n-1)}$

7. $u(x, t) = T_0 + \sum b_n \sin \lambda_n x \exp(-\lambda_n^2 kt), \qquad \lambda_n = (2n-1)\pi/2a,$

$$b_n = \frac{8T(-1)^{n+1}}{\pi^2(2n-1)^2} - \frac{4T_0}{\pi(2n-1)}$$

9. The steady-state solution is $v(x) = T_0 + T(x(x-2a)/2a^2)$. The transient satisfies Eqs. (5)–(8) with

$$g(x) = T_0 - v(x) = \frac{Tx(x-2a)}{2a^2}.$$

11. $u(x, t) = T_0 + \sum c_n \cos \lambda_n x \exp(-\lambda_n^2 kt),$

$$\lambda_n = (2n-1)\pi/2a, \qquad c_n = \frac{4(T_1 - T_0(-1)^{n+1})}{\pi(2n-1)}$$

Section 6, p. 117

1. The graph of $V(x)$ is a straight line from T_0 at $x=0$ to T^* at $x = a$, where

$$T^* = T_0 + \frac{ha}{k + ha}(T_1 - T_0).$$

In all cases, T^* is between T_0 and T_1.

3. Negative solutions provide no new eigenfunctions.

7. $b_m = \dfrac{2(1 - \cos \lambda_m a)}{\lambda_m[a + (\kappa/h)\cos^2 \lambda_m a]}$

9. $b_m = \dfrac{-2(\kappa + ah)\cos \lambda_m a}{\lambda_m(ah + \kappa \cos^2 \lambda_m a)}$

Section 7, p. 121

1. $\lambda_n = n\pi/\ln 2, \; \phi_n = \sin(\lambda_n \ln x)$

3. **a.** $\sin \lambda_n x, \; \lambda_n = (2n-1)\pi/2a$
   b. $\cos \lambda_n x, \; \lambda_n = (2n-1)\pi/2a$
   c. $\sin \lambda_n x, \; \lambda_n$ a solution of $\tan \lambda a = -\lambda$
   d. $\lambda_n \cos \lambda_n x + \sin \lambda_n x, \; \lambda_n$ a solution of $\cot \lambda a = \lambda$
   e. $\lambda_n \cos \lambda_n x + \sin \lambda_n x, \; \lambda_n$ a solution of $\tan \lambda a = 2\lambda/(\lambda^2 - 1)$

5. The weight functions in the orthogonality relations, and limits of integration are:
   a. $1 + x$, 0 to a
   b. e^x, 0 to a
   c. $1/x^2$, 1 to 2
   d. e^x, 0 to a.
7. Because λ appears in a boundary condition.

Section 8, p. 124

1. $x = \sum c_n \phi_n$
   $c_n = 2n\pi(1 - b\cos n\pi)/(n^2 \pi^2 + \ln^2 b)$
3. $1 = \sum c_n \phi_n$
   $c_n = 2n\pi(1 - e^{a/2} \cos n\pi)/(n^2 \pi^2 + a^2/4)$
   (Hint: find the sine series of $e^{x/2}$.)
5. $b_n = \int_l^r f(x) \psi_n(x) p(x) \, dx$
7. 1 and $\sqrt{2} \cos n\pi x$

Section 9, p. 129

1. a. $v(x) = \text{constant}$
   b. $v(x) = AI(x) + B$
3. If $\partial u / \partial x = 0$ at both ends, then the steady-state problem is indeterminate, but Eqs. (1)–(3) are homogeneous, so separation of variables applies directly. Note that $\lambda_0 = 0$ and $\phi_0 = 1$. The constant term in the series for $u(x, t)$ is

$$a = \int_l^r p(x) f(x) \, dx \Big/ \int_l^r p(x) \, dx.$$

Section 10, p. 132

1. $u(x, t) = \dfrac{2}{\pi} \int_0^\infty \dfrac{1 - \cos \lambda b}{\lambda} \sin \lambda x \exp(-\lambda^2 kt) \, d\lambda$

3. $u(x, t)$ is given by Eq. (6) with $B(\lambda) = \dfrac{2T_0 \lambda}{\pi(\alpha^2 + \lambda^2)}$.

5. $u(x, t) = \displaystyle\int_0^\infty A(\lambda) \cos \lambda x \exp(-\lambda^2 kt) \, d\lambda$

   $A(\lambda) = \dfrac{2}{\pi \lambda} \sin \lambda b$

7. $u(x, t) = T_0 + \int_0^\infty B(\lambda) \sin \lambda x \exp(-\lambda^2 kt) \, d\lambda$

$B(\lambda) = \dfrac{2}{\pi} \int_0^\infty (f(x) - T_0) \sin \lambda x \, d\lambda$

Section 11, p. 135

1. a. $u(x, t) = \int_0^\infty \dfrac{2 \sin \lambda a}{\pi \lambda} \cos \lambda x \exp(-\lambda^2 kt) \, d\lambda$

b. $u(x, t) = \dfrac{1}{\sqrt{4\pi kt}} \int_{-a}^a \exp\left(\dfrac{-(\xi - x)^2}{4kt}\right) d\xi$

3. The function $u(x, t)$, as a function of x, is the famous "bell-shaped" curve. The smaller t is, the more sharply peaked the curve.

5. In Eq. (3) replace both $f(\xi)$ and $u(x, t)$ by 1.

7. $u(x, t) = \mathrm{erf}(x/\sqrt{4kt})$

Miscellaneous exercises, p. 138

1. SS: $v(x) = T_0, \quad 0 < x < a$

EVP: $\phi'' = \lambda^2 \phi = 0, \quad \phi(0) = 0, \quad \phi(a) = 0, \quad \lambda_n = n\pi/a, \quad \phi_n = \sin \lambda_n x$

$u(x, t) = T_0 + \sum_1^\infty b_n \sin \lambda_n x e^{-\lambda^2_n kt}$

$b_n = \dfrac{2}{a} \int_0^a (T_1 - T_0) \sin \dfrac{n\pi x}{a} \, dx$

3. SS: $v(x) = T_0 + \dfrac{r}{2} x(x - a), \quad 0 < x < a$

EVP: $\phi'' + \lambda^2 \phi = 0, \quad \phi(0) = 0, \quad \phi(a) = 0, \quad \lambda_n = n\pi/a, \quad \phi_n = \sin \lambda_n x$

$u(x, t) = T_0 + \dfrac{r}{2} x(x - a) + \sum_1^\infty b_n \sin \lambda_n x e^{-\lambda^2_n kt}$

$b_n = \dfrac{2}{a} \int_0^a [T_1 - T_0 - \dfrac{r}{2} x(x - a)] \sin \dfrac{n\pi x}{a} \, dx$

5. SS: $v(x) = 0$, $0 < x < a$

   EVP: $\phi'' + (\lambda^2 - \gamma^2)\phi = 0$, $\phi'(0) = 0$, $\phi'(a) = 0$

   $$\lambda_n = \sqrt{\left(\frac{n\pi}{a}\right)^2 + \gamma^2}, \qquad n = 0, 1, 2, \ldots$$

   $$\phi_0 = 1, \qquad \phi_n = \cos\frac{n\pi x}{a}$$

   $$u(x, t) = a_0\, e^{-\lambda^2_0 kt} + \sum_1^\infty a_n \cos\frac{n\pi x}{a}\, e^{-\lambda^2_n kt}$$

   $$a_0 = T_1/2, \qquad a_n = \frac{2}{a}\int_0^a T_1(x/a) \cos\frac{n\pi x}{a}\, dx$$

7. $u(x, t) = T_0$

9. $u(x, t) = T_0 + \sum c_n \sin \lambda_n x \exp(-\lambda_n^2 kt)$,

   $$\lambda_n = \frac{(2n-1)\pi}{2a}, \qquad c_n = \frac{(T_1 - T_0)\cdot 4}{(2n-1)\pi}$$

11. $u(x, t) = T_0 + \displaystyle\int_0^\infty B(\lambda) \sin \lambda x \exp(-\lambda^2 kt)\, d\lambda$,

    $$B(\lambda) = \frac{-2\lambda T_0}{\pi(\alpha^2 + \lambda^2)}$$

13. $u(x, t) = \displaystyle\int_0^\infty A(\lambda) \cos \lambda x \exp(-\lambda^2 kt)\, d\lambda$,

    $$A(\lambda) = \frac{2T_0 \sin \lambda a}{\pi\lambda}$$

15. $u(x, t) = \displaystyle\int_0^\infty (A(\lambda)\cos \lambda x + B(\lambda)\sin \lambda x) \exp(-\lambda^2 kt)\, d\lambda$,

    $$A(\lambda) = \frac{T_0 \sin \lambda a}{\pi\lambda}, \qquad B(\lambda) = \frac{T_0(1 - \cos \lambda a)}{\pi\lambda}$$

    or

    $$u(x, t) = \frac{T_0}{\sqrt{4\pi kt}}\int_0^a \exp\left(-\frac{(\xi - x)^2}{4kt}\right)\, d\xi$$

17. $u(\rho, t) = \dfrac{1}{\rho} \sum b_n \sin \lambda_n \exp(-\lambda_n^2 kt),$

$$\lambda_n = n\pi/a, \qquad b_n = \dfrac{2}{a} \int_0^a \rho T \sin \lambda_n \rho \, d\rho$$

19. $v(x) = T_0 + \Delta T \dfrac{x}{a} - \dfrac{\Delta T \sinh \gamma x}{a\gamma \cosh \gamma a}$

21. $w(x, t) = -\dfrac{2}{u} \dfrac{\partial u}{\partial x}$ where $u(x, t) = a_0 + \sum a_n \cos n\pi x \exp(-n^2 \pi^2 t)$

where $a_0 = 2(1 - e^{-1/2})$ and $a_n = \dfrac{1 - e^{-1/2} \cos n\pi}{\tfrac{1}{4} + (n\pi)^2}$

23. $u_2 = \dfrac{\beta_2 V}{\beta_1 + \beta_2}, \qquad u_1 = 1 - \dfrac{\beta_1 V}{\beta_1 + \beta_2}$

where $V = 1 - \exp(-(\beta_1 + \beta_2)t)$ and $\beta_i = h/c_i$

Chapter 3

Section 1, p. 146

1. $[u] = L, \qquad [c] = L/t$

3. $v(x) = \dfrac{(x^2 - ax)g}{2c^2}$

Section 2, p. 151

1. $u(x, t) = \sum B_n \sin\left(\dfrac{n\pi x}{a}\right) \sin\left(\dfrac{n\pi ct}{a}\right)$

$$B_n = \dfrac{2a(1 - \cos n\pi)}{n^2 \pi^2 c}$$

3. a. $\sin \dfrac{n\pi x}{a}$ **b.** $\sin \dfrac{2n - 1}{2} \dfrac{\pi x}{a}$

5. Product solutions are $\phi_n(x) T_n(t)$ where

$$\phi_n = \sin\left(\dfrac{n\pi x}{a}\right)$$

$$T_n(t) = \exp(-kc^2t/2) \begin{Bmatrix} \sin \\ \cos \end{Bmatrix} \sqrt{\lambda_n^2 c^2 - \tfrac{1}{4}k^2 c^4}\, t.$$

7. Product solutions are $\phi_n(x)T_n(t)$ where

$$\phi_n(x) = \sin\left(\frac{n\pi x}{a}\right)$$

$$T_n(t) = \sin \quad \text{or} \quad \cos\left(\frac{n\pi}{a}\right)^2 ct.$$

Frequencies $n^2\pi c/2a^2$ Hz.

9. Yes.

Section 3, p. 155

1. Table shows $u(x, t)/h$.

t	0	$0.2a/c$	$0.4a/c$	$0.8a/c$	$1.4a/c$
$x = 0.25a$	0.5	0.5	0.2	-0.5	-0.2
$x = 0.5a$	1.0	0.6	0.2	-0.6	-0.2

3. $u(0, 0.5a/c) = 0$; $u(0.2a, 0.6a/c) = 0.2\alpha a$; $u(0.5a, 1.2a/c) = -0.2\alpha a$. (Hint: $G(x) = \alpha x, 0 < x < a$.)

5. $G(x) = \begin{cases} 0, & 0 < x < 0.4a \\ 5(x - 0.4a), & 0.4a < x < 0.6a \\ a, & 0.6a < x < a \end{cases}$

Notice that G is a continuous function whose graph is composed of line segments.

7.

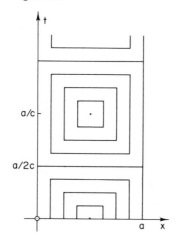

9. $u(x, t) = -c^2 \cos t + \phi(x - ct) + \psi(x + ct)$

Section 4, p. 159

1. If f and g are sectionally smooth, and f continuous.
3. The frequency is $c\lambda_n$ rads/sec, and the period is $2\pi/c\lambda_n$ sec.
5. Separation of variables leads to the following in place of Eqs. (11) and (12):

$$T'' + \gamma T' + \lambda^2 c^2 T = 0 \tag{11'}$$

$$(s\phi')' - q\phi + \lambda^2 p\phi = 0. \tag{12'}$$

The solutions of Eq. (11') all approach 0 as $t \to \infty$, if $\gamma > 0$.

Section 5, p. 163

1. If $q \geq 0$, the numerator in Eq. (3) must also be greater than or equal to 0, since $\phi_1(x)$ cannot be identically 0.
3. $2\pi^2/3$ is one estimate from $y = \sin \pi x$.

Section 6, p. 168

1. $u(x, t) = \frac{1}{2}[f(x + ct) + G(x + ct)] + \frac{1}{2}[\bar{f}(x - ct) - \bar{G}(x - ct)]$, where $\bar{f}$ is the even extension of f and $\bar{G}$ the odd extension of G.

3.

5.

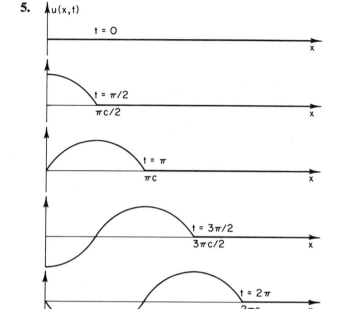

7. $u(x, t) = \dfrac{1}{2} \left[f(x + ct) + f(x - ct) \right] + \dfrac{1}{2c} \displaystyle\int_{x-ct}^{x+ct} g(y)\, dy$

Miscellaneous exercises, p. 170

1. $u(x, t) = \displaystyle\sum_{1}^{\infty} b_n \sin \lambda_n x \cos \lambda_n ct$

$b_n = 2(1 - \cos n\pi)/n\pi$

$\lambda_n = n\pi/a$

3.

5.

7.

9.

11.

13.

15. $\lambda_1{}^2 \leq 8.324$, using $y(x) = (x - \frac{1}{4})(\frac{5}{4} - x)$. Also $\lambda_1{}^2 \leq 8.265$ from $y(x) =$ $\sin \pi(x - \frac{1}{4})$. For this case, one integral had to be calculated numerically (trapezoidal rule).

17. $f(q) = 12a^2 \operatorname{sech}^2(aq)$, $\quad c = 4a^2$

21. $v(x, t) = \sum (a_n \cos \lambda_n ct + b_n \sin \lambda_n ct) \sin \lambda_n x$

$$\lambda_n = (2n - 1)\pi/2a$$

$$a_n = \frac{8aU_0(-1)^{n+1}}{\pi^2(2n - 1)^2}, \qquad b_n = \frac{16a^2k(-1)^n}{\pi^3(2n - 1)^3}$$

23. $\dfrac{Y''}{Y} = \dfrac{2V}{k}\dfrac{\psi'}{\psi}$

The function $\phi(x - Vt)$ cancels from both sides.

25. $\phi_n(-Vt) = T_0 \exp(\lambda_n{}^2 kt/2)b_n$, $\quad t > 0$

$\quad\quad \phi_n(x) = T_1 \exp(\lambda_n{}^2 kx/2V)b_n$, $\quad x > 0$

where $\sum b_n \sin \lambda_n y = 1$, $\quad 0 < y < b$

27. $\dfrac{\partial^2 i}{\partial x^2} = CR\dfrac{\partial i}{\partial t}$, $\quad \dfrac{\partial^2 e}{\partial x^2} = CR\dfrac{\partial e}{\partial t}$

29. $e(x, t) = \sum a_n \cos \lambda_n ct \sin \lambda_n x$,

$$a_n = \frac{2V(1 - \cos n\pi)}{n\pi}, \qquad \lambda_n = n\pi/a$$

31. $e(x, t) = V_0\dfrac{(a - x)}{a} + \sum b_n \sin \lambda_n x \exp(-\lambda_n{}^2 kt)$

$$b_n = -\frac{2V_0 \cos n\pi}{n\pi}, \qquad \lambda_n = n\pi/a, \quad k = 1/Rc$$

33. $e(x, t) = V + \frac{1}{2}(\bar{f}(x + ct) + \bar{f}(x - ct))$

where $\bar{f}$ is the odd extension of

$$f(x) = -V(1 - e^{-\alpha x}), \qquad 0 < x$$

Chapter 4

Section 1, p. 178

1. $f + d = 0$

3. $Y(y) = A \sinh \pi y$, $A = 1/\sinh \pi$

5. $v(r) = a \ln r + b$

7. $\dfrac{\partial u}{\partial x} = \dfrac{\partial v}{\partial r} \cos \theta - \dfrac{\partial v}{\partial \theta} \dfrac{\sin \theta}{r}$

$\dfrac{\partial u}{\partial y} = \dfrac{\partial v}{\partial r} \sin \theta + \dfrac{\partial v}{\partial \theta} \dfrac{\cos \theta}{r}$

Section 2, p. 183

1. Show by differentiating and substituting that both are solutions of the differential equation. The Wronskian of the two functions is

$$\begin{vmatrix} \sinh \lambda y & \sinh \lambda(b - y) \\ \lambda \cosh \lambda y & -\lambda \cosh \lambda(b - y) \end{vmatrix} = -\lambda \sinh \lambda b \neq 0$$

3. In the case $b = a$, use two terms of the series: $u(a/2, a/2) = 0.32$.

5. $u(x, y) = \displaystyle\sum_1^\infty b_n \sin\left(\dfrac{n\pi x}{a}\right) \dfrac{\sinh(n\pi y/a)}{\sinh(n\pi b/a)}$

$b_n = \dfrac{8}{n^2 \pi^2} \sin\left(\dfrac{n\pi}{2}\right)$

7. a. $u(x, y) = 1$, but the form found by applying the methods of this section is

$$u(x, y) = \sum a_n \dfrac{\sinh \lambda_n y + \sinh \lambda_n(b - y)}{\sinh \lambda_n b} \cos \lambda_n x$$

$$+ \sum b_n \dfrac{\cosh \mu_n x}{\cosh \mu_n a} \sin \mu_n y$$

where

$$\lambda_n = \dfrac{(2n - 1)\pi}{2a}, \quad a_n = \dfrac{4 \sin \dfrac{(2n - 1)\pi}{2}}{\pi(2n - 1)}, \quad \mu_n = \dfrac{n\pi}{b}, \quad b_n = \dfrac{2(1 - \cos n\pi)}{n\pi}.$$

b. $u(x, y) = y/b$, and this is found by the methods of this section. In this case, 0 is an eigenvalue.

c. $\dfrac{4}{b} \displaystyle\sum_1^\infty \dfrac{(-1)^{n+1} \cos \lambda_n y \sinh \lambda_n(a - x)}{(2n - 1) \sinh \lambda_n a}$

$\lambda_n = \left(\dfrac{2n - 1}{2} \dfrac{\pi}{b}\right)$

Section 3, p. 187

1. $a_n = \dfrac{2}{a} \displaystyle\int_0^a f(x) \sin\left(\dfrac{n\pi x}{a}\right) dx$

3. $A(\mu) = \dfrac{2}{\pi} \displaystyle\int_0^\infty g_2(y) \sin \mu y \, dy$

5. $u(x, y) = \displaystyle\int_0^\infty (A(\mu) \cos \mu y + B(\mu) \sin \mu y) \dfrac{\sinh \mu x}{\sinh \mu a} \, d\mu$

   $\qquad + \displaystyle\int_0^\infty (C(\mu) \cos \mu y + D(\mu) \sin \mu y) \dfrac{\sinh \mu(a - x)}{\sinh \mu a} \, d\mu$

7. $u(x, y) = \dfrac{2}{\pi} \displaystyle\int_0^\infty \dfrac{\lambda}{1 + \lambda^2} (e^{-\lambda y} \sin \lambda x + e^{-\lambda x} \sin \lambda y) \, d\lambda$

9. $u(x, y)$ is given by Formula (5) with $B(\mu) = 0$ and $A(\mu) = 2\mu/\pi(1 + \mu^2)$.

11. $u(x, y) = \displaystyle\sum_1^\infty b_n \sin \lambda_n x \exp(-\lambda_n y)$

   $\qquad + \displaystyle\int_0^\infty \left(A(\mu) \dfrac{\sinh \mu x}{\sinh \mu a} + B(\mu) \dfrac{\sinh \mu(a - x)}{\sinh \mu a}\right) \sin \mu y \, d\mu$

   $\lambda_n = n\pi/a, \qquad b_n = 2(1 - \cos n\pi)/n\pi, \qquad A(\mu) = B(\mu) = 2\lambda/\pi(\lambda^2 + 1)$

   Also see Exercise 12.

13. a. $u(x, y) = \dfrac{2}{\pi} \displaystyle\int_0^\infty \dfrac{1 - \cos \lambda a}{\lambda} \sin \lambda x \dfrac{\sinh \lambda y}{\sinh \lambda b} \, d\lambda$

   b. $u(x, y) = \dfrac{2}{\pi} \displaystyle\int_0^\infty \dfrac{\lambda}{1 + \lambda^2} \sin \lambda x \dfrac{\sinh \lambda(b - y)}{\sinh \lambda b} \, d\lambda$

Section 4, p. 191

1. $v(r, \theta)$ is given by Formula (10) with $b_n = 0$, $a_0 = \pi/2$,
   $a_n = -2(1 - \cos n\pi)/\pi n^2 c^n$.

3. Exercise 1, $v(0, \theta) = \pi/2$; Exercise 2, $v(0, \theta) = 0$.

5. Convergence is uniform in θ.

7. $a_0 = \dfrac{1}{2\pi} \displaystyle\int_{-\pi}^\pi f(\theta) \, d\theta$

   $a_n = \dfrac{c^n}{\pi} \displaystyle\int_{-\pi}^\pi f(\theta) \cos n\theta \, d\theta, \qquad b_n = \dfrac{c^n}{\pi} \displaystyle\int_{-\pi}^\pi f(\theta) \sin n\theta \, d\theta$

9. $\dfrac{2}{\pi}\displaystyle\sum_{n=1}^{\infty}\dfrac{1-\cos n\pi}{nc^{2n}}r^{2n}\sin 2n\theta = v(r,\theta)$

Section 5, p. 193

1. Hyperbolic (a) and (e); elliptic (b) and (c); parabolic (d).

3. Only (e).

5. a. $u(x,y) = \displaystyle\sum_{1}^{\infty} a_n \sin n\pi x\, e^{-n\pi y}$

   b. $u(x,y) = \displaystyle\sum_{1}^{\infty} a_n \sin n\pi x \cos n\pi y$

   c. $u(x,y) = \displaystyle\sum_{1}^{\infty} a_n \sin n\pi x \exp(-n^2\pi^2 y)$

   $$a_n = 2\int_0^1 f(x)\sin n\pi x\, dx$$

Miscellaneous exercises, p. 195

1. $u(x,y) = \displaystyle\sum_{1}^{\infty} b_n \dfrac{\sinh \lambda_n(a-x)}{\sinh \lambda_n a}\sin \lambda_n y$

   $\lambda_n = n\pi/b, \qquad b_n = 2(1-\cos n\pi)/n\pi$

3. $u(x,y) = 1$. Note that 0 is an eigenvalue.

5. $u(x,y) = \displaystyle\sum \dfrac{a_n \sinh \lambda_n x + b_n \sinh \lambda_n(a-x)}{\sinh \lambda_n a}\cos \lambda_n y$

   $\lambda_n = (2n-1)\pi/2b, \qquad a_n = b_n = 4(-1)^{n+1}/\pi(2n-1)$

7. $u(x,y) = w(x,y) + w(y,x)$, where

   $w(x,y) = \displaystyle\sum b_n \dfrac{\sinh \lambda_n(a-y)}{\sinh \lambda_n a}\sin \lambda_n x$

   $\lambda_n = n\pi/a, \qquad b_n = 8h \sin\dfrac{n\pi}{2}\bigg/n^2\pi^2$

9. $u(x,y) = \displaystyle\int_0^{\infty} A(\lambda)\dfrac{\sinh \lambda(b-y)}{\sinh \lambda b}\cos \lambda x\, d\lambda$

   $A(\lambda) = (2\sin \lambda a)/\lambda\pi$

11. $u(x,y) = \displaystyle\int_0^{\infty} A(\lambda)\cos \lambda x e^{-\lambda y}\, d\lambda$

    $A(\lambda) = 2\alpha/\pi(\alpha^2 + \lambda^2)$

13. $u(x, y) = 1$

15. $u(r, \theta) = a_0 + \sum \left(\dfrac{r}{c}\right)^n (a_n \cos n\theta + b_n \sin n\theta)$

$a_0 = \frac{1}{2}$, $\quad a_n = 0$, $\quad b_n = (1 - \cos n\pi)/n\pi$

17. Same form as Exercise 15, but $a_0 = 2/\pi$, $a_n = 2(1 + \cos n\pi)/(n^2 - 1)$, $b_n = 0$ (and $a_1 = 0$).

19. $u(r, \theta) = (\ln r - \ln b)/(\ln a - \ln b)$

21. $u(r, \theta) = \sum_1^\infty b_n \left(\dfrac{r}{c}\right)^{n/2} \sin(n\theta/2)$

$b_n = \dfrac{1}{\pi} \int_0^{2\pi} f(\theta) \sin(n\theta/2)\, d\theta$

23. $2(d + f) = -1$

25. $u(x, y) = \dfrac{y(b - y)}{2} + \sum_1^\infty \dfrac{a_n \sinh \mu_n x + b_n \sinh \mu_n(a - x)}{\sinh \mu_n a} \sin \mu_n y$

$\mu_n = \dfrac{n\pi}{b}$, $\quad a_n = b_n = \dfrac{2b^2}{\pi^3} \dfrac{1 - \cos n\pi}{n^3}$

27. The equations become

$$\frac{\partial^2 \phi}{\partial y\, \partial x} = \frac{\partial^2 \phi}{\partial x\, \partial y}, \qquad (1 - M^2)\frac{\partial^2 \phi}{\partial x^2} + \frac{\partial^2 \phi}{\partial y^2} = 0.$$

29. $\phi(x, y) = \displaystyle\int_0^\infty (A(\alpha) \cos \alpha x + B(\alpha) \sin \alpha x)e^{-\beta y}\, d\alpha + c$

where $\beta = \alpha\sqrt{1 - M^2}$, c is an arbitrary constant, and

$$\left.\begin{matrix} A(\alpha) \\ B(\alpha) \end{matrix}\right\} = -\frac{U_0}{\beta\pi} \int_{-\infty}^\infty f'(x)\left\{\begin{matrix} \cos \alpha x \\ \sin \alpha x \end{matrix}\right\} dx$$

Chapter 5

Section 1, p. 203

1. $\dfrac{\partial^2 u}{\partial x^2} + \dfrac{\partial^2 u}{\partial y^2} = \dfrac{1}{c^2}\dfrac{\partial^2 u}{\partial t^2}$, $\quad 0 < x < a$, $\quad 0 < y < b$, $\quad 0 < t$

$u(x, 0, t) = 0$, $\quad u(x, b, t) = 0$, $\quad 0 < x < a$, $\quad 0 < t$

$u(0, y, t) = 0$, $\quad u(a, y, t) = 0$, $\quad 0 < y < b$, $\quad 0 < t$

$u(x, y, 0) = f(x, y)$, $\quad \dfrac{\partial u}{\partial t}(x, y, 0) = g(x, y)$, $\quad 0 < x < a$, $\quad 0 < y < b$

3. $\dfrac{\partial^2 u}{\partial x^2} + \dfrac{\partial^2 u}{\partial y^2} + \dfrac{\partial^2 u}{\partial z^2} = \dfrac{1}{c^2}\dfrac{\partial^2 u}{\partial t^2}$

Section 2, p. 205

1. $\nabla^2 u = \dfrac{1}{k}\dfrac{\partial u}{\partial t}$, $\qquad 0 < x < a, \;\; 0 < y < b, \;\; 0 < t$

$\dfrac{\partial u}{\partial x}(0, y, t) = 0$, $\qquad u(a, y, t) = T_0$, $\qquad 0 < y < b, \;\; 0 < t$

$u(x, 0, t) = T_0$, $\qquad \dfrac{\partial u}{\partial y}(x, b, t) = 0$, $\qquad 0 < x < a, \;\; 0 < t$

3. $\kappa\left(\dfrac{\partial^2 u}{\partial x^2} + \dfrac{\partial^2 u}{\partial y^2} + \dfrac{\partial^2 u}{\partial z^2}\right) = \rho c \dfrac{\partial u}{\partial t} - r$

Section 3, p. 209

1. If $a = b$, the lowest eigenvalues are those with indices (m, n) in this order: $(1, 1); (1, 2) = (2, 1); (2, 2); (3, 1) = (1, 3); (3, 2) = (2, 3); (1, 4) = (4, 1); (3, 3)$.

3. Frequencies are $\lambda_{mn} c/2\pi$ (Hz) where λ^2_{mn} are the eigenvalues found in the text.

5. The $\lambda^2_{mn} = (m\pi/a)^2 + (n\pi/b)^2$, for $m = 0, 1, 2, \ldots, n = 1, 2, 3, \ldots$.

7. a. $u(x, y, t) = 1$

For **(b)** and **(c)** the solution has the form:

$$u(x, y, t) = \sum_{m, n} a_{mn} \cos\left(\dfrac{m\pi x}{a}\right) \cos\left(\dfrac{n\pi y}{b}\right) \exp(-\lambda^2_{mn} kt)$$

where $\lambda^2_{mn} = (m\pi/a)^2 + (n\pi/b)^2$, and m and n run from 0 to ∞.

b. $a_{00} = \dfrac{(a + b)}{2}$, $\qquad a_{m0} = -\dfrac{2b(1 - \cos m\pi)}{m^2\pi^2}$

$a_{0n} = -\dfrac{2a(1 - \cos n\pi)}{n^2\pi^2}$, $\qquad a_{mn} = 0 \quad$ otherwise

c. $a_{00} = \dfrac{ab}{4}$, $\qquad a_{m0} = -\dfrac{ab(1 - \cos m\pi)}{m^2\pi^2}$

$a_{0n} = -\dfrac{ab(1 - \cos n\pi)}{n^2\pi^2}$

$a_{mn} = \dfrac{4ab(1 - \cos n\pi)(1 - \cos m\pi)}{m^2 n^2 \pi^4}$

if m and n are greater than zero.

9. The choice of a positive constant for either X''/X or Y''/Y, under the boundary conditions in Eqs. (9) and (10), will lead to the trivial solution.

Section 4, p. 212

1. The partial differential equations are the same, the boundary conditions become homogeneous, and in the initial conditions, $g(r, \theta)$ is replaced by $g(r, \theta) - v(r, \theta)$.
3. In the heat problem, $T' + \lambda^2 kT = 0$. In the wave problem, $T'' + \lambda^2 c^2 T = 0$.
5. The boundary conditions Eqs. (10) and (11) would be replaced by

$$\Theta(0) = 0, \qquad \Theta(\pi) = 0$$

Solutions are $\Theta(\theta) = \sin n\theta, \; n = 1, 2, \dots$.
7. Taking the hint, and using the fact that $\nabla^2 \phi = -\lambda^2 \phi$, the left-hand side becomes

$$(\lambda_k{}^2 - \lambda_m{}^2) \iint_R \phi_k \, \phi_m$$

while the right-hand side is zero, due to the boundary condition.

Section 5, p. 217

1. $\lambda_n = \alpha_n/a$, where α_n is the nth zero of the Bessel function J_0. The solutions are $\phi_n(r) = J_0(\lambda_n r)$, or any constant multiple thereof.
3. This is just the chain rule.
5. Rolle's theorem says that if a differentiable function is zero in two places, its derivative is zero somewhere between. From Exercise 4 it is clear that J_1 must be zero between consecutive zeros of J_0. Check Fig. 5.9 and Table 5.1.
7. Use the second formula of Exercise 6, after replacing $\mu + 1$ on both sides.

Section 6, p. 221

1. $v(0, t)/T_0 \cong 1.602 \exp(-5.78\tau) - 1.065 \exp(-30.5\tau)$ where $\tau = kt/a^2$
3. In Exercise 7 of Section 5, set $\mu = 0$ and make the substitution $x = \lambda r$.
5. Integration leads to the equality

$$\int_0^a (r\phi'^2)' \, dr + \lambda^2 \int_0^a r^2 (\phi^2)' \, dr = 0.$$

The first integral is evaluated directly. The second must be integrated by parts.

Section 7, p. 228

1. Use

$$\frac{1}{r}\frac{d}{dr}\left(r\frac{d}{dr}J_0(\lambda r)\right) = -\lambda^2 J_0(\lambda r).$$

3. The frequencies of vibration are $\lambda_{mn}c = \alpha_{mn}c/a$. The five lowest values of α_{mn}, in order, have subscripts $(0, 1)$, $(1, 1)$, $(2, 1)$, $(0, 2)$, and $(3, 1)$. See Table 5.1.

5. $\phi(a, \theta) = 0$ and $\phi(r, -\pi) = \phi(r, \pi)$, $\quad\dfrac{\partial\phi}{\partial\theta}(r, -\pi) = \dfrac{\partial\phi}{\partial\theta}(r, \pi)$

7. Set $J_m(\lambda_{mn}r) = \phi_n$. Then $(r\phi_n')' = -\lambda_{mn}r\phi_n$ and $(r\phi_q')' = -\lambda_{mq}r\phi_q$ are the equations satisfied by the functions in the integrand. Follow the proof in Chapter 2, Section 7.

Section 8, p. 233

1. $\phi(x) = x^\alpha[AJ_p(\lambda x) + BY_p(\lambda x)]$, where $\alpha = (1 - n)/2$, $p = |\alpha|$.
3. For $\lambda^2 = 0$, $Z = A + Bz$.
5. $\phi(\rho + ct) = \bar{F}(\rho + ct) + \bar{G}(\rho + ct)$
   $\psi(\rho - ct) = \bar{F}(\rho - ct) - \bar{G}(\rho - ct)$
   where $\bar{F}(x)$ is the odd periodic extension with period $2a$ of $\frac{1}{2}xf(x)$, and $\bar{G}(x)$ is the even periodic extension with period $2a$ of $\int (x/2c)g(x)\,dx$.
7. The weight function is ρ^2 and the interval is 0 to a.
9. $v(x) = (b - x)(x - a)/(a + b)x^2$
11. No. The idea is to find a solution of the partial differential equation that depends on only one variable. That is impossible if f depends on both x and y.

13. $a_n = \dfrac{-\displaystyle\int_a^b v(x)X_n(x)x^3\,dx}{\displaystyle\int_a^b X_n^2(x)x^3\,dx}$

Section 9, p. 240

1. $[k(k + 1) - \mu^2]a_{k+1} - [k(k - 1) - \mu^2]a_{k-1} = 0$, valid for $k = 1, 2, \dots$.

3. $P_5 = \dfrac{1}{8}(63x^5 - 70x^3 + 15x)$

5. $y = A \ln\left(\dfrac{1 + x}{1 - x}\right)$

7. Product solutions are

$$u_{mn}(\rho, \phi) = J_{m+1/2}(\lambda_{mn}\rho)P_m(\cos \phi)$$

$m = 0, 1, \ldots,$ and λ_{mn} are chosen to satisfy the boundary condition at $\rho = c$.

9. Leibnitz's rule states that

$$(uv)^{(k)} = \sum_{r=0}^{k} \binom{k}{r} u^{(k-r)}v^{(r)}.$$

(A superscript k in parentheses means kth derivative.) The right-hand side looks like the binomial theorem. In the case at hand, at most two terms are not zero.

11. $u(\rho, \phi) = \sum_{0}^{\infty} b_n \rho^n P_n(\cos \phi)$ where $b_0 = \frac{1}{2}$, and

$$b_n = \frac{1}{2}[P_{n-1}(0) - P_{n+1}(0)]$$

Miscellaneous exercises, p. 242

1. $u(x, y, t) = \sum a_m \sin \mu_m y \exp(-\mu_m^2 kt)$

$$+ \sum a_{mn} \cos \lambda_n x \sin \mu_m y \exp(-(\mu_m^2 + \lambda_n^2)kt)$$

$$\mu_m = m\pi/b, \qquad \lambda_n = n\pi/a$$

$$a_m = T \frac{1 - \cos m\pi}{m\pi}$$

$$a_{mn} = \frac{4T}{\pi^3} \frac{(\cos n\pi - 1)(1 - \cos m\pi)}{n^2 m}$$

3. $u(a/2, b/2, t) = \sum b_{mn} \sin \frac{n\pi}{2} \sin \frac{m\pi}{2} \exp(-(\lambda_n^2 + \mu_m^2)kt)$

where $\lambda_n = n\pi/a$, $\mu_m = m\pi/b$, and

$$b_{mn} = \frac{4T}{\pi^2} \frac{(1 - \cos m\pi)(1 - \cos n\pi)}{mn}$$

The first three nonzero terms are, for $a = b$, those with $(m, n) = (1, 1)$, $(1, 3) = (3, 1), (3, 3)$.
All terms with an even index are 0.

$$u(a/2, a/2, t) \cong \frac{16t}{\pi^2} \left(e^{-2\tau} - \frac{2}{3} e^{-10\tau} + \frac{1}{9} e^{-18\tau} \right)$$

where $\tau = kt/\pi^2 a^2$

5. $u(r) = (a^2 - r^2)/2$ and $u(r) = \sum_1^\infty C_n J_0(\lambda_n r)$, with

$$C_n = \frac{2a^2}{\alpha_n^3 J_1(\alpha_n)}$$

7. $w(x, t) = a_0 + \sum a_n \cos \lambda_n x \exp(-\lambda_n^2 kt)$

$v(y, t) = \sum b_m \sin \mu_m y \exp(-\mu_m^2 kt)$ where $\mu_m = m\pi/b$, $\lambda_n = n\pi/a$, and initial conditions are

$$v(y, 0) = 1, \quad 0 < y < b; \qquad w(x, 0) = Tx/a, \quad 0 < x < a.$$

9. $J_0(\lambda r) \exp(-\lambda^2 kt)$

11. $B_k = b_k/k(k + 1)$ for $k = 1, 2, \ldots$; b_0 must be 0, and B_0 is arbitrary.

13. $((1 - x^2)y')' - \dfrac{m^2}{1 - x^2} y + \mu^2 y = 0$

15. $u(r, z) = \sum a_n \dfrac{\sinh \lambda_n z}{\sinh \lambda_n b} J_0(\lambda_n r)$ where $\lambda_n = \alpha_n/a$ and $a_n = \dfrac{2U_0}{\alpha_n J_1(\alpha_n)}$

17. $u(r, z, t) = \sin \mu z J_0(\lambda z) \sin vct$ is a product solution if $\mu = m\pi/b$, $\lambda = \alpha_n/a$, and $v = \sqrt{\mu^2 + \lambda^2}$. The frequencies of vibration are therefore vc or

$$c\sqrt{\left(\frac{m\pi}{b}\right)^2 + \left(\frac{\alpha_n}{a}\right)^2}.$$

Chapter 6

Section 1, p. 249

1. c. $\dfrac{s^2 + 2\omega^2}{s(s^2 + 4\omega^2)}$ **d.** $\dfrac{\omega \cos \phi - s \sin \phi}{s^2 + \omega^2}$

e. $\dfrac{e^2}{s - 2}$ **f.** $\dfrac{2\omega^2}{s(s^2 + 4\omega^2)}$

3. a. $\dfrac{e^{-as}}{s}$ **b.** $\dfrac{(e^{-as} - e^{-bs})}{s}$ **c.** $\dfrac{(1 - e^{-as})}{s^2}$

5. a. $e^{-t} \sinh t$ **b.** e^{-2t} **c.** $e^{-at} \dfrac{\sin \sqrt{b^2 - a^2} t}{\sqrt{b^2 - a^2}}$

7. a. $\dfrac{(e^{at} - e^{bt})}{(a - b)}$ **b.** $\dfrac{t}{2a} \sinh at$ **d.** $\dfrac{t^2 e^{at}}{2}$

e. $f(t) = 1, \quad 0 < t < 1; \quad = 0, \quad 1 < t$

Section 2, p. 257

1. **a.** e^{2t} **b.** e^{-2t} **c.** $\dfrac{(3e^{-t} - e^{-3t})}{2}$ **d.** $\frac{1}{3}\sin 3t$

3. **a.** $\dfrac{(1 - e^{-at})}{a}$ **b.** $t - \sin t$ **c.** $\dfrac{(\sin t - \frac{1}{2}\sin 2t)}{3}$

 d. $\dfrac{(\sin 2t - 2t\cos t)}{8}$ **e.** $-\frac{3}{4} + \frac{1}{2}t + e^{-t} - \frac{1}{4}e^{-2t}$

5. **a.** $\dfrac{(e^{2t} - e^{-2t})}{4}$ **b.** $\frac{1}{2}\sin 2t$

 c. $\dfrac{3}{2} + \dfrac{i\sqrt{2} - 3}{4}\exp(-i\sqrt{2}t) - \dfrac{i\sqrt{2} - 3}{4}\exp(i\sqrt{2}t)$ **d.** $4(1 - e^{-t})$

7. **a.** $1 - \cos t$ **b.** $\dfrac{(e^{t} - \cos \omega t + \omega \sin \omega t)}{(\omega^2 + 1)}$ **c.** $t - \sin t$

Section 3, p. 264

1. **a.** $s = -\left(\dfrac{2n - 1}{2}\pi\right)^2,\qquad n = 1, 2, \ldots$

 b. $s = \pm i\dfrac{2n - 1}{2}\pi,\qquad n = 1, 2, \ldots$

 c. $s = \pm in\pi,\qquad n = 0, 1, 2, \ldots$

 d. $s = i\eta,\qquad \text{where } \tan \eta = \dfrac{-1}{\eta}$

 e. $s = i\eta,\qquad \text{where } \tan \eta = \dfrac{1}{\eta}$

3. **a.** $\dfrac{\sinh \sqrt{s}x}{s^2 \sinh \sqrt{s}}$ **b.** $\dfrac{1}{s} - \dfrac{\cosh \sqrt{s}(\frac{1}{2} - x)}{s(s + 1)\cosh(\sqrt{s}/2)}$

5. **a.** $u(x, t) = x + \sum \dfrac{2 \sin n\pi x}{n\pi \cos n\pi}\exp(-n^2\pi^2 t)$

 b. $u(x, t)$ is 1 minus the solution of Example 3.

Section 4, p. 270

1. $t + \dfrac{x^2}{2}$

3. $v(x, t) = \dfrac{4}{\pi^2} \sum_{1}^{\infty} \dfrac{\cos(2n-1)(\frac{1}{2}-x)\sin(2n-1)\pi t}{(2n-1)^2 \sin\left(\dfrac{2n-1}{2}\pi\right)}$

5. **a.** $\dfrac{\omega}{\omega^2 - \pi^2}\left(\dfrac{1}{\pi}\sin \pi t - \dfrac{1}{\omega}\sin \omega t\right)\sin \pi x$

    b. $\dfrac{1}{2\pi^2}(\sin \pi t - \pi t \cos \pi t)\sin \pi x$

7. **a.** $u(x, t) = x - \dfrac{\sin\sqrt{a}x}{\sin\sqrt{a}}e^{-at} + \dfrac{2a}{\pi}\sum_{1}^{\infty}\dfrac{\sin n\pi x \exp(-n^2\pi^2 t)}{n(a - n^2\pi^2)\cos n\pi}$

    b. The term $-\dfrac{x \cos n\pi x}{\cos n\pi}\exp(-n^2\pi^2 t)$ arises.

Miscellaneous exercises, p. 272

1. $U(s) = \dfrac{T_0}{\gamma^2 + s} + \dfrac{\gamma^2 T}{s(\gamma^2 + s)}$

    $u(x, t) = T_0 \exp(-\gamma^2 t) + T(1 - \exp(-\gamma^2 t))$

3. $U(s) = \dfrac{\cosh\sqrt{s}x}{s^2 \cosh\sqrt{s}}$

    $u(x, t) = t - \dfrac{1 - x^2}{2} - \sum_{n=1}^{\infty}\dfrac{2\cos \rho_n x}{\rho_n \sin \rho_n}\exp(-\rho_n^2 t)$

    where $\rho_n = (2n-1)\pi/2$

5. $u(x, t) = \dfrac{x(1-x)}{2} - \sum\dfrac{4\cos \rho_n(x - \frac{1}{2})}{\rho_n \sin(\rho_n/2)}\exp(-\rho_n^2 t)$

    where $\rho_n = (2n-1)\pi$

7. $u(x, t) = x + \sum_{1}^{\infty}\dfrac{2\sin n\pi x}{n\pi \cos n\pi}\exp(-n^2\pi^2 t)$

9. $U(x, s) = \dfrac{1}{s}(1 - e^{-\sqrt{s}x})$

11. $f(t) = \dfrac{x}{\sqrt{4\pi t^3}}\exp(-x^2/4t)$

13. $u(x, t) = \sum_{n=0}^{\infty} \left[\text{erfc}\left(\frac{2n + 1 - x}{\sqrt{4t}}\right) - \text{erfc}\left(\frac{2n + 1 + x}{\sqrt{4t}}\right) \right]$

15. $F(s) = 2 \sum_{n=1}^{\infty} \frac{1}{s^2 + n^2}$

17. $f(t) = \sum_{-\infty}^{\infty} \frac{1}{2a} G\left(\frac{in\pi}{a}\right) e^{in\pi t/a}$

19. $F(s)$ must be of the form $F(s) = G(s)/H(s)$ where $G(s)$ is never infinite. The solutions of $H(s) = 0$ must form an arithmetic sequence of purely imaginary numbers, and $H'(s) \neq 0$ if $H(s) = 0$.

21. $F(s) = \dfrac{(1 - e^{-\pi s})^2}{s(1 - e^{-2\pi s})} = \dfrac{1 - e^{-\pi s}}{s(1 + e^{-\pi s})}$

23. $F(s) = \dfrac{1 + e^{-\pi s}}{(s^2 + 1)(1 - e^{-\pi s})}$

Chapter 7

Section 1, p. 282

1. Exact solution $\dfrac{3x - x^2}{2}$.

3. $\dfrac{\Delta x_2}{\delta} u(x + \Delta x_1) - 2u(x) + \dfrac{\Delta x_1}{\delta} u(x - \Delta x_2),$ where $\delta = \dfrac{(\Delta x_1 + \Delta x_2)}{2}$

5. $\dfrac{[u(x + 2\Delta x) - 4u(x + \Delta x) + 6u(x) - 4u(x - \Delta x) + u(x - 2\Delta x)]}{(\Delta x)^4}$

7. $u_1 = -0.02,$ $u_2 = -0.08,$ $u_3 = -0.18,$ $u_4 = -0.32,$

    $u_5 = -0.50.$

9. $u_1 = \frac{4}{81},$ $u_2 = \frac{10}{81},$ $u_3 = \frac{26}{81}$

Section 2, p. 286

1. Line m of the solution should be exactly the same as line $m + 1$ of Table 7.4.

3. $r = \dfrac{2}{5},$ $\Delta t = \dfrac{1}{40}$

i	0	1	2	3	4
m					
0	0	0	0	0	0
1	1	0.	0.	0.	0.
2	1	0.4	0.	0.	0.
3	1	0.48	0.16	0.	0.
4	1	0.56	0.224	0.064	0.
5	1	0.6016	0.2944	0.1024	0.0512

5. $\Delta t = \frac{1}{32}$. Remember that $u_4(m) = u_0(m) = m\,\Delta t$.

i	0	1	2	3	4
m					
0	0	0	0	0	0
1	1	0	0	0	1
2	2	$\frac{1}{2}$	0	$\frac{1}{2}$	2
3	3	1	$\frac{1}{2}$	1	3
4	4	$\frac{7}{4}$	1	$\frac{7}{4}$	4
5	5	$\frac{5}{8}$	$\frac{7}{4}$	$\frac{5}{8}$	5

All numbers in this table should be multiplied by Δt.

7. $\Delta t = \frac{1}{32}$

i	0	1	2	3	4
m					
0	0	0	0	0	0
1	0	1	1	1	0
2	0	$\frac{3}{2}$	2	$\frac{3}{2}$	0
3	0	2	$\frac{5}{2}$	2	0
4	0	$\frac{9}{4}$	3	$\frac{9}{4}$	0
$\vdots$					
∞	0	3	4	3	0

All numbers in this table should be multiplied by Δt.

9. $\Delta t = \frac{1}{32}$. Remember $u_{-1} = u_1$.

i / m	0	1	2	3	4
0	0	1	2	3	4
1	1	1	2	3	4
2	1	$\frac{3}{2}$	2	3	4
3	$\frac{3}{2}$	$\frac{3}{2}$	$\frac{9}{4}$	3	4
4	$\frac{3}{2}$	$\frac{15}{8}$	$\frac{9}{4}$	$\frac{25}{8}$	4
5	$\frac{15}{8}$	$\frac{15}{8}$	$\frac{9}{4}$	$\frac{25}{8}$	4

All entries in this table should be multiplied by $\Delta x = \frac{1}{4}$.

Section 3, p. 291

1. Using symmetry, with $r = \frac{1}{4}$, the replacement equations are

$$u_1(m + 1) = \tfrac{1}{4}(u_2(m) + u_5(m))$$
$$u_2(m + 1) = \tfrac{1}{4}(u_1(m) + u_2(m) + u_6(m))$$
$$u_5(m + 1) = \tfrac{1}{4}(2u_1(m) + u_6(m))$$
$$u_6(m + 1) = \tfrac{1}{4}(2u_2(m) + u_5(m) + u_6(m)).$$

3. $\Delta t \le 1 \Big/ 2\left(\dfrac{1}{\Delta x^2} + \dfrac{\gamma}{\Delta y^2}\right)$

Section 4, p. 295

1. At $(\frac{1}{4}, \frac{1}{4})$, $11/256$; at $(\frac{1}{2}, \frac{1}{4})$, $14/256$; at $(\frac{1}{2}, \frac{1}{2})$, $18/256$.

3 and 4. In both cases the exact solution is $u(x, y) = xy$, and the numerical solutions are exact.

5. Coordinates and values of the corresponding u_i are: $(\frac{1}{7}, \frac{1}{7})$, 5α; $(\frac{2}{7}, \frac{1}{7})$, 10α; $(\frac{3}{7}, \frac{1}{7})$, 14α; $(\frac{1}{7}, \frac{2}{7})$, 21α; $(\frac{2}{7}, \frac{2}{7})$, 32α. Here $\alpha = 19/1159$.

7. Coordinates and values of the corresponding u_i are: $(\frac{1}{4}, \frac{1}{4})$, 165β; $(\frac{1}{2}, \frac{1}{4})$, 218β; $(\frac{3}{4}, \frac{1}{4})$, 176β; $(\frac{1}{4}, \frac{1}{2})$, 174β; $(\frac{1}{2}, \frac{1}{2})$ 263β. Here $\beta = 1/268$.

Section 5, p. 302

1.

m \ i	0	1	2	3	4
0	0	0	0	0	0
1	0	$\frac{1}{4}$	$\frac{1}{4}$	$\frac{1}{4}$	0
2	0	$\frac{1}{4}$	$\frac{1}{2}$	$\frac{1}{4}$	0
3	0	$\frac{1}{4}$	$\frac{1}{4}$	$\frac{1}{4}$	0
4	0	0	0	0	0
5	0	$-\frac{1}{4}$	$-\frac{1}{4}$	$-\frac{1}{4}$	0

3.

m \ i	0	1	2	3	4
0	0	0	0	0	0
1	0	$\alpha/4$	$\frac{1}{4}$	$\alpha/4$	0
2	0	$\frac{1}{4}$	$\alpha/2$	$\frac{1}{4}$	0
3	0	$\alpha/4$	$\frac{1}{4}$	$\alpha/4$	0
4	0	0	0	0	0
5	0	$-\alpha/4$	$-\frac{1}{4}$	$-\alpha/4$	0

In this table, $\alpha = 1/\sqrt{2}$.

5.

t_m	m \ i	0	1	2	3	4
0	0	0	$\frac{1}{2}$	1	$\frac{1}{2}$	0
0.177	1	0	$\frac{1}{2}$	$\frac{3}{4}$	$\frac{1}{2}$	0
0.354	2	0	$\frac{3}{8}$	$\frac{1}{4}$	$\frac{3}{8}$	0
0.530	3	0	0	$-\frac{1}{8}$	0	0
0.707	4	0	$-\frac{7}{16}$	$-\frac{3}{8}$	$-\frac{7}{16}$	0
0.884	5	0	$-\frac{5}{8}$	$-\frac{11}{16}$	$-\frac{5}{8}$	0

7.

i \ m	0	1	2	3	4
0	0	0	0	0	0
1	0	0	0	0	1
2	0	0	0	1	1
3	0	0	1	1	1
4	0	1	1	1	0
5	0	1	1	0	-1
6	0	0	0	-1	-1
7	0	-1	-2	-1	-1
8	0	-2	-2	-2	0

9. Subscript i is assigned according to Fig. 7.6.

i \ m	1	2	5
0	1	1	1
1	$\frac{1}{2}$	$\frac{3}{4}$	1
2	$-\frac{1}{4}$	0	$\frac{1}{2}$
3	$-\frac{1}{2}$	$-\frac{3}{4}$	$-\frac{3}{2}$
4	$-\frac{1}{2}$	$-\frac{5}{4}$	-2
5	$-\frac{3}{4}$	$-\frac{3}{4}$	-1

11. See figure for numbering of points.

i \ m	1	2	3	4	5	6
0	1	0	0	0	0	0
1	0	$\frac{1}{4}$	$\frac{1}{4}$	0	0	0
2	$-\frac{3}{4}$	0	0	$\frac{1}{4}$	$\frac{1}{8}$	0
3	0	$-\frac{1}{2}$	$-\frac{7}{16}$	0	0	$\frac{3}{16}$

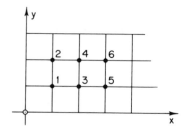

Index

A
B 8
C 9
D 0
E 1
F 2
G 3
H 4
I 5
J 6